Developments in Agricultural Engineering 9

Trickle Irrigation for Crop Production

OTHER TITLES IN THIS SERIES

1. Controlled Atmosphere Storage of Grains
by J. Shejbal (Editor)
1980 viii + 608 pp.

2. Land and Stream Salinity
by J.W. Holmes and T. Talsma (Editors)
1981 vi + 392 pp.

3. Vehicle Traction Mechanics
by R.N. Yong, E.A. Fattah and N. Skiadas
1984 xi + 307 pp.

4. Grain Handling and Storage
by G. Boumans
1984 xiii + 436 pp.

5. Controlled Atmosphere and Fumigation in Grain Storages
by B.E. Ripp et al. (Editors)
1984 xiv + 798 pp.

6. Housing of Animals
by A. Maton, J. Daelemans and J. Lambrecht
1985 xii + 458 pp.

7. Soil Cutting and Tillage
by E. McKyes
1985 viii + 218 pp.

8. Mechanics of Agricultural Materials
by G. Sitkei

Developments in Agricultural Engineering 9

Trickle Irrigation for Crop Production

Design, Operation and Management

Edited by

F.S. NAKAYAMA and D.A. BUCKS

U.S. Department of Agriculture, Agricultural Research Service, U.S. Water Conservation Laboratory, Phoenix, Arizona, U.S.A.

ELSEVIER

Amsterdam — Oxford — New York — Tokyo 1986

ELSEVIER SCIENCE PUBLISHERS B.V.
Sara Burgerhartstraat 25
P.O. Box 211, 1000 AE Amsterdam, The Netherlands

Distributors for the United States and Canada:

AMERICAN SOCIETY OF AGRICULTURAL ENGINEERS
2950 Niles Road
St. Joseph, MI 49085-9659 U.S.A.

Library of Congress Cataloging-in-Publication Data

Trickle irrigation for crop production.

(Development in agricultural engineering ; 8)
Includes index.
1. Trickle irrigation. I. Nakayama, F. S.,
1930- . II. Bucks, D. A. III. Series.
S619.T74T75 1986 616.5'87 86-2109
ISBN 0-444-42615-9

ISBN 0-444-42615-9 (Vol. 9)
ISBN 0-444-41940-3 (Series)

Printed in The Netherlands

PREFACE

An entirely new agricultural technology, trickle or drip irrigation, began its development in the early 1960's. Initial progress was sporadic even though the advantages in water management with trickle systems were recognized. Operators were reluctant to use the system because of its high initial cost and questions regarding its reliability. Once the main problems were isolated and solutions developed to make the system reliable, rapid acceptance by the growers resulted. Today, trickle irrigation is being used on crops that were earlier considered to be uneconomical. To achieve reliability, a multidisciplinary approach was needed to look critically at all aspects of the trickle system because, like other organized but complex operational systems, various types of expertise were required. The excellent interaction among the research, manufacturing, distributing and operational personnel made trickle irrigation successful. Many greenhouse and field experiments are being conducted using trickle irrigation as a tool incidental to the primary objectives of the particular research. This alone proves the acceptance of the trickle system because researchers are selective in using properly functioning equipment.

This multipurpose textbook/handbook is an outgrowth of cooperative efforts by workers in the private and public sectors. The various educational and research groups, particularly in the Western United States, where most of the irrigated agriculture is located, recognized the need to make a concerted effort to investigate trickle irrigation. Practically all phases of the system were studied, both basic and applied. This lead to a better understanding of trickle irrigation, and the results from our work and others have been combined in this book. Other long-term experiments are continuing, particularly the ones related to specific crops and will be reported in another volume directed toward crop production.

Trickle irrigation is still growing rapidly in terms of land area and economic impact. Continuous progress is being made in system design and operation so that some of the discussion may need revision by the time this book is published. Chapter leaders were encouraged to included the latest development. They were also given the greatest flexibility in the organization and presentation of their materials. Each chapter was to be essentially independent of another so that some repetition in subject material may be present, but such topics are seen from different viewpoints depending upon the observer's training.

Trickle irrigation truly involves interdisciplinary participation by the agricultural and hydraulic engineers, and soil and plant scientists. The design engineer may look at the system in terms of the hydraulics, water distribution and flow patterns, and the soil scientist in terms of water and salt distribution and flow rates, and the plant scientist in terms of water and nutrient use and crop behavior, but each in his or her own way has contributed to the success of trickle irrigation. Special problems, such as emitter clogging required the involvement of chemists and microbiologists, who were already aware of similar problems in tile drainage. Besides the research inputs, we recognize the participation and fruitful contributions of industrial personnel who improved on the materials and equipment components which make up the trickle system.

We have concentrated mainly on the trickle irrigation system and not on the crop in this book, although in many instances it is difficult to separate one from the other. Material is being prepared to consider the crop, with the trickle system as a vehicle for water application as it is with sprinkler or flood irrigation. We have tried to assemble in this volume the latest information available on the design, operation, maintenance, and management of trickle systems. Hopefully, the material can be used by the on-farm operator, and as a teaching tool and reference source. We anticipate that others will build upon our presentation to make further improvements in trickle irrigation.

Acknowledgment is made to the Western Regional Research Committee, W-128, Trickle Irrigation to Improve Crop Production and Water Management, under whose guidance this textbook/handbook developed. The Editors also give special thanks to Mrs. Frieda Bell, who single-handedly undertook the difficult task of preparing the many drafts and the final camera-ready copies. Her perseverance truly was a moral support to all in the completion of this book.

And to our wives, Mitzi and Betsy, we extend our heartfelt appreciation for their patience and encouragement during the hectic periods when the book was being prepared and finalized.

PHOENIX, ARIZONA 1985 F.S.N. and D.A.B.

CONTENTS

CONTRIBUTORS

R.S. Bowman, Soil Scientist, U.S. Department of Agriculture, Agricultural Research Service, U.S. Water Conservation Laboratory, Phoenix, Arizona.

V.F. Bralts, Agricultural Engineer, Michigan State University, East Lansing, Michigan.

D.A. Bucks, Irrigation Engineer, U.S. Department of Agriculture, Agricultural Research Service, U.S. Water Conservation Laboratory, Phoenix, Arizona.

S. Davis, Agricultural Engineer (retired), U.S. Department of Agriculture, Agricultural Research Service, Riverside, California.

H.W. Ford, Horticulturalist (retired), University of Florida, Agricultural Research and Education Center, Lake Alfred, Florida.

R.G. Gilbert, Microbiologist, U.S. Department of Agriculture, Agricultural Research Service, Irrigated Agricultural Research and Extension Center, Prosser, Washington.

H.M. Gitlin, Agricultural Engineer (retired), University of Hawaii, Honolulu, Hawaii.

D.A. Goldhamer, Extension Specialist, University of California, San Joaquin Valley Agricultural and Research and Extension Center, Parlier, California.

G.J. Hoffman, Agricultural Engineer, U.S. Department of Agriculture, Agricultural Research Service, Water Management Research Laboratory Fresno, California.

T.A. Howell, Agricultural Engineer, U.S. Department of Agriculture, Agricultural Research Service, Conservation and Production Laboratory, Bushland, Texas.

L.G. James, Agricultural Engineer, Washington State University, Pullman, Washington.

J.M. Lima, Agricultural Engineer, Superior Farming Co., Bakersfield, California.

R.J. Miller, Water Specialist (retired), University of California, Davis, California.

F.S. Nakayama, Chemist, U.S. Department of Agriculture, Agricultural Research Service, U.S. Water Conservation Laboratory, Phoenix, Arizona.

C.J. Phene, Soil Physicist, U.S. Department of Agriculture, Agricultural Research Service, Water Management Research Laboratory, Fresno, California.

D.E. Rolston, Soil Physicist, University of California, Davis, California.

C.A. Saruwatari, Agricultural Engineer, University of Guam, Mamgilao, Guam.

H. Schulbach, Extension Specialist, University of California, Colusa, California.

W.M. Shannon, Graduate Assistant, University of Colorado, Fort Collins, Colorado; formerly Design Engineer, Boyle Engineering, Bakersfield, California.

K.H. Solomon, Agricultural Engineer, U.S. Department of Agriculture, Agricultural Research Service. U.S. Salinity Laboratory, Riverside, California.

R.D. von Bernuth, Agricultural Engineer, University of Tennessee, Knoxville, Tennessee.

A.W. Warrick, Soil Physicist, University of Arizona, Tucson, Arizona.

I.P. Wu, Agricultural Engineer, University of Hawaii, Honolulu, Hawaii.

Chapter 1

INTRODUCTION

1.1 HISTORICAL DEVELOPMENT

D. A. BUCKS AND S. DAVIS

Trickle irrigation is the slow application of water on, above, or beneath the soil by surface trickle, subsurface trickle, bubbler, spray, mechanical-move, and pulse systems. Water is applied as discrete or continuous drops, tiny streams, or miniature spray through emitters or applicators placed along a water delivery line near the plant. The shape or design of the emitter reduces the operating pressure from the supply line, and a small volume of water is discharged at the emission point. Water flows from the emission points through the soil by capillarity and gravity. The trickle irrigation method is characterized by the following features: (1) water is applied at a low rate; (2) water is applied over a long period of time; (3) water is applied at frequent intervals; (4) water is applied near or into the plant's root zone; and (5) water is applied by a low-pressure delivery system.

Advances in trickle irrigation systems and practices have been rapid with user interest developing principally in the 1970's. Trickle irrigation, like other irrigation methods, will not fit every agricultural crop, land situation, or user objective. However, trickle irrigation does offer many unique agronomic, agrotechnical, and economic advantages for the present and future irrigation technologies.

1.1.1 *History*

Historical and archaeological findings show that irrigation has played a major role in the development of ancient civilizations. The oldest civilization with irrigation developed along the Nile, Tigris, Euphrates, Indus, and Yellow rivers; for example, gravity irrigation began along the Nile about 6,000 B.C. The dominant methods of irrigation from these early times have been surface or gravity and sprinkler irrigation. Trickle irrigation is a considerably new approach compared to these methods and developed from subirrigation where irrigation water is applied by raising the water table.

1.1.1.a Early history worldwide

Although simple in its concept, the widespread use of trickle irrigation has not been practical until very recently because of the lack of suitable, economical materials. Beginning in 1860, German researchers experimented with a combination irrigation and drainage system using clay pipes. These subirrigation and drainage tiles lasted for more than 20 years, where irrigation water was pumped into the underground drainage system. The first work in the subsurface trickle irrigation in which water was applied to the root zone without raising the water table was conducted in the United States at Colorado State University in 1913 by E.B. House, who concluded that it was too expensive for practical use.

An important breakthrough was made around 1920 in Germany when perforated pipe was introduced; and in the United States, porous pipe or canvas was used for subsurface irrigation at Michigan State University by O.E. Robey. Thereafter, research and development centered around using perforated and porous pipe made of various materials to determine whether water flow from these pipes into the soil could be controlled by the soil moisture tension rather than by water pressure in the system. At the same time, subirrigation experiments in Germany, the United Soviet Socialist Republic (1923), and France resulted in an irrigation method in which water was applied through closely-spaced channels to raise the ground water level to near the root zone. Various other forms of subirrigation were also used in a number of other countries, including the United States, Netherlands and United Kingdom.

With the development of plastics during and after World War II, the idea of using plastic pipe for irrigation became feasible. In the late 1940's in the United Kingdom, plastic pipe trickle systems were used to irrigate greenhouse plants. About the same time, K. Dorter (1962) and others in Germany began extensive work on subsurface irrigation (called underground irrigation at this time) with plastic pipe; over 100 publications were listed on the concept of underground irrigation. Publications on the modern-day surface trickle system began to appear from Israel in 1963 and the United States in 1964, while research and development from both countries started some years before (figure 1.1.1). The observation of S. Blass, an engineer who developed the first patented surface trickle irrigation emitter, has been often quoted describing greater vigor of a large tree near a leaking faucet over other trees in the area. From Israel the concept of surface trickle (often called drip) irrigation spread to Australia, North America, and South Africa by the late 1960's, and eventually throughout the world. In 1971, the First International Drip Irrigation Meeting was held in Tel Aviv, Israel, where 24 papers were presented.

1.1.1.b Early History in United States

In the mid 1950's, a small irrigation manufacturing firm in Watertown, New York, began to supply polyethylene tubing (called spaghetti) to water plants and flowers grown in greenhouses. By the early 1960's, plastic-pipe trickle irrigation systems were extensively used in greenhouse research and most commercial enterprises. S. Davis installed the first field experiment with a subsurface trickle irrigation system on a lemon orchard at Pomona, California, in 1963 and on

Fig. 1.1.1 The first surface trickle irrigated tomato field in the United States (New York) in 1965 which includes plastic tunnels and heating to prevent frost damage.

oranges near Riverside, California, in 1964. With the success stories of surface trickle irrigation coming from Israel, D. Gustafson visited Israel in 1968, and returned to install the first research and demonstration study on trees on a private grower's avocado orchard in San Diego, California, in 1969. About the same time, B. Hall began to conduct trials using surface trickle irrigation on strawberries and tomatoes along with plastic mulches also in and around San Diego.

From 1968 through early 1970, numerous inventors and companies began to develop trickle irrigation emitters that totaled well over 250 devices before the mid 1970's (figure 1.1.2). Extensive research and development began in the United States and other countries on all aspects of trickle irrigation in the early 1970's. In 1974, eighty-three papers were presented at the Second International Drip Irrigation Congress in San Diego, California, with over 1000 registrants from 26 countries.

1.1.1.c Development in plastics

As often happens, the initial breakthrough in plastic materials came accidentally when polyethylene (PE) plastic was produced in a British laboratory in 1935. PE plastic is produced when ethylene gas, a component of natural gas and other fossil fuels, is subjected to

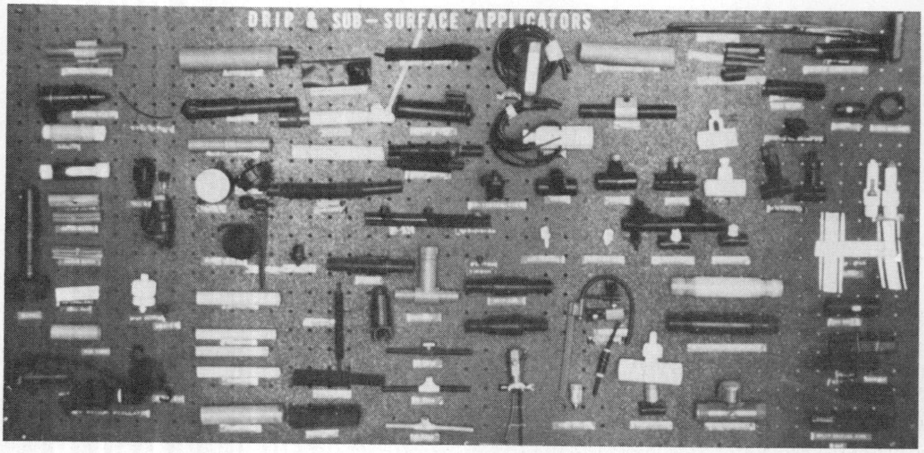

Fig. 1.1.2 Examples of numerous surface and subsurface trickle emitters which have been manufactured.

heat and pressure. In the early years of its development, only the high density form of PE was known. High-density PE was produced at relatively low pressures (2,000 kPa, 300 lb/in^2), was rigid, and has been used to make bottles and other rigid plastic products. In 1948, the discovery of low-density PE, which was created at high pressures (20,000 kPa, or more), spurred rapid growth in the plastics industry, and produced a suitable, economical material for trickle irrigation lateral lines.

In 1977, the plastic industry introduced a new low-density PE that was produced at a lower pressure, which lowered the cost and increased the quality of polyethylene resins. The development of the linear, low-density PE (LLDPE) has greatly improved the strength and stress-crack resistance of plastics. LLDPE is a remarkably tough and flexible material, with a wide range of physical properties, depending on the size and arrangement of ethylene molecules. LLDPE manufactures currently produce resins which are tailored to specific uses. Suitable additives, such as antioxidants, stabilizers, and carbon black, must be added to the clear plastic before a durable and economical trickle irrigation lateral line can become a reality. The development of new LLDPE plastic for trickle irrigation lateral lines and other plastic materials for emitters has resulted in increased research and developments. Further details on materials used for emitter construction is presented in chapter 2.1.

1.1.1.d Expansion in land area

Irrigated land development has kept pace with the world population since about 1800. In 1977, the Food and Agriculture of the United

Nations Organization estimated that the total global irrigated area was 223 million ha (550 million ac) and that this would increase to about 273 million ha (675 million ac) by 1990. The 1978 Census of Agriculture indicated that there were 20.3 million ha (50.2 million ac) irrigated in the United States. Of these acreages, about 12.6 million ha (31.2 million ac, 62.2 percent) were irrigated by gravity irrigation, 7.4 million ha (18.4 million ac, 36.7 percent) by sprinkler irrigation, and 0.2 million ha (0.6 million ac, 1.1 percent) by trickle or subirrigation.

A recent survey conducted by the International Commission on Irrigation and Drainage (ICID) indicated that about 417,000 ha (1.0 million ac) were under trickle irrigation throughout the world. The major use of trickle irrigation was in the United States, where the area has expanded from approximately 4,000 ha (9,800 ac) in 1972 to 185,300 ha (458,000 ac) in 1982. There has also been extremely rapid growth in Israel where the area has expanded from 10,000 ha in 1975 to 81,700 ha in 1982. Table 1.1.1 shows the irrigated area and major crops for the countries with over 1,000 ha (2,500 ac). The seven countries with over 10,000 ha (25,000 ac) were the United States, Israel, South Africa, France, Australia, United Soviet Socialist Republic, and Italy. The ICID survey also showed that fruit trees were the principal crops irrigated with trickle and that stone fruit was the most widely irrigated crop in the world. All types of soils were being trickle irrigated, although sandy soils predominated. The main reasons for choosing trickle irrigation were as follows: (1) water and labor were expensive; (2) the water supply was limited; (3) the water supply was saline (although periodic leaching was still required); (4) the use of other methods was difficult (example, hillside orchards); and (5) landscaping or greenhouse irrigation was required.

In the United States, the states of California, Florida, Georgia, Hawaii, Michigan, and Texas account for over 85 percent of the trickle irrigated acreage in 1982 (figure 1.1.3). Although the area under trickle irrigation is presently small compared to the total irrigated area, the higher valued crops and land are being utilized. The list of trickle irrigated agricultural crops includes avocado, grape, citrus, strawberry, tomato, sugarcane, cotton, and others. Application of trickle irrigation for landscaping, greenhouses, and nurseries has also increased tremendously and includes ornamental trees and shrubs, ground covers on highway roadsides and residential properties, avocado and citrus nurseries, forestry trees, and others.

1.1.2 *System components*

Many significant advancements have been made in the design of trickle irrigation components. The basic components of a trickle irrigation system include a pump, fertilizer injector, filters, distribution lines, emitters, and other control and monitoring equipment (figure 1.1.4).

TABLE 1.1.1

Trickle irrigated land area and principal crops throughout the world in 1982 (adapted from International Commission on Irrigation and Drainage Working Group survey on micro irrigation, Abbott, 1984).

Country	Land area (ha)	Principal crops
United States	185,300	Orchard, vine, vegetable, sugarcane
Israel	81,700	Citrus, cotton, fruit
South Africa	44,000	Vine, orchard
France	22,000	Orchard, vine, glasshouse, vegetable
Australia	20,050	--
United Soviet Socialist Republic	11,200	Orchard, glasshouse, vine, tea
Italy	10,300	Orchard, vegetable, glasshouse
China	8,040	Orchard
Cyprus	6,600	Orchard, glasshouse, vine
Mexico	5,500	
Canada	4,985	Orchard, berry, glasshouse
Morocco	3,600	Orchard
United Kingdom	3,150	Orchard, glasshouse
Hungary	2,500	Orchard, vine
Brazil	2,000	Orchard, vine, nursery
Jordan	1,020	Glasshouse, fruit
New Zealand	1,000	--
Others	3,715	Orchard, vine, vegetable, nursery
Total	416,660	

1.1.2.a Emitters

Emitters are used to dissipate pressure and discharge water in a trickle irrigation system. Ideally, an emitter permits a small uniform flow or trickle of water at a constant discharge that does not vary significantly throughout the field or subunit. Many different emitters have been devised and manufactured with the concept that the emitters should be inexpensive, reliable (not clog), and compact as well as provide a uniform discharge of water. Some of the more distinctive emitter designs are the short-path, long-path, short-orifice, vortex, pressure-compensating, self-flushing, perforated single- and double-chamber tubings, as well as the aerosol emitters, foggers, misters, or miniature sprinklers used in spray irrigation.

The point on or beneath the soil surface at which water is discharged from the emitter is called an emission point. Emitter designs can be classified into two types, point-source and line-source. Point-source emitters discharge water from individual or multiple outlets

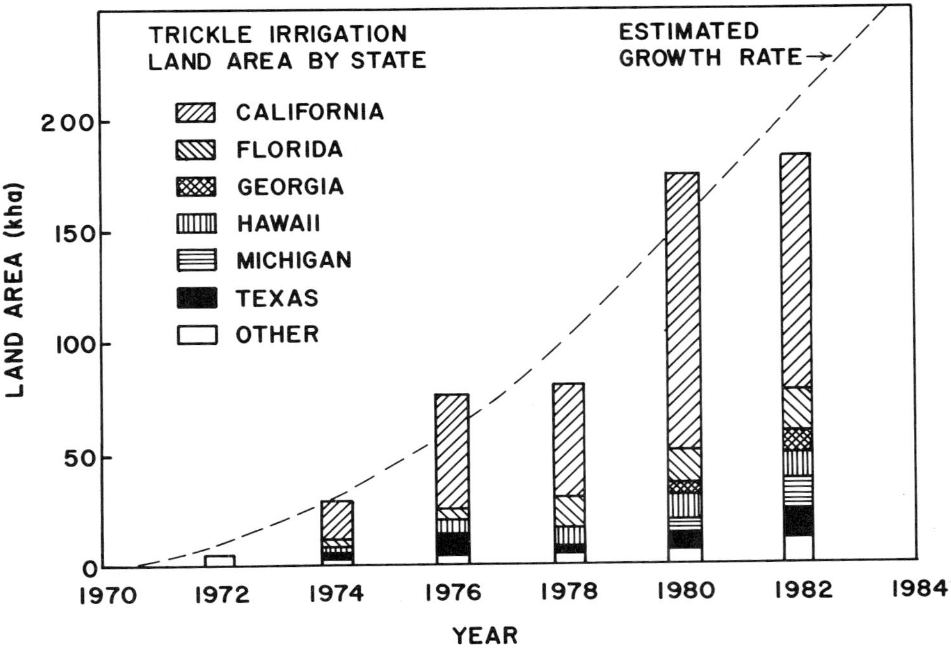

Fig. 1.1.3 Trickle irrigation land area and distribution in the United States from 1970 through 1982 (based on Irrigation Journal surveys).

that are spaced at least 1 m (3 ft) apart. Typically, point-source emitters are used for widely spaced plants such as trees, vines, ornamentals, and shrubs; but some point-source emitters are being used for closely spaced row crops. We are classifying bubbler and spray irrigation emitters as point-source systems although this may not be the case under all situations (See section 1.1.3 on types of trickle systems). Line-source emitters have perforations, holes, or porous walls in the irrigation tubing that discharge water at close spacings, or even continuously along a lateral. Line-source emitter systems are used primarily on small fruits, vegetables, or other closely spaced row crops. Better materials and manufacturing have improved extrusion and molding of point-source emitters; and coextrusion, laser technology, and plastic formulation techniques have improved the reliability of line-source emitters.

Emitter construction and performance requirements are very important because all emitters are subjected to exposure from sunlight and chemicals applied through or on the emitters as the crop is grown,

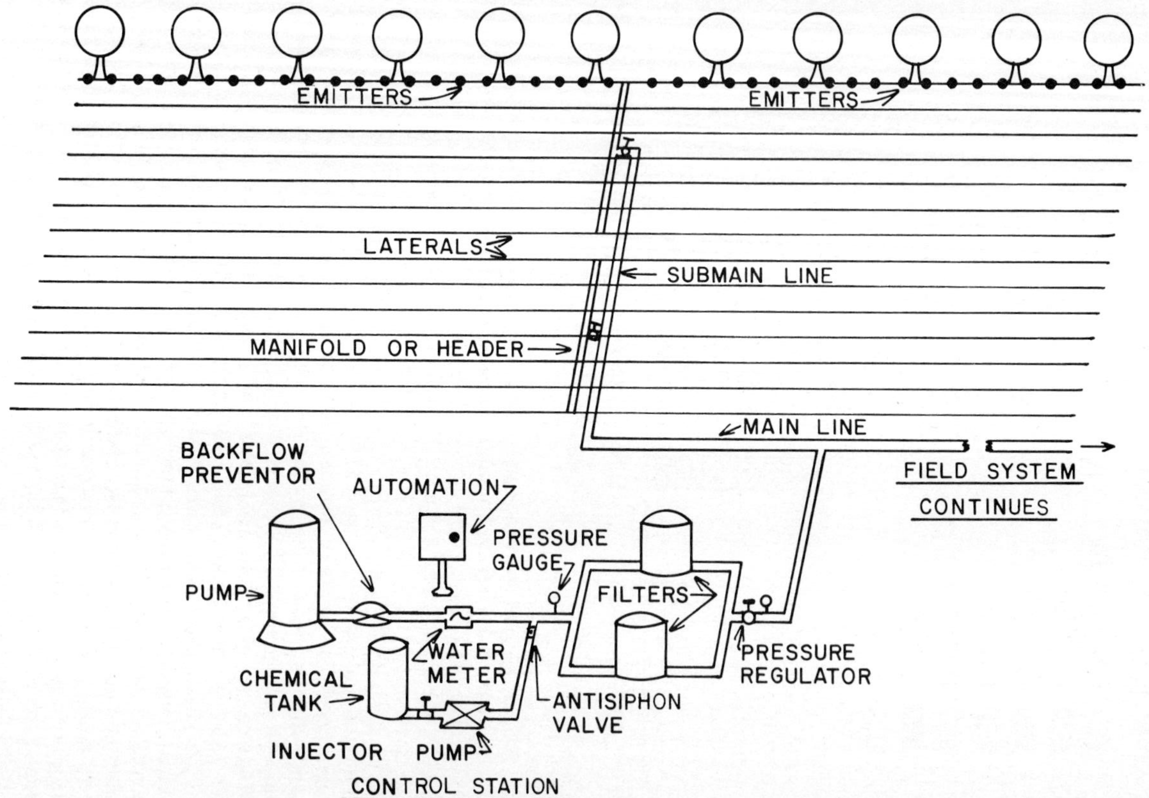

Fig. 1.1.4 An example of a basic trickle irrigation system.

extremes in environmental changes, and physical abuse. Details on principles of discharge, emitter hydraulics, type of emitters, and construction methods and materials are presented in chapter 2.1, Emitter Construction. Emitter performance will also be discussed and is an important aspect of chapters 3.1 through 3.4 on Operational Principles.

1.1.2.b Distribution lines and fittings

Distribution lines consist of a network of graduated pipe sizes starting with a single, large main line followed by smaller submain and lateral lines (figure 1.1.4). The buried main and submain lines are normally permanent and made of polyvinyl chloride (PVC), asbestos-cement or PE plastic pipe. Main and submain lines often range from 40 to 200 mm (1.5 to 8.0 in) inside diameter depending upon the required water flow to the laterals and the economics tradeoff between power costs and pipe installation costs. Both main and submain lines should have valved outlets for periodic pipeline flushing. The submain line may also contain pressure regulators, flow control valves, manual or automatic control valves, secondary filters, and other safety devices. Lateral lines are usually made of PE plastic and range from 10 to 26 mm (0.380 to 1.050 in) inside diameter, with the most commonly used size of 16 mm (0.5 in nominal size with an inside diameter of 0.625 in)

In most cases, the submain line connects the main line to the lateral; however, in larger installations, there may also be a manifold or header line that is coupled between the submain and lateral line. The size, length, and maximum allowable pressure loss depend on topography, pressure loss in the laterals, and total pressure variation allowed for the emitters along the laterals. The hydraulic design of main, submain, and lateral lines will be presented in chapter 2.2, System Design.

As part of the design of a trickle irrigation system, the appropriate fittings or parts must be selected to connect together the main, submain, and lateral lines. Appropriate fittings are available for connecting both PVC and PE distribution lines. PVC fittings are solvent cemented using specifically designed primers and cements, whereas PE equipment is noncementable and must be connected with barbed or compression fittings. A barbed fitting is inserted inside the pipeline, and a compression fitting is fitted over the outside of the pipeline. The PE compression fittings are becoming more popular since they have less friction loss and fewer stress-cracking problems than the barbed fitting. Examples of common fittings used in trickle irrigation are shown in figure 1.1.5. All these fittings are available in couplers (slip by slip, slip to thread, thread to insert, etc.), elbows (90 and 45 deg.), tees, crosses, unions, and plugs or caps. Standards and recommendations are being written to insure proper sizing of fittings in both metric and English units.

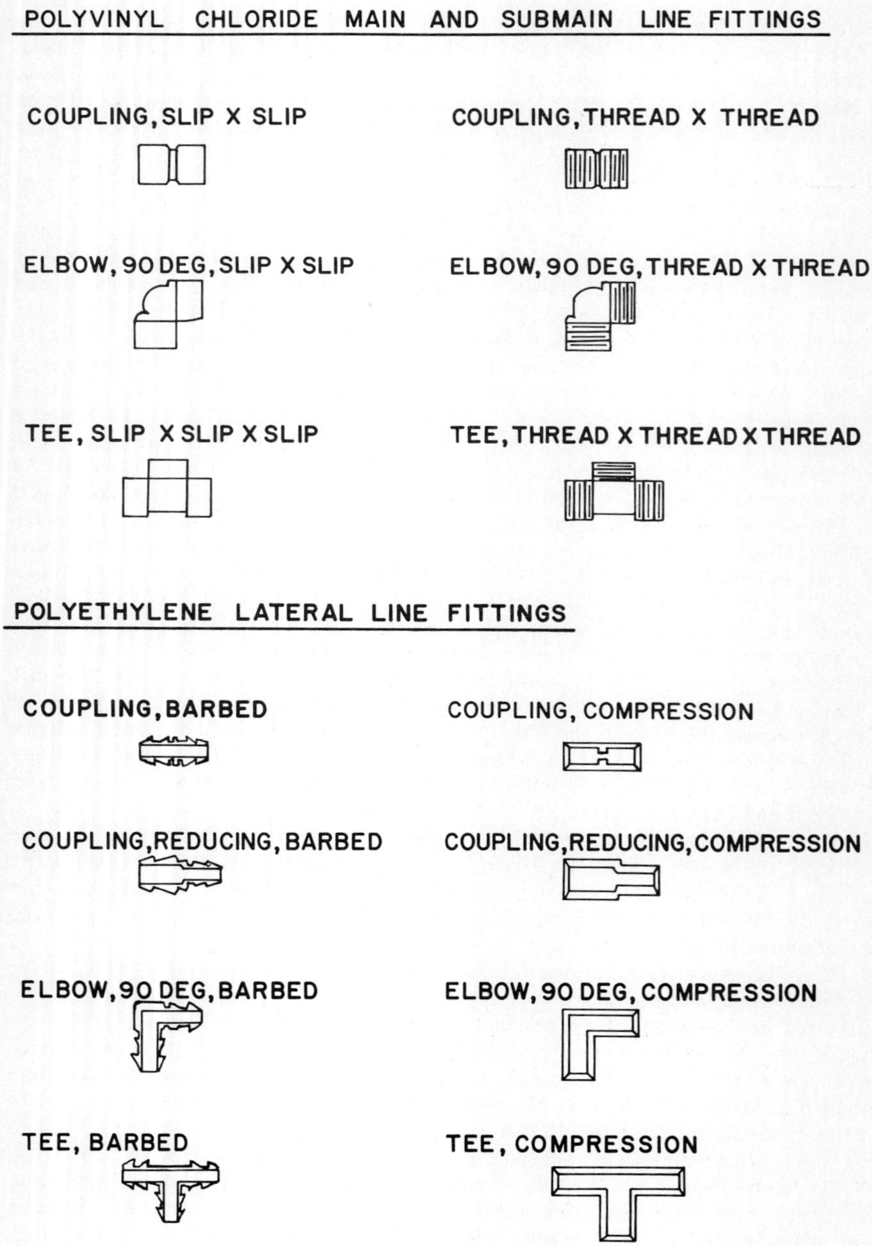

Fig. 1.1.5 Examples of plastic pipe and tubing fittings used in trickle irrigation.

1.1.2.c Control station and automation

The main control station for the trickle irrigation system is organized to measure, filter or screen, and treat the water, and to regulate pressure and time of water application. The control station includes the pump, backflow-prevention device, primary filter, pressure regulator (automatic or mechanical flow control valves), pressure gauge, water meter, and usually automation and chemical injection equipment (figure 1.1.4). Prevention of emitter clogging is important in the successful operation of a trickle irrigation system. Details on emitter clogging and water treatment are presented in chapters 3.1 and 3.2, respectively. Proper water filtration is essential and achieved by using either screen, media (fine gravel and sand), cartridge, or centrifugal filters either individually or in combination. The selection of the type, size, and capacity of the filtration unit depends on the initial water quality and emitter design. Chemical injectors are used to apply fertilizer, acid, bacteriacide, or other chemicals through trickle systems. The different types of injectors, their operating principles, and safety precautions needed for injecting chemicals will be discussed in chapters 3.2 on water treatment and 4.3 on fertilization.

The mechanical pressure regulator is used to maintain the design operation pressure of at the emitter, which may range from 27 to 205 kPa (4 to 30 lb/in^2) for the different types of emitter systems. The water meter is required to monitor flow volume, to check the initial design, to schedule irrigations, and to provide information on possible maintenance problems. Trickle systems are readily automated, as explained in chapter 3.3, using single and multistation timers or controllers and related solenoid valves to eliminate the need to operate, otherwise manual flow control values. Filter backwashing and lateral line flushing for system maintenance can also be automated. Controllers are available that can operate on electrical outlet, battery, or solar power sources. Automation can be partial or total to include sequential operations from a few minutes in an hour, an entire day, or any number of cycles per day. Automation can be accomplished on a volumetric or timeclock basis, or by soil plant stress sensors which actuate the controllers.

1.1.3 *System types*

Trickle irrigation systems are typically defined in terms of installation method, emitter discharge rate, wetted soil surface area, or mode of operation. The five basic types of trickle systems are surface, subsurface, bubbler, spray, mechanical-move, and pulse.

1.1.3.a Surface trickle

A surface trickle irrigation system is one where the emitters and lateral lines are laid on the soil surface. It is sometimes called drip irrigation and is the most prevalent type of trickle system. Surface trickle has been primarily used on widely spaced plants, but can also be used for row crops. Generally, discharge rates are less than 12 L/hr (3 gal/hr) for single-outlet, point-source emitters and less than 12 L/hr/m (1 gal/hr/ft) for line-source emitters. Advantages of surface trickle irrigation include the ease of installation, inspection, changing and cleaning emitters, plus the possibility of checking soil surface wetting patterns and measuring individual emitter discharge rates. An example of a surface trickle irrigation emitter and lateral line is shown in figure 1.1.6.

1.1.3.b Subsurface trickle

In subsurface trickle irrigation, water is applied slowly below the surface through emitters with discharge rates in the same range as those for a surface trickle system. A typical soil surface wetting pattern from a subsurface trickle system is illustrated in figure 1.1.7. This method of application is not to be confused with subirrigation, in which the root zone is irrigated through or by water table control. Lately, subsurface trickle systems have gained wider acceptance since earlier problems of emitter clogging have been reduced. These systems are now being used on small fruit and vegetable crops. Experience has shown that emitter outlets should be pointed upwards and that maintenance requirements are similar to surface trickle systems. Advantages of subsurface trickle irrigation include freedom from anchoring of the lateral lines at the beginning and removing them at the end of the growing season, little inference with cultivation or other cultural practices, and possibly a longer operational life. In

Fig. 1.1.6 Point-source emitter used for surface trickle irrigation.

Fig. 1.1.7 Line-source emitter system used for subsurface trickle irrigation showing the wetting pattern on the soil surface.

addition, combination subsurface and surface trickle irrigation systems have been tried where the lateral lines are buried and the emitters are located on or above the soil surface through the use of riser tubing.

1.1.3.c Bubbler

In bubbler irrigation, water is applied to the soil surface in a small stream or fountain from an opening with a point-source discharge rate greater than surface or subsurface trickle irrigation, but usually less than 225 L/hr (60 gal/hr). Because the emitter discharge rate normally exceeds the infiltration rate of the soil, a small basin is required to control the distribution of water (figure 1.1.8). Advantages of the bubbler system include reduced filtration, maintenance or repair, and energy requirements compared with other trickle systems. However, larger size lateral lines are usually required with the bubbler systems to reduce the pressure loss associated with the higher discharge rates.

1.1.3.d Spray

In spray irrigation, water is applied to the soil surface as a small spray, jet, fog, or mist. The air is instrumental in distributing the water; whereas in surface trickle, subsurface trickle, and bubbler irrigation, the soil is primarily responsible for water distribution. Spray systems have discharge rates typically less than 175

Fig. 1.1.8 Bubbler system with a small basin around each tree to contain the high water application rate.

L/hr (46 gal/hr) and are used to irrigate tree or other widely spaced crops (figure 1.1.9). Spray systems can be vulnerable to high wind and evaporation losses, particularly when plants are young having a limited crop canopy. However, the spray and bubbler systems normally have minimal filtration and other maintenance requirements.

1.1.3.e Mechanical-move

Mechanically moving trickle irrigation systems that expand the bubbler concept to large-scale row crops include traveling trickle or spray and drag or hose-reel trickle systems. First, a traveling trickle system can utilize a linear-move sprinkler line except that drop tubes, miniature sprayers, or attached trickle lateral lines rather than conventional sprinkler devices are used to deliver water as a continuously moving stream to each row. Operating pressures are lower than those of most conventional sprinkler systems, and the uniformity of water application over the field is usually good. Discharge rates from the traveling or spray trickle systems often exceed the soil infiltration rate so that check dams are required in the furrow to prevent soil erosion or water runoff. Potential advantages of traveling or spray trickle systems include a possible reduction in clogging problems and a less expensive pipe network, compared to a solid-set surface or subsurface trickle system.

Fig. 1.1.9 Spray system to deliver water in a full- or part-circle pattern.

Second, hose-reel or drag-type trickle irrigation systems use surface trickle lateral lines that are either pulled as a set across the field, moved singly from one row to the next, or wound onto a reel before being moved by a tractor. Discharge rates are sometimes similar to conventional trickle irrigation systems, and the main possible advantage of these systems is their adaptability to supplemental irrigation practices. These mechanical-move trickle systems have the potential disadvantage of high initial cost, limited water application rates, and extensive maintenance requirements.

1.1.3.f Pulse

Pulse trickle systems use high discharge rate emitters and consequently have short water application times. Pulse systems have application cycles of 5, 10, or 15 minutes in an hour, and discharge rates for pulse emitters are typically 4 to 10 times larger than conventional surface trickle systems. The primary advantage of this system is a possible reduction in clogging problems, whereas a possible disadvantage is the need for reliable, inexpensive pulse emitters and automatic controllers. Adapting standard trickle emitters to pulse irrigation systems can increase startup and shutdown inefficiences by increasing the number of application cycles.

1.1.4 *General principles*

Many commercial companies and government agencies have invested large sums of money and time to foster the advancement of trickle irrigation. But, like anything that is new, trickle irrigation had few supporters in its early conception, and initially had many unanticipated design, management, and maintenance problems. The development of better plastics and water filtration systems reduced much of the maintenance problems, and research and experience solved many of the design and management problems. However, the user must recognize that trickle irrigation has advantages and disadvantages, and to maximize efficiency, the system must be tailored to specific field and water conditions before success can be achieved by proper design, installation, maintenance, and management. This includes the proper design which will be discussed briefly in this chapter and in more detail in succeeding chapters.

1.1.4.a Potential advantages

Numerous reports (Bucks et al., 1983; Davis and Bucks, 1983; Goldberg, 1976; Howell et al., 1981; and Shoji, 1977) have listed and summarized potential advantages of trickle compared to other irrigation methods. However, we will attempt to discuss only the more important benefits.

(i) Increased beneficial use of available water. General agreement exists that irrigation water requirements can be less with trickle irrigation than with traditional irrigation methods. The savings, of course, depend on the crop, soil, and environmental conditions, and attainable on-farm efficiency. Primary reasons given for the water savings include irrigation of a smaller portion of the soil volume, decreased surface evaporation, reduced irrigation runoff from the field, and controlled deep percolation losses below the crop root zone. Direct evaporation from the soil surface, and water uptake by weeds could be reduced by not wetting the entire soil surface between rows or trees, especially on young tree or row crops. Water loss by wind drift or by increased evaporation due to poor water application uniformity under strong winds such as from sprinkler irrigation can be controlled. Water loss from runoff on steep hills or lack of water penetration on low-intake permeability or crusting soils can also be minimized (figure 1.1.10).

(ii) Enhanced plant growth and yield. Under trickle irrigation, the soil water content in a portion of the plant root zone should remain fairly constant because water can be applied slowly and frequently at a predetermined rate. Generally, wide fluctuations in the soil water content that traditionally occur under gravity and some sprinkler irrigation methods will not take place under the trickle method. Frequent trickle irrigations should improve plant growth and increased yields assuming that no problems occur such as those related to soil aeration, plant disease, or restricted plant rooting. The

Fig. 1.1.10 Trickle irrigation system being used to deliver water on hillside orchards in San Diego, California.

major causes of improved plant growth are probably related to improved water distribution along the row and to reduced effects of variation in texture and water holding capacity in heterogenous soils. Published data on crop response has shown that yields were equal or better in most cases and the amounts of water applied were equal or less with trickle compared with other irrigation methods.

(iii) Reduced salinity hazard to plants. Considerable evidence exists indicating that waters of higher salinity can be used with trickle than other irrigation methods without greatly reducing crop yields. Minimizing the salinity hazard to plants irrigated by trickle irrigation can be attributed to (1) keeping salts in the soil water more diluted because the high frequency irrigations maintain a stable soil moisture condition; (2) eliminating leaf damage caused by foliar salt absorption with sprinkler irrigation; and (3) moving salts beyond the active plant root zone. On the other hand, the improper placement of the system in relation to the crop could increase the salinity hazard.

(iv) Improved fertilizer and other chemical application. Trickle irrigation can maximize flexibility in fertilizer scheduling. Frequent or nearly continuous application of plant nutrients with the irrigation water is feasible and appears to be beneficial for many crop production situations. Various reasons have been proposed for increased efficiency of fertilization (1) decreased quantity of applied fertilizer because fertilizer is applied only to the root zone; (2) improved timing of fertilization because the more frequent applications make it possible to match plant requirements at various growth stages; and (3) improved distribution of fertilizer with minimum leaching beyond the root zone or runoff. Besides fertilizer, other material, such as herbicides, insecticides, fungicides, nematicides, growth regulators, and carbon dioxide can be efficiently supplied to improve crop production.

(v) Limited weed growth. Weed infestation may be reduced under trickle irrigation because only a fraction of the soil surface is wetted. Also, water filtration for trickle irrigation may result in delivery of fewer weed seeds to the field in comparison with other methods. Numerous investigators have mentioned the overall reduction in weed control; however, others have experienced an increase in weed growth and control problems within the small wetted portion of the soil surface, primarily when the crop is young. Selective herbicides have been applied through trickle systems with mixed results. Further development and water soluble products are needed.

(vii) Decreased energy requirements. Energy costs for pumping irrigation water may be reduced with trickle irrigation since the operating pressures are lower than with other types of pressurized systems. However, most of the energy conservation should come from reducing the quantity of water pumped. Trickle systems can save energy over gravity of flood irrigation only when irrigation efficiencies are significantly increased.

(viii) Improved cultural practices. Cultural operations such as spraying, weeding, thinning, and harvesting are possible on a continuous basis without interrupting the normal irrigation cycle. More hedge plantings for tree crops, higher plant populations for row crops, and increased use of plastic, natural, and synthetic mulches are examples of improved cultural practices that are adaptable to trickle irrigation. Minimum tillage practices can be easily utilized along with trickle irrigation, since dry areas are available for controlled traffic.

1.1.4.b Potential disadvantages

Despite observed successes and possible advantages, several problems have been encountered in the mechanics of applying water with trickle systems for some soils, water qualities, and environmental conditions. Again, the most important possible disadvantage of trickle irrigation will be compared with other irrigation methods.

(i) Persistent maintenance requirements. Complete or partial clogging of emitters represents the most serious problem encountered with trickle irrigation. Clogging will adversely affect the uniformity of water and fertilizer application, increase maintenance costs, and result in crop damage and decreased yield if not detected and corrected early. Other maintenance problems, not as serious as emitter clogging, are pipeline and component damage as leaks or flow restrictions caused by rodent or other animals, personnel, or machinery.

(ii) Salt accumulation near plants. Where high salinity waters are used in arid regions, salts tend to accumulate at the soil surface and along the periphery of the wetted soil volume. Rain water may move harmful amounts of salts into the plant root zone and cause injury to the plants. Salt accumulation from a prior trickle operation can be a problem if seeds are located in areas of such high salt concentration.

(iii) Restricted plant root development. Because trickle irrigation normally supplies water to a concentrated portion of the total soil volume, root development is limited to the wetted soil volume near each emitter or along each lateral line. Excessive restriction of plant root development has the potential to decrease plant growth and yields. Also, good root development may be needed to anchor the plant against strong winds.

(iv) High system costs. Supporting equipment requirements make trickle systems expensive initially although costs are generally less than solid-set sprinklers, but more than center pivot systems. Actual costs may vary considerably depending on the type of crop, trickle system, filtration equipment, automation, etc., selected. Most economists attribute the annual operation costs for trickle to be comparable with other irrigation methods provided that maintenance costs are not excessive.

1.1.4.c Design and installation considerations

The main goal when designing a trickle system is to insure that an acceptable uniformity of water and fertilizer application is obtained throughout the field. A trickle system designer must consider the emitter type, emitter uniformity, hydraulics, topography, desired water distribution uniformity, crop salt tolerance, water requirement, water quality, fertilizer injection, soil salinity, cultural practice, and other site-specific variables in setting up an appropriate system. The trickle irrigation installer must not only be aware of minimum design recommendations, but communicate with the users on the proper operation of the trickle system. Specifics on design and operation will be covered in chapters 2.1 through 3.4; however, as you explore these chapters, a listing of key design and installation suggestions is provided for future discussion as follows:

(1) Lateral lines should run flat, downhill, or along the contour.

(2) Generally, lateral lines of 13-to-16-mm (0.5-to-0.625-in) inside diameter should be less than 150 m (500 ft) for tree and vineyard crops and less than 200 m (660 ft) for row crops, but may vary depending upon the desired uniformity of water distribution.
(3) Usually, submain length should be less than 200 m (660 ft) and in some cases 100 m (330 ft), but may vary depending upon terrain, pipe size, and economics.
(4) System capacity must meet peak crop evapotranspiration requirements.
(5) Filtration units must meet water quality, flow capacity, and backwash capacity.
(6) Flushing valves should be installed at the ends of main, submain, and lateral lines.
(7) Normally, emitters should be installed facing upwards along lateral lines whenever possible.
(8) A backflow prevention device should be installed after the pump or well.
(9) Air or vacuum relief valves should be installed where needed.
(10) Chemical injection points should be provided before and after the main filter.
(11) A water meter should be included in the design.

1.1.4.d Maintenance considerations

The primary goal of a maintenance schedule is to control emitter clogging to assure a suitable economical life for the trickle system. A maintenance schedule varies with water quality, depending on three factors: (1) physical--the suspended inorganic (sand, silt, and clay) and organic materials and plastic particles; (2) chemical--the precipitation of calcium or magnesium carbonate, calcium sulfate, heavy metal hydroxides, and some fertilizers; and (3) biological--the bacterial and algal filaments, slimes, and microbial-chemical deposits. A listing of possible maintenance requirements depending upon the emitter design and water quality is as follows:

(1) Filters should be cleaned, backwashed or inspected regularly with no more than a 14 to 34 kPa (2 to 5 lb/in^2) increase in operating pressure across the filter before cleaning.
(2) Automatic flushing devices should be used where the water is high in silt and clay.
(3) Chemical injectors, time clocks, pressure regulators, water meters, and main pump should be checked at least weekly.
(4) Field inspections for malfunctioning emitters and pipeline leaks will be required at least monthly (sometimes weekly).
(5) Flush lateral lines by hand every six months for tree crops, or at least three times per season for row crops.
(6) Chemical water treatment is needed when the chemical or biological hazards are moderate or severe.

(7) Inject only approved or tested chemicals that are tested in a separate container for chemical reactions prior to injection.
(8) Follow safety requirements for injecting chemicals.

1.1.4.e Management considerations

The purpose of the total management scheme is to insure optimum crop response. A listing of key management suggestions which you will be exposed to more fully in chapter 4 on Management is as follows:

(1) Water measurement is an important tool in irrigation scheduling.
(2) Automation can save labor in water and fertilizer applications, but may increase malfunction problems.
(3) Soil water penetration and storage should be checked regularly.
(4) Use as many field measurements as possible to assist in irrigation scheduling.
(5) Irrigation water applications should closely match the crop evapotranspiration rate.
(6) Irrigation frequency can be flexible for many soil, water, crop conditions.
(7) More frequent irrigations can be helpful for salinity management.
(8) Fertilizer applications should be more frequent during the early stages of plant growth.
(9) Cultural practices must be modified where trickle systems are used.

1.1.5 *Economic considerations*

Because yields are higher or more consistent on irrigated than rainfed lands, irrigation plays a major role in stabilizing food and fiber production. Irrigation technology has advanced significantly during the past two decades, but many existing on-farm irrigation systems have not kept pace with technology in over 30 years. Problems of improving on-farm irrigation efficiencies will need to be addressed as the competition for limited water supplies increases. Trickle irrigation systems have the potential to obtain high efficiency (85 to 95%) and may become even more economical in the future. An economic analysis of an irrigation system should stress the total system rather than the individual parts of the system.

1.1.5.a System costs

Trickle irrigation systems are usually costly and will require better management skills and farming practices. Initial costs of a trickle irrigation system in 1985 dollars may average about $1500 to $3500/ha ($600 to $1400/ac), and maintenance costs could range from about $50 to $200/ha/yr ($20 to $80/ac/yr) in the United States.

These costs will undoubtedly vary considerably for other countries. In a recent economic study on trickle irrigation for cotton, operation costs of both a trickle and conventional-furrow system were within 5 percent, but annual fixed costs per land area were 2.3 times higher for trickle than surface irrigation ($865 compared with $370/ha). Implications were that trickle irrigated cotton could be profitable under many conditions in the southwestern United States if yields were increased by at least 1.2 to 2.5 bales/ha (0.5 to 1.0 bales/ac). Other studies on tomatoes in California and sugarcane in Hawaii have indicated that operating and ownership costs can be comparable for both trickle and surface irrigation. In some situations, the initial equipment and maintenance costs for a trickle system have become less in the last few years and may become even lower or remain unchanged in the future. With the increased need to conserve water, public agencies may wish to subsidize new or improved irrigation systems through cost-sharing or tax incentive programs, similar to those presently available for soil conservation. Also, public and private groups could encourage additional research and development that enhance the economic attractiveness of trickle irrigation and other efficient irrigation methods.

1.1.5.b Land costs

Land quality is an important factor in selecting an irrigation system. Consider an analysis of crop production in a region with two irrigation technologies (figure 1.1.11). In this example, it is more profitable to continue surface or gravity irrigation on all land of a quality greater than C before introducing a new technology. With the adoption of trickle irrigation, more land qualities would become profitable for farming; below a certain land quality, a farmer should change from the traditional to trickle irrigation methods. Land qualities between A and B could be added to the agricultural land base; those between B and C could be converted from traditional to trickle methods; and those greater than C should continue with former methods. This type of analysis leads one to the conclusion that trickle irrigation should be adopted and used first on lower quality land, where the relative profit potential is greater. Later, if the cost of the system should continue to decline, its adoption on higher quality land may become profitable.

1.1.5.c Water costs

Profitability of trickle irrigation will obviously change as the multitude of economic factors involved in crop production vary. Farmers depending on groundwater are affected by escalating pumping costs as the result of dropping water tables. Figure 1.1.12 shows the cost conditions under which a grower in the southwestern United States operates and must consider before switching irrigation technologies. Under some circumstances, neither method would be profitable (shaded area). If the farmer had a shallow well (less than 60 m, 200 ft), energy prices would have to go above $1.60/m ($0.50/ft) of lift before

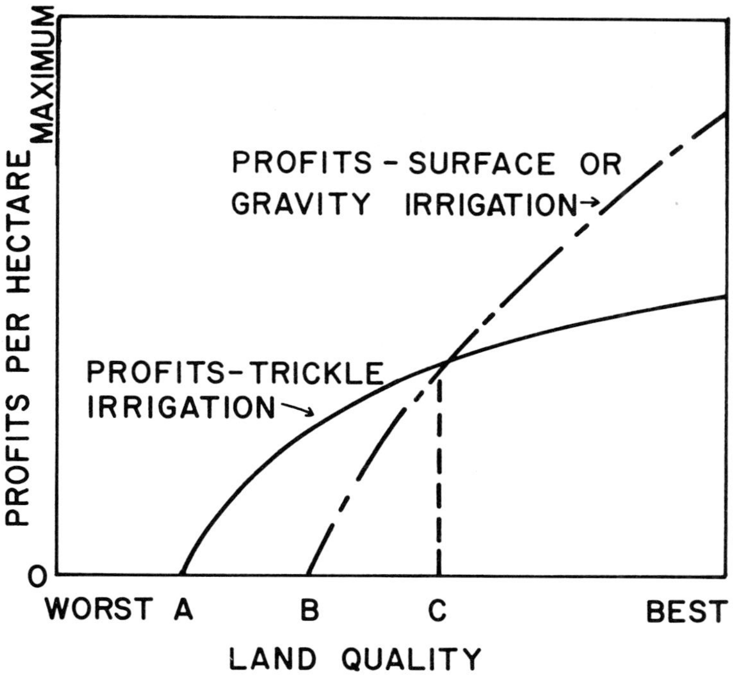

Fig. 1.1.11 Selection of an irrigation method based on land quality (after Caswell et al., 1984).

a technology change would be recommended. With wells deeper than 150 m (500 ft), any energy price greater than \$0.60/m (\$0.20/ft) of lift would prompt a change from gravity to trickle irrigation. Obviously, this type of economic analysis leads one to conclude that rising energy prices and declining water tables may automatically justify a switch to trickle irrigation, but other things must be taken into account.

Although the land and water cost factors can be the two most important features in an economic comparison of different irrigation technologies, other costs (labor, tillage, weed control, fertilization, harvesting, etc.) and profits (higher yield, earlier ripening, price, product quality, etc.) need to be considered in a complete economic analysis for the selection of the proper irrigation method for each location.

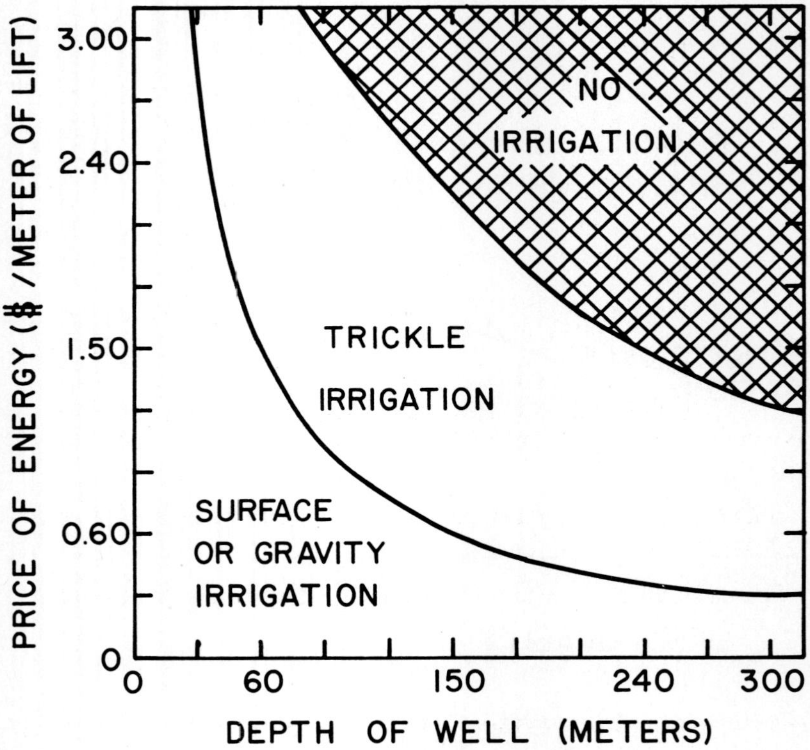

Fig. 1.1.12 Selection of an irrigation method based on energy prices pumping depths (after Caswell et al., 1984).

PROBLEMS

1. Describe the early history of trickle irrigation and some of the the reasons why this method did not develop until the 1960's.

2. Discuss some of the early developments in plastics and the importance of durable and economical materials.

3. What is the present land area under trickle irrigation worldwide, and how does this compare with the total irrigated land area?

4. List and describe the major components of a trickle irrigation system along with some of the fittings used to connect the network of pipelines.

5. List the six major types of trickle irrigation systems, and discuss field situations where each type could best be used.

6. Describe the three most important potential advantages of the trickle irrigation method and situations where these advantages would most likely occur.

7. Describe the three most important potential disadvantages of the trickle irrigation method and situations where these disadvantages could be the most problem.

8. Discuss the importance of proper design, installation, maintenance, and management to achieve success with all types of irrigation systems.

9. Discuss why maintenance and management principles are sometimes more important with the trickle method than other irrigation methods.

10. Discuss briefly how the selection of an irrigation method can depend on land quality.

11. Discuss briefly how the selection of an irrigation method can depend on water costs.

REFERENCES AND SUPPLEMENTAL READING

Abbott, J.S. (Chairman), 1984. Micro irrigation - world wide usage. ICID Working Group on Micro Irrigation, International Commission on Irrig. and Drain., New Delhi, India, 33(1): 4-9.

American Society of Agricultural Engineers, 1982. Design, installation, and performance of trickle irrigation systems. ASAE EP 405, Agricultural Engineers Yearbook, St. Joseph, Michigan, pp. 507-510.

Boswell, M.S., 1984. Micro-irrigation Design Manual. James Hardie Irrigation, El Cajon, California, 234 pp.

Bresler, E., 1977. Trickle-drip irrigation: principles and applications to soil-water management. In Brady, N.C. (Ed.), Advances in Agronomy 29: 343-393, Academic Press, New York.

Bucks, D.A., Nakayama, F.S. and Warrick, A.W., 1982. Principles, practices and potentialities of trickle (drip) irrigation. In Hillel, D. (Ed.), Advances in Irrigation 1: 219-298, Academic Press, New York.

Bureau of the Census, 1982. 1979 Farm and Ranch Survey. U.S. Department of Commerce, 1978 Census of Agriculture, U.S. Government Printing Office, Washington, DC, 5(8): 1-163.

Caswell, M., Zilberman, D. and Goldman, G.E., 1984. Economic implications of drip irrigation. Calif. Agric. 38: 4-5.

Davis, S., 1974. History of irrigation. Agribusiness News 10 (7): 1.
Davis, S. and Bucks, D.A., 1983. Drip Irrigation. In Pair, C.H., Hinz, W.W., Sneed, R.E., Frost, K.R., and Schiltz, T.J., (Eds.), Irrigation (formerly Sprinkler Irrigation), Fifth Edition, Irrigation Association, Silver Spring, Maryland, pp. 528-546.
Dorter, K., 1962. Study on solution of existing problems with underground irrigation. Report of the Inst. for Soil Conservation of the Martin Luther University, W. Germany, 35 pp.
Fereres, E. (Ed.), 1981. Drip Irrigation Management. Univ. of Calif. Leaflet 21259, Berkeley, California, 39 pp.
Goldberg, D., Gornat, B. and Rimon, D., 1976. Drip Irrigation: Principles, Design, and Agricultural Practices. Drip Irrig. Scientific Publ., Kfar Shumaryahu, Israel, 296 pp.
Howell, T.A., Bucks, D.A. and Chesness, J.L., 1981. Advances in trickle irrigation. Proc. 2nd Nat. Irrig. Symposium, Lincoln, Nebraska, Amer. Soc. Agric. Eng., St. Joseph, Michigan, pp. 69-94.
House, E.B., 1918. Irrigation by Means of Underground Porous Pipe. Colorado Agricultural Exp. Sta. Bull. 240, Ft. Collins, Colorado, 42 pp.
Karmeli, D. and Keller, J., 1975. Trickle Irrigation Design. Rain Bird Sprinkler Manufacturing Co., Glendora, California, 133 pp.
Shoji, K., 1977. Drip Irrigation. Scientific Amer. 237 (5): 62-68.
Soil Conservation Service, 1983. Trickle Irrigation. Chapter 5, Sec. 15, National Engineering Handbook, U.S. Dept. of Agric., Washington, DC, 129 pp.
Robey, O.E., 1934. Porous Hose Irrigation. Cooperative Extension Bull. 133, Michigan State College, East Lansing, Michigan, 6 pp.
Wilson, P., Ayer, H. and Snider, G., 1984. Drip Irrigation for Cotton: Implications for Farm Profits. National Resource Economics Division, Economics Research Ser., U.S. Dept. of Agric., Agric. Econ. Report 517, Washington, DC, 29 pp.

Chapter 2

DESIGN PRINCIPLES

2.1 EMITTER CONSTRUCTION

R. D. VON BERNUTH AND K. H. SOLOMON

2.1.1 *Introduction*

Trickle irrigation offers several potential advantages over other irrigation systems, the primary one being the precise application of water from the emitter system. Many factors contribute to the overall precision of water application; however, the most critical component of the system in this regard is probably the emitter.

2.1.1.a Performance requirements for emitters

With trickle irrigation, water from a central source is distributed throughout the field to be irrigated in a system of pressurized pipe lines. Adequate pressure must be maintained in the lines to overcome friction losses and elevation differences. The difference in pressure between the distribution line and atmospheric pressure must be dissipated in the emitter, which is the last component of the system through which water passes before reaching the soil. It is possible to dissipate some or all of the pressure into velocity as in a small spray, but many advantages of drip irrigation are negated if that is done.

Precise water application requires that the discharge from the emitter must remain relatively constant. Pressure variations in a drip system are unavoidable due to topography of the land and distribution system hydraulics.

Environmental and seasonal factors affect water and lateral line temperatures and, consequently, emitter flow characteristics. When emitter flow rate depends upon water temperature, then this sensitivity will become an emitter design factor.

2.1.1.b Modes of operation

The emitter must dissipate the pressure difference between internal line pressure and external atmospheric pressure. The process through which this is accomplished can be explained by examining the Darcy-Weisbach head loss equation

$$h_L = \left(\frac{fl}{d}\right)\left(\frac{v^2}{2g}\right) \qquad (2.1.1)$$

where h_L = head loss, f = friction coefficient, l = length of flow channel, d = diameter of flow channel, v = velocity of flow, and g = gravitational constant.

The pressure differential is to be dissipated as head loss. The factors over which the designer has control are f (to some extent), l, and d. An emitter with large l is a long path type emitter, and one with very small l is an orifice type emitter. It is also possible to introduce losses due to abrupt changes in flow direction, and that principle is used in the tortuous path or labyrinth type emitter. The possible modes of operation for dissipating the pressure differential are: (1) long path, (2) orifice, (3) labyrinth, or (4) combination of the three.

2.1.2 *Discharge principles*

Theoretically, two basic methods can be used to control flow in emitters, but in practice most emitters incorporate a combination of methods. These methods depend upon the flow regime and are discussed in the following sections.

2.1.2.a Flow regimes

The flow regime in fluid mechanics is characterized by the Reynolds number which is the ratio of inertial forces to viscous forces. The inertial forces are represented by the product of mass and acceleration (ma) and the viscous forces are represented by νd. The inertial force can be reduced to $\rho(v^2)d^2$, so that the ratio of inertial to viscous forces, R_e, can be written as $\frac{\rho v^2 d^2}{\mu v\ d}$ which reduces to

$$R_e = \frac{dv}{\nu} \qquad (2.1.2)$$

where d = representative emitter diameter, v = velocity, μ = viscosity (N-sec/m^2), ρ = density (kg/m^3), and ν = kinematic viscosity (μ/ρ, m^2/sec). The Reynolds number can be calculated by equation 2.1.2.

Flow is usually categorized by the region or range of Reynolds numbers within which the flow occurs. The primary reason for doing so is because the friction factor f can be estimated when the region of Reynolds numbers can be determined. Typical flow regimes and equations for f are given in table 2.1.1.

2.1.2.b Methods of controlling flow

Emitter flow can be controlled by altering the design of the flow length, l, and the diameter, d. The sensitivity of the control achieved with l and d depends upon the flow regime and can be affected if orifice discharge occurs. The method of controlling flow is often referred to as the emitter type as shown in parentheses.

<u>Laminar flow without orifice discharge (Laminar)</u>

If flow is laminar, $f = \frac{64}{R_e} = \frac{64}{dv/\nu} = \frac{64\nu}{dv}$ (2.1.7)

TABLE 2.1.1

Equations for friction factor f for different flow regimes.

Flow regime	Reynolds number, R_e	"f" equation	
Laminar	$R_e \leq 2{,}000$	$f = 64/R_e$	(2.1.3)
Unstable	$2{,}000 < R_e \leq 4{,}000$	$f = 3.42 \times 10^{-5} R_e^{0.85}$	(2.1.4)
Partially turbulent	$4{,}000 < R_e \leq 10{,}000$	$f = \frac{0.316}{R_e^{0.25}}$	(2.1.5)
Fully turbulent	$R_e > 10{,}000$	$f = \frac{0.316}{R_e^{0.25}}$	(2.1.6)

By substituting equation 2.1.7 into equation 2.1.1, the following relationship can be obtained:

$$h_L = \frac{(64\nu/dv) \cdot 1}{d} \frac{v^2}{2g} = \frac{64\nu}{2g} 1 v$$

Since $q_e = Av = \frac{\pi d^2 v}{4}$, $v = \frac{4q_e}{\pi d^2}$

and $h_L = \frac{64\nu}{2g} 1 \frac{4q_e}{\pi d^2}$

then

$$h_L = \frac{128\nu}{\pi g} \frac{1}{d^2} q_e$$

where $\pi = 3.1416$, q_e = emitter flow rate, and A = emitter cross sectional flow area.

Solving for q_e and assuming the pressure differential (dP) is dissipated as head loss,

$$q_e = \frac{\pi g}{128\nu} \frac{d^2}{1} dP \qquad (2.1.8)$$

Thus, in a laminar flow condition, flow from an emitter varies with the square of the passage diameter, with dP, and inversely with kinematic viscosity and length of flow passage.

<u>Turbulent flow without orifice discharge (Turbulent)</u>

In a manner similar to that developed for laminar flow, turbulent flow can be related to a flow equation where the pressure differential is dissipated as head loss,

$$q_e = K\ \nu^{-1/7}\ (\frac{dP}{l})^{4/7}\ d^{19/7} \tag{2.1.9}$$

where K is a constant.

Flow in this type of emitter is relatively insensitive to viscosity (-1/7 power), and is not as sensitive to changes in pressure differential, or length as with laminar flow. However, it would be somewhat more sensitive to diameter.

<u>Turbulent flow with orifice discharge controlling (Orifice)</u>

Discharge from an orifice is given by

$$q_e = \frac{\pi C_d\ (2g)^{1/2}}{4}\ d^2(dP)^{0.5} \tag{2.1.10}$$

where C_d = coefficient of discharge.

In this case, flow does not depend upon the viscosity, but does vary with the square of the diameter and square root of the pressure differential.

2.3.3 *Characterizing hydraulic performance*

Relationships describing the dependence of emitter discharge upon passage length, passage diameter, pressure differential, and viscosity (water temperature) have been derived theoretically for the three principal types of emitters. However, a simpler and more universal form to describe flow variation with key dependent factors is desirable. The simplest forms which adequately characterize the flow are presented in the following sections.

2.1.3.a Emission exponent

The emission exponent is the parameter which characterizes the flow from an emitter as a function of operating pressure (pressure differential). The equation is given by

$$q_e = kH^x \tag{2.1.11}$$

where q_e = emitter flow, k = an emitter constant which includes factors to make units consistent, H = operating pressure, and x = emission exponent.

The most common method of determining the values of k and x is to perform the linear regression on the logarithms of flow and operating pressures. Using equation 2.1.11

$$\ln(q_e) = x\ln H + \ln k$$

which is of the linear form

$$y = mZ + b$$

where $y = \ln(q_e)$, $Z = \ln H$, $m = x$, and $b = \ln k$.

A linear regression of lnH on $\ln(q_e)$ will yield the values for m and b. Recalling that $b = \ln k$, then $k = e^b$, and equation 2.1.11 can be expressed as

$$q_e = e^b H^m. \qquad (2.1.12)$$

The emission exponent, x (or m from the regression equation), is a measure of the variation of flow from the emitter due to pressure change. It is also possible to estimate x from the discharges from two different operating pressures.

$$x = \frac{\ln(q_1/q_2)}{\ln(H_1/H_2)} \qquad (2.1.13)$$

where q_1 and q_2 are the emitter flows at pressures H_1 and H_2, respectively.

For laminar flow, x would be expected to be 1.0 (refer to equation 2.1.8). In the turbulent flow regime without orifice discharge x would be expected to be 4/7 = 0.57 (equation 2.1.9), and for orifice controlled discharge x would be 0.50 (equation 2.1.10). Reasonable values of x lie between 0.0 (no flow variation with pressure) and 1.0. Values of x greater than 1.0 or less than 0.0 are possible where elastomeric or moving parts are used in the flow passages. It is also possible that a rather poor regression fit would be obtained when the emitter operates in the transition regime somewhere in its operating pressure range.

2.1.3.b Coefficient of manufacturing variation

The coefficient of variation, CV, is a statistical parameter expressed as $CV = s/q_{avg}$ where s is the standard deviation of flow and q_{avg} the mean flow for a sampled number of emitters of the same type tested at a fixed pressure and temperature (20°C).

A parameter which can be used as a measure of emitter flow variation caused by variation in manufacturing of the emitter is called the coefficient of manufacturing variation and CV serves as that parameter. Common causes of manufacturing variation are the inability to hold dimensional tolerances due to molding pressure and temperature, variation in material used, mold parting line flash, welding and gluing flash, and mold wear. The extent to which the manufacturer is able to control variation depends not only upon manufacturing and materials quality control but also on emitter design. The mode of operation of the emitter is also quite significant. Generally, the most critical dimension of an emitter to maintain is the flow passage diameter, d. With laminar and orifice emitters, flow varies with d^2, but turbulent emitter flow varies with $d^{19/7}$. Consequently a 2% change in d results in a 4% change in flow in laminar and orifice emitters, whereas, a 6% change occurs in the turbulent emitter. As a result it would be expected that the manufacturing coefficient of variation would be the greatest for turbulent emitters. Reasonable ranges of CV run from 0.01 to 0.15, and an interpretation of values is given in table 2.1.2.

TABLE 2.1.2

Interpretation of manufacturing variation values (after Solomon, 1979, and American Society of Agricultural Engineers, 1984).

Coefficient of variation CV	Interpretation
0.05 or less	Good
0.05 - 0.10	Average
0.10 - 0.15	Marginal
0.15 or more	Unacceptable

2.1.3.c Temperature sensitivity

The dependency of emitter flow on temperature of the water is known as temperature sensitivity and is caused by the change in viscosity of water with temperature. The kinematic viscosity of water at various temperatures is shown in table 2.1.3, and it can be seen that viscosity decreases with increasing temperature.

Equations 2.1.8, 2.1.9, and 2.1.10 show that flow can vary with viscosity (and temperature). The manner in which flow varies with viscosity depends upon the way flow is controlled. Orifice emitter flow is insensitive to viscosity (and temperature). Turbulent emitter flow increases as viscosity decreases (temperature increases), and laminar emitter flow also increases as viscosity decreases, but at a much greater rate than that of turbulent emitter flow.

		laminar	turbulent	orifice
$\left(\frac{q_{70}}{q_{20}}\right)_c$	=	$(1.005)^2$	$(1.005)^{19/7}$	$(1.005)^2$
	=	1.010	1.014	1.010

Flow change due to viscosity change varies with ν^{-1}, $\nu^{-1/7}$, and ν^0.

$\left(\frac{q_{70}}{q_{20}}\right)_\nu$	=	$\left(\frac{0.413}{1.003}\right)^{-1}$	$\left(\frac{0.413}{1.003}\right)^{-1/7}$	$\left(\frac{0.413}{1.003}\right)^{0}$
	=	2.429	1.135	1.000

Expansion final factor

$\left(\frac{q_{70}}{q_{20}}\right)_c \left(\frac{q_{70}}{q_{20}}\right)_\nu$	=	2.453	1.151	1.010

Contraction final factor

$\left(\frac{q_{70}}{q_{20}}\right)_\nu / \left(\frac{q_{70}}{q_{20}}\right)_c$	=	2.405	1.119	0.990

The parameter m can be estimated by

$$\frac{100\ (\text{final factor} - 1)}{dT}$$

m (expansion)	2.906	0.302	0.020
m (contraction)	2.810	0.238	-0.020

2.1.4 *Types*

Emitters are usually categorized by their modes of operation, but because of the numerous variations in how emitters can be constructed further differentiation is necessary. Since clogging is a major problem with emitters, many manufacturers have decided to modify one of the three basic modes of operation to allow flushing of foreign matter. In order to accomplish flushing, moving parts, flaps, and elastomeric materials are used. Other manufacturers have attempted to make the emitter flow less sensitive to pressure change by using a series of orificies, moving parts, or elastomers.

2.1.4.a Laminar (long path) emitters

The long path emitter is the simplest of the emitter devices but the most difficult to utilize in a trickle system design. Typical long path emitters are either microtubes or spiral long path emitters as shown in figures 2.1.1a, b, and c. In some cases, an elastomeric flap on the pressurized side gives flow compensation by reducing the effective flow passage diameter with increasing pressure.

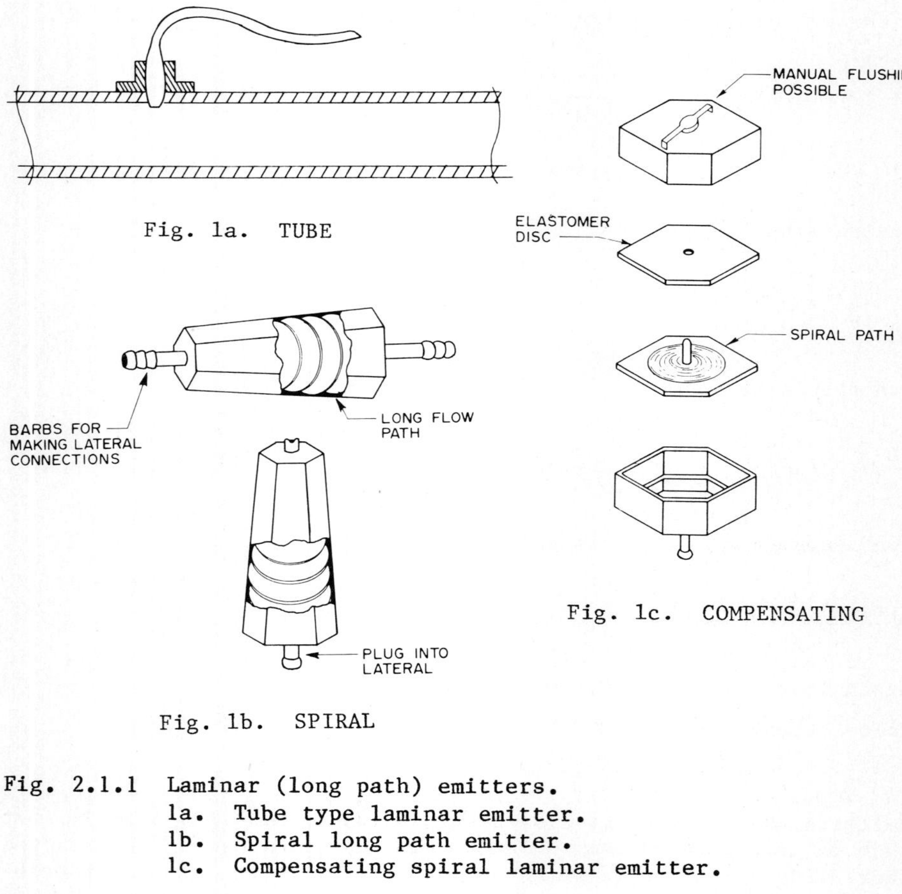

Fig. 1a. TUBE

Fig. 1b. SPIRAL

Fig. 1c. COMPENSATING

Fig. 2.1.1 Laminar (long path) emitters.
1a. Tube type laminar emitter.
1b. Spiral long path emitter.
1c. Compensating spiral laminar emitter.

2.1.4.b Turbulent emitters

There are two primary types of turbulent emitters. One is a modification of a long path emitter with a shortened maze-like flow passage. This emitter, often called a labyrinth or tortuous long path emitter maintains turbulence with continuously changing flow direction. The second type of turbulent flow emitter is the short path type incorporating elastomeric flaps which decrease the flow passage diameter under increasing pressure. Examples of both types are presented in figures 2.1.2a, b, and c.

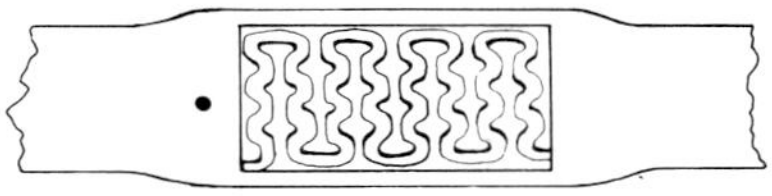

Fig. 2a. LABYRINTH

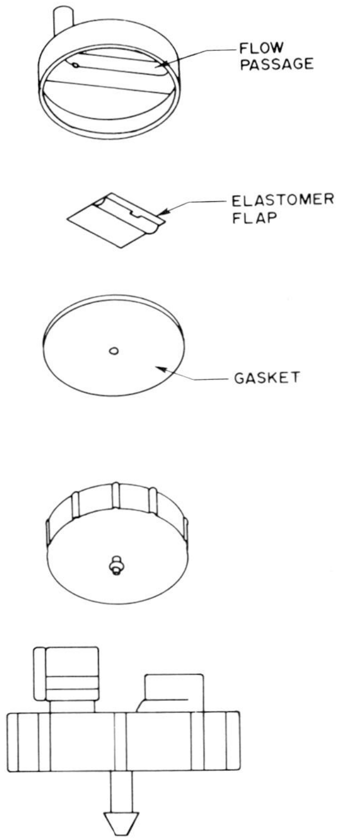

Fig. 2b. GROOVE AND FLAP

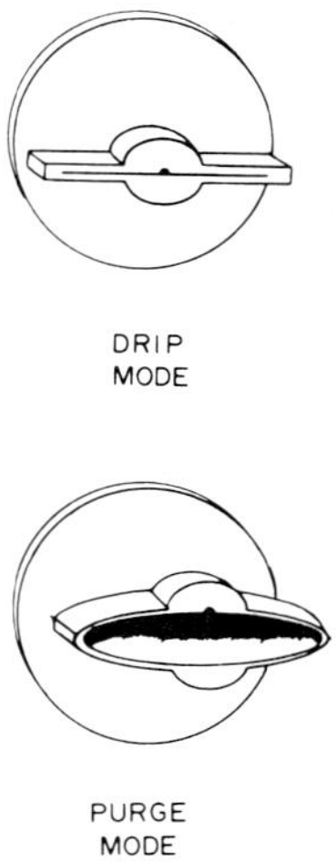

Fig. 2c. FLAPPER

Fig. 2.1.2 Turbulent emitters.
2a. Labyrinth turbulent emitter.
2b. Groove and flap compensating turbulent emitter.
2c. Flapper flushing turbulent emitter.

2.1.4.c Orifice emitters

The most common orifice emitters are the simple orifice, the vortex orifice, and multiple orifices in series. Examples of each of the orifice emitters are shown in figures 2.1.3a, b, c, and d. The multiple flexible orifice emitter has the advantage of being self flushing when an orifice becomes plugged, but has the disadvantage of a high discharge coefficient because the orifice expands and enlarges as the operating pressure is increased.

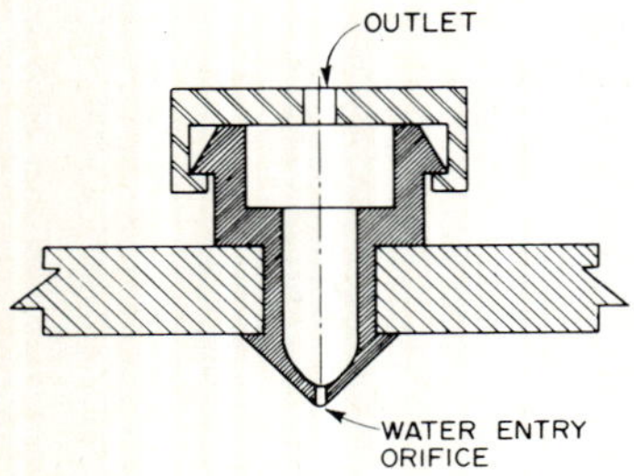

Fig. 3a. ORIFICE

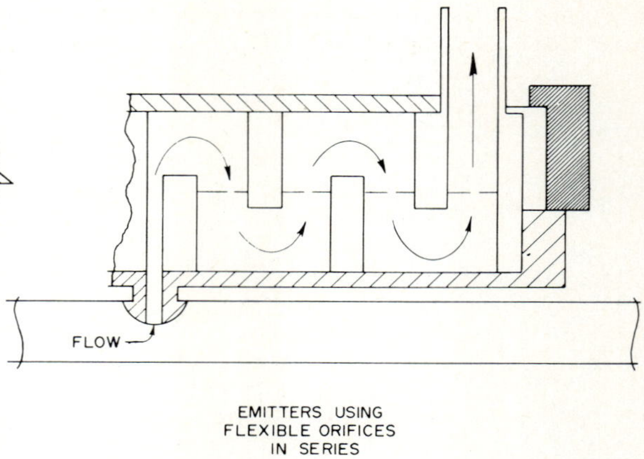

Fig. 3c. MULTIPLE ORIFICE

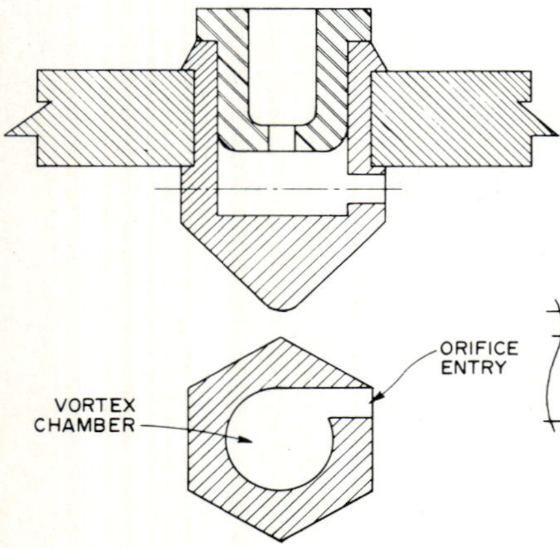

Fig. 3b. ORIFICE-VORTEX

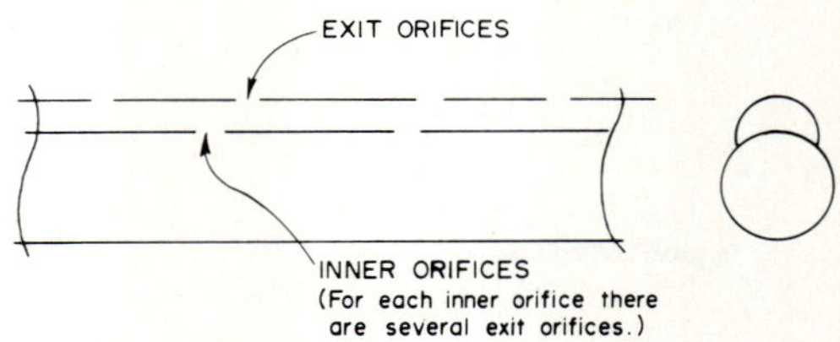

Fig. 3d. TWIN-WALL LATERAL

Fig. 2.1.3 Orifice emitters.
3a. Simple orifice type emitter.
3b. Vortex chamber orifice emitter.
3c. Multiple orifice emitter.
3d. Twin wall (two chambers) emitter.

2.1.5 *Molding methods*

The molding or forming process is important in producing the desirable characteristics of the emitter. Plastics have been found to be the best construction materials for emitters. There are many different kinds of plastics from which to choose, and the forming method also depends upon the material chosen. Nearly all emitters are now being molded from the family of plastics known as thermoplastics which are defined to be plastics capable of being repeatedly softened by increases in temperature and hardened by decreases in temperature with the changes being physical rather than chemical. The two practical methods of molding thermoplastics are discussed in the following sections.

2.1.5.a Injection molding

Injection molding is the process where a powdered or granular plastic resin is heated to plasticity in a cylinder at controlled temperature, forced under controlled pressure through sprues and runners into cavities of a mold. The resin solidifies rapidly and the mold is opened and the parts removed. It is adaptable to rapid production rates, requires little finishing, gives good dimensional accuracy, and hence leads to low part cost. It does require relatively expensive tools and dies and is therefore not suitable to small runs. An illustration of the injection molding process is given in figure 2.1.4.

To maintain uniformity of parts, temperature and pressure controls are important in the molding process. The speed of the cycle is determined by how rapidly the material can be cooled, and varies from one material to another with the thermal conductivity of the material. The speed at which the mold can be cycled is a very significant part of the cost of producing parts.

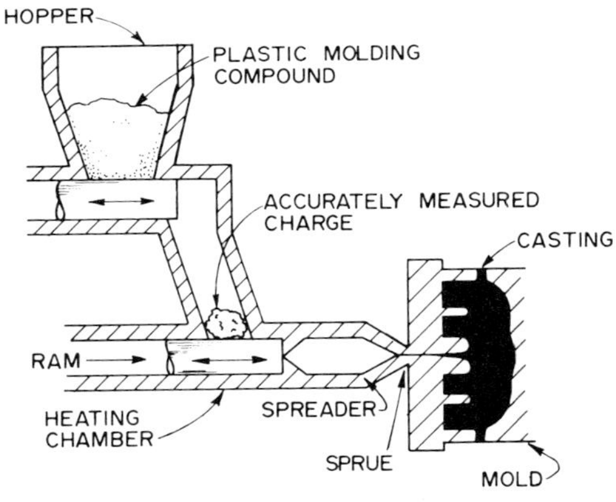

Fig. 2.1.4 Diagrammatic representation of injection molding process.

2.1.5.b Compression molding

Compression molding with thermoplastic materials involves placing a preformed piece of material into a heated mold cavity, closing the cavity, and applying heat and pressure. The material flows and fills the cavity and it is reopened for part removal. Although little finishing is necessary, fabricating extremely intricate parts with this

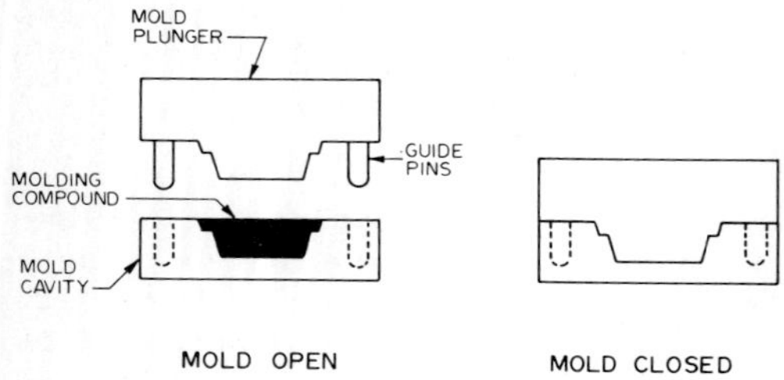

Fig. 2.1.5. Diagrammatic representation of compression molding process.

method is not practical. Furthermore, it is difficult to hold tolerances. An illustration of this method is presented in figure 2.1.5.

The capital investment to produce parts with compression molding is generally less than required with injection molding, but the inability to hold tolerances as well may limit the applications of compression molding.

2.1.5.c Other molding methods

Other forming or molding methods are available for emitter construction but are not presently being used. Calendaring is the process of working a mass of material into a uniform sheet by passing it over and through a series of rollers. This process, often used to form elastomeric parts, requires that the parts then be cut to size.

Blow molding involves placing a heated thermoplastic material between two halves of an open split mold and expanding the material against the sides with air pressure. This process is limited to hollow or tubular parts and tolerances are often hard to hold.

Cold forming is similar to compression molding but does not require heat. The parts are formed by pressure. Parts are cured in an oven in a separate process. This process is limited to rather simple shapes and it is often difficult to hold the exact shape.

2.1.6 ***Materials***

Emitters are subjected to extremes in environmental conditions including light and temperature, to chemicals run through them or applied to them as the crop is chemically treated, and physical abuse. They must be resistant to their environment and not change chemically or physically or the flow rate and life expectancy of the emitters will be adversely affected.

The materials chosen must maintain the same characteristics over their lifetime. This is especially true with elastomers where the performance of the emitter depends upon the consistent maintenance of properties and dimensions. All thermoplastics and elastomers have a tendency to deform under load. The dimensional change with time of a material under load is known as creep. Creep at room temperature is known as cold flow. If a material has a tendency to creep under small loads, it will be less useful as an emitter material. Hardness is a property which describes the local deformation (indentation) of a material under load and is measured either by a Rockwell hardness number or shore hardness using a durometer. The American Society for Testing and Materials (ASTM, 1982) has established numerous standards for measuring properties of plastics and should be referred to for specific details.

Plastic materials degrade by photo-oxidation with the partial breakdown of long molecules into shorter less strong components. This is largely due to ultraviolet (UV) radiation and water. This degradation is reflected in a tendency for plastics to become brittle. As a result, nearly all plastics have a UV stabilizer (UV absorbing agent) such as carbon black.

2.1.6.a Acetals, polyethylenes, and polypropylenes

These three types of materials are the most frequently used to mold the body or rigid parts of emitters. All three exhibit good chemical resistance characteristics and are highly adaptable to injection molding.

Acetals

Acetal copolymers are prepared by copolymerization of trioxane ($C_3H_6O_3$) with small amounts of a comonomer which randomly distributes C-C bonds in the polymer chain. They are strong, hard, highly crystalline thermoplastics used where metals have formerly been used and are known as engineering plastics. Other common acetals are the acetal homopolymers composed of linear polymers of formaldehyde, CH_2O. They are highly crystalline engineering thermoplastics. Both types of the acetals are strong, have good resistance to fatigue, and have high melting points and can be rapidly cycled in injection molding. They also have excellent creep resistance and are highly chemical resistant.

Polyethylenes

Polyethylene (PE) is produced by addition polymerization of ethylene (C_2H_4). There are many kinds of polyethylenes available with the most common ones used in emitter construction being Low Density Polyethylene (LDPE) and High Density Polyethylene (HDPE). The characteristics of PE are chemical resistance to solvents, acids and alkalines, toughness and flexibility, and low cost. Most drip applications are of the LDPE type, especially for lateral tubing. The amount of crosslinking is affected by radiation, so some manufacturers irradiate the tubing to improve its strength. PE is susceptible to stress cracking and the rate is dependent upon the amount of stress and temperature. Agents such as metallic soaps and sulfated or sulfonated alcohols accelerate cracking. This cracking can be minimized by correct composition of polyethlene and design to reduce stress.

Polypropylenes

Polypropylene is a linear polymer of propylene (C_3H_6). Their properties are quite similar to those of polyethylene except that they have greater rigidity and heat resistance and are less resistant to environmental stress cracking than PE. They also have low moisture absorbency. (Any absorption of water is undesirable as it leads to swelling and changes in critical dimensions.). The properties of these thermoplastics are summarized in table 2.1.5 and their polymers are shown in figure 2.1.6.

Strengthening Fillers

The structural strength of themoplastics can often be increased by adding fillers which reinforce the plastics. Two fillers commonly used are glass fibers and talc. The property changes of glass filled materials are listed in table 2.1.6.

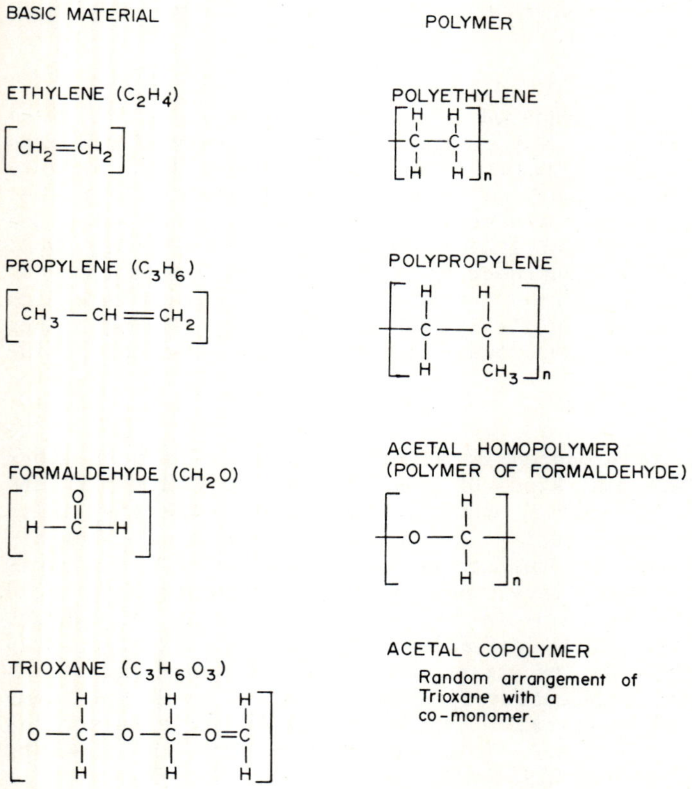

Fig. 2.1.6 Chemical composition of polymers of thermoplastics.

TABLE 2.1.5

Properties of thermoplastics used in the construction of trickle emitters (adapted from Modern Plastics Encyclopedia, 1982).

	Acetal				Polyethylene		Polypropylene			
	Homopolymer	Copolymer	20% Glass Homopolymer	25% Glass Copolymer	Low Density	Medium Density	Homopolymer	Copolymer	40% Talc-Filled	40% Glass-Filled
Melting Temp, °F	181	175	181	175	95-130	120-140	168	160-168	158-168	168
Injection Molding Temp Range, °F	380-470	360-450	350-480	380-480	300-450	300-450	400-550	400-550	410-550	500-550
Molding Press, 10^3 psi	10-20	10-20	10-20	10-20	5-10	5-15	10-20	10-20	10-20	10-20
Shrinkage, in/in	.020-.025	.020	.009-.012	.004-018	.015-.050	.015-050	.010-.025	.020-.025	.008-.015	.003-.005
Tensile Strength at break, psi	—	—	8,500	18,500	600-2300	1200-3500	4500-6000	4000-5000	4300-5000	8500-15000
Elongation at break, %	25-75	40-75	7	3	90-800	40-600	100-600	200-700	3-8	2-4
Flexural Strength, psi	14,000	13,000	15,000	28,000	—	—	6000-8000	5000-7000	8500-9000	10500-22000
Impact Strength, psi	1.3-2.3	1.0-1.5	.08	1.8	No break	.5-16	.4-1.0	1.0-20.0	.4-.6	1.4-2.0
Coeff. of thermal expansion 10^{-6} in/in/°C	100	85	36-81	—	100-220	140-160	81-100	68-95	55-80	27-32
Thermal conductivity, 10^{-4} cal-in/sec-cm^2 °C	5.5	5.5	—	—	8	8-10	2.8	3.5-4.0	7.6	8.4-8.8
Water absorption, %,24 hr	.24-.40	.22	.25	.27	<.01	<.01	.01-.03	03	.01-.03	.05-.06

TABLE 2.1.6

Important properties of raw and glass-filled thermoplastics (adapted from Harper, 1975).

Material	Percent loading	Tensile strength at 73°F $lb/in^2 \times 10^{-3}$	Elongation at 73°F %	Flexural strength at 73°F $lb/in^2 \times 10^{-3}$	Impact strength at 73°F notched Izod	Rockwell hardness	Specific gravity
Polyacetal:							
Raw	—	10.0	15	14	1.4	M94-R120	1.425
Short fiber	20	10-13.5	2-3	14-15	0.8-1.4	M70-75-95	1.55
Long fiber	20	10.5	2.3	15	2.2	M75-80 (Shore)	1.55
Polyethylene:							
Raw	—	1.2	50-600	4.8	0.5-16	D50-60	0.92-0.94
Short fiber	20	6	3.0	7	1.1	R60	1.10
Long fiber	20	6.5	3.0	8	2.1	R60	1.10
Polypropylene:							
Raw	—	4.3	200-700	6	—	R85-110	0.90-0.91
Short fiber	20	6.0	3.0	7.5	1.0	M40	1.05
Long fiber	20	8	2.2	10	3.5	M50	1.05

2.1.6.b Elastomers

Elastomers are defined as materials which at room temperature can be stretched repeatedly to at least twice their original length and upon release will return to their approximate original length. They are most commonly used in emitter parts which deform under pressure to reduce the flow passage diameter and to give some degree of pressure compensation. Emitters with these operational characteristics are illustrated in figures 2.1.1c, 2.1.2b, 2.1.2c, and 2.1.3d which all use elastomers.

Properties of elastomers which are important to emitter construction are hardness, mechanical strength, resistance to chemicals, abrasion, and tear. Past experiences of emitter manufacturers have shown difficulties with several elastomeric materials such as natural rubber (chemical resistance), Buna N (tensile strength and tear resistance), and butyl rubber (chemical resistance). Most manufacturers have now selected one of several types of silicone elastomers primarily because of their chemical resistance. Some of the key properties of elastomers are listed in table 2.1.7.

2.1.7 ***Bonding and assembly methods***

Emitters generally are designed so that some inner flow passages (perhaps with an elastomer incorporated) are contained within a general body. Because the flow passage must be molded inside, most emitters are molded in parts and assembled. An important part of the design is how the emitter parts are held together. Several methods of bonding can be used, and some are discussed in the following sections.

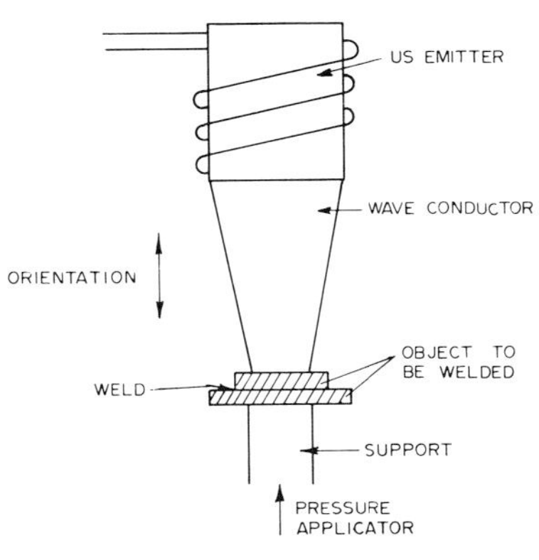

Fig. 2.1.7 Diagrammatic representation of ultrasonic welding.

TABLE 2.1.7

Properties of elastomers used in the construction of trickle emitters (adapted from Harper, 1975).

Property	Natural rubber (cis-polyisoprene)	Butadiene-acrylonitrile (nitrile) (Buna N)	Chloroprene) (neoprene)	Butyl (isobutyleneisoprene	Silicone (polysiloxane)
Physical Properties:					
Specific gravity (ASTM D 792)	0.93	0.93	1.25	0.90	1.1-1.6
Thermal conductivity, Btu/(h)(ft^2)(°F/ft) (ASTM C 177)	0.082	0.143	0.112	0.053	0.13
Coefficient of thermal expansion (cubical), 10^{-5} per °F (ASTM D 696)	37	39	34	32	45
Electrical insulation	Good	Fair	Fair	Good	Excellent
Flame resistance	Poor	Poor	Good	Poor	Good
Min recommended service temp, °F	-60	-60	-40	-50	-178
Max recommended service temp, °F	180	300	240	300	600
Mechanical properties:					
Tensile strength, lb/in^2:					
Pure gum (ASTM D 412)	3,500	900	4,000	3,000	600-1,300+
Black (ASTM D 412)	3,500-	3,000-	3,000-	2,500-	———
Elongation, %:					
Pure gum (ASTM D 412)	750-850	300-700	800-900	750-950	100-500+
Black (ASTM D 412)	550-650	300-650	500-600	650-850	———
Hardness (durometer)	A30-90	A40-95	A20-95	A40-90	A30-90
Rebound:					
Cold	Excellent	Good	Very good	Bad	Very Good
Hot	Excellent	Good	Very good	Very good	Very Good
Tear resistance	Excellent	Good	Fair to good	Good	Fair
Abrasion resistance	Excellent	Good to Excellent	Good	Good to Excellent	Poor
Chemical resistance:					
Sunlight aging	Poor	Poor	Very good	Very good	Excellent
Oxidation	Good	Good	Excellent	Excellent	Excellent
Heat aging	Good	Excellent	Excellent	Excellent	Excellent
Solvents:					
Aliphatic hydrocarbons	Poor	Excellent	Good	Poor	Fair
Aromatic hydrocarbons	Poor	Good	Fair	Poor	Poor
Oxygenated, alcohols	Good	Good	Very good	Very good	Excellent
Gasoline	Poor	Excellent	Good	Poor	Poor
Animal, vegetable oils	Poor to good	Excellent	Excellent	Excellent	Excellent
Acids:					
Dilute	Fair to good	Good	Excellent	Excellent	Very Good
Concentrated	Fair to good	Good	Good	Excellent	Good
Permeability to gases	Low	Very low	Low	Very low	High
Water-swell resistance	Fair	Excellent	Fair to Excellent	Excellent	Excellent

2.1.7.a Ultrasonic welding

This method is the most commonly used, and involves application of a high frequency sound vibration transmitted through a metal horn. The vibrations generate friction at the bond area melting the plastics enough to form a bond. An illustration of the method is shown in figure 2.1.7. The method works well for most thermoplastics and can be done quickly. Strong bonds are achievable. In order to get strong bonds, care must be taken in the design to allow for a ridge to be melted to form the joint and to ensure that weld flash does not interfere with a flow passage. Several types of joint designs are given in figure 2.1.8.

2.1.7.b Spin welding

Parts are bonded in spin welding by spinning one part against the other developing friction at the bond area in much the same fashion as ultrasonic welding. When the spinning stops, the parts cool and are bonded. The method is fast and forms strong bonds, but the bond area must be circular. Capital setup cost for such an operation can be very expensive.

2.1.7.c Solvent welding and adhesives

Use of solvent welding is not recommended for polyethylenes, polypropylenes, and acetal homopolymers and is not currently practiced in the trickle industry. Acetals, nylons, polyethylenes, and polypropylenes can bond with adhesives, but it is rarely done with emitters. Hot melt adhesives to bond similar materials are used extensively in the line source type emitters.

2.1.7.d Interlocking parts

Design of emitters so that the parts snap together is a rather common assembly technique. The major disadvantages arise from difficulty in maintaining a water-tight seal and from localized stresses on the molded fastener parts.

2.1.8 ***Attachment to lateral pipe and hydraulic implications***

Most discrete emitters are not part of the lateral tubing and must be attached to it. The method of attachment affects the hydraulics of the lateral line and consequently affects the flow from emitters along the line.

2.1.8.a On-line emitter attachment

On-line attachment is accomplished by punching a hole in the wall of the lateral tubing and inserting the inlet barb of the emitter into the hole. The lip of the barb protrudes (see figure 2.1.9) into the lateral tubing and causes some friction loss. The amount of friction loss depends upon the flow in the lateral, the size of the lateral, and the projected area of the barb. The barbs are generally one of two sizes. The larger projects an area of about 40 mm^2 and the smaller projects an area of about 20 mm^2. In order to account for the

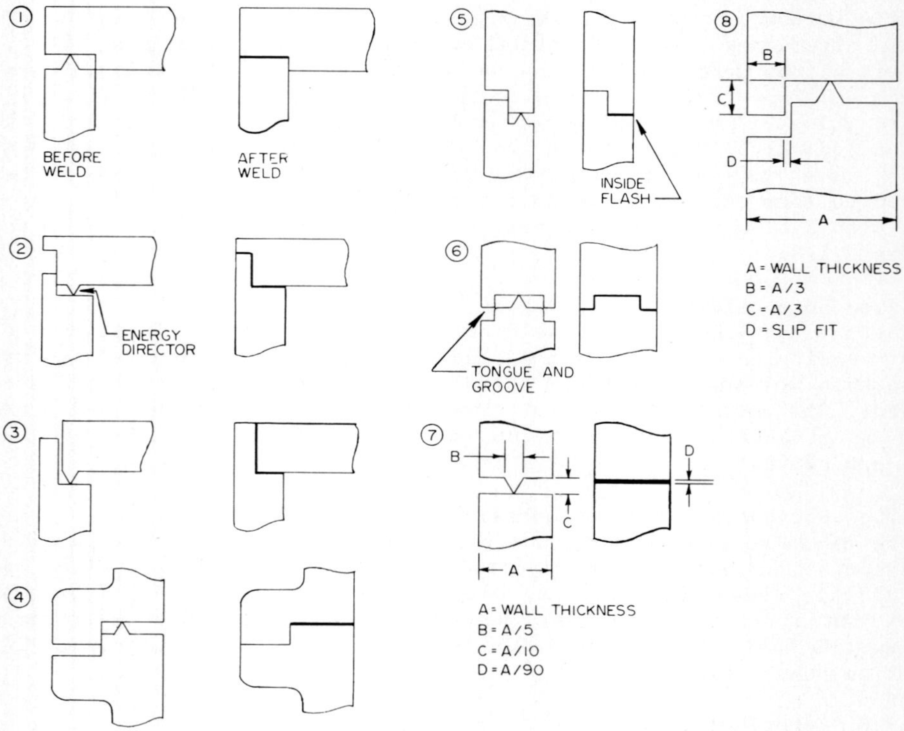

Fig. 2.1.8 Common joint designs for ultrasonic welding.

presence of barbs when calculating friction loss, an equivalent length of pipe must be added so that a new length includes the actual length and the new length. The larger barb results in about 50% more length being added than for the smaller barb.

2.1.8.b In-line attachment

Some emitters are attached directly into the lateral tubing, and in some cases the lateral tubing forms the outer wall of the emitter (see figure 2.1.2a). An usual method of in-line attachment is with barbed ends on the emitter which can be used as a coupling. Although the in-line attachment offers an advantage in that it does not protrude out from the lateral, the friction loss due to in-line emitters has been measured to be as much as 10 times that of the small barb on line emitters.

TABLE 2.1.8

Characteristics of hydraulic performance of selected trickle emitters.

Type	Figure Reference	Exponent X	Manufacturing Variability V	Temperature Sensitivity m	Minimum Flow Passage (mm)	Barb Size	Body Material	Flap Material	Bonding
Theoretical									
Long Path (Laminar)	2.3.1a,b,c	1.00	--	2.86					
Turbulent	2.3.2a,b,c	0.57	--	-0.192					
Orifice	2.3.3a,b,c,d	0.50	--	0.0					
Measured Data									
Long Path (Laminar)									
Small Tube	2.3.1a	0.70	0.05	0.290	1.0	NA	PE	NA	PE
		0.80	0.05	0.495	1.0	NA	PE	NA	
Spiral Long Path	2.3.1b	0.75	0.06	0.760 to 0.415[a]	0.8	--	--	--	
Lake 1000		0.81	--	1.20 to 0.003[a]	0.9	M	PP	NA	TH
Turbulent									
Short Path									
Groove and Flap	2.3.2b	0.33	0.02	0.0	0.3	S	AH	EPDM	TH
Global STF1		0.07	0.10	0.001	--	S	PPG	SI	USW
Labyrinth									
Olson 1500		0.46	0.02	0.0		S	PP	NA	USW
Lake 1500									
Drip In		0.54	0.01	0.001	1.0	IL	LDPE	NA	--
Agrifim		0.50	0.02	0.0	--	IL	PP	NA	--
Orifice									
Vortex	2.3.3b	0.41	0.05	--					
		0.24	0.11						

[a] Range of values indicating emitter operation in transition regime.

AH = Acetal homopolymer
EPDM = Ethylene-propylene elastomer
IL = In-line
LDPE = Low density polyethylene
M = Medium
NA = Not applicable
PE = Polyethylene
PP = Polypropylene
PPG = Polypropylene glassed filled
S = Small
SI = Silicone
TH = Threaded
USW = Ultrasonically welded

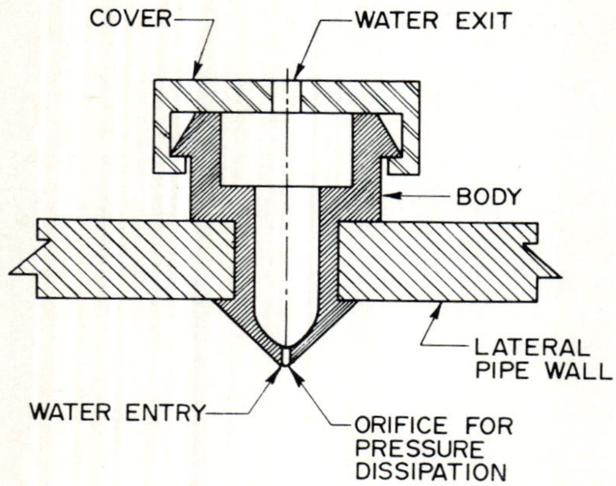

Fig. 2.1.9 Emitter barb protrusion into lateral line.

2.1.9 *Hydraulic performance*

The hydraulic performance of an emitter is the composite result of its design, materials, and manufacturing. The effects of mode of operation, type of emitter, characteristics of materials, and methods of molding and bonding have been discussed. The characteristics of hydraulic performance which take into account all the variables are listed in table 2.1.8. It is worthy of note that many emitters do not perform according to their classification. The laminar emitters tend to perform more like the theoretical turbulent emitters, and the turbulent emitters perform more like the theoretical orifice emitters. Many of the laminar emitters do become turbulent with temperature and/or pressure increase, and some turbulent emitters may have some limiting flow passage which cause them to act as orifice emitters.

PROBLEMS

1. Derive equation 2.1.9.

2. Determine the emitter constant for a turbulent flow emitter with a 4 L/hr flow rate and an emitter exponent of 0.57 operating at 75 kPa. Answer: 0.34

3. Determine the pressure needed for a laminar flow emitter with passage length of 10m, inside diameter of 1.0 mm flowing 4 L/hr at 30°C. Answer: 36.9 m or 179 kPa

4. Compute the coefficient of manufacturing variation for the following set of emitter flow rates. How would you interpret the results?

Emitter #	Emitter flow rate (L/hr)
1	4.3
2	4.0
3	3.2
4	3.8
5	4.1
6	4.0
7	3.8
8	4.2
9	3.9
10	3.7

Answer: CV = 0.079
Interpretation: Average

5. Compare the temperature sensitivity of an orifice type emitter made from medium density polyethylene with that of a 40% glass filled polypropylene. Assume the temperature varies from 20°C to 40°C and that the orifice contracts with increasing temperature.

6. Discuss some of the advantages and disadvantages of laminar, turbulent, and orifice type emitters.

7. Discuss the theoretical dependence of emitter discharge with respect to flow passage length, flow passage diameter, pressure differential and viscosity (water temperature).

8. Discuss the importance of a low coefficient of manufacturing variation and the causes of manufacturing variation in emitter construction.

9. Compare the relative importance of manufacturing and water temperature variations for the three main types of emitters.

10. Describe and discuss briefly the differences among the three types of materials that emitters are commonly manufactured.

REFERENCES AND SUPPLEMENTAL READINGS

American Society of Agricultural Engineers, 1984. Design, Installation, and Performance of Trickle Irrigation Systems. ASAE EP405, Amer. Soc. Agric. Eng., St. Joseph, Michigan, pp. 502-505.

American Society of Testing and Materials, 1982. Annual Book of Standards, part 36, Plastics. ASTM, Philadelphia, Pennsylvania.

Anonymous, 1972. How to Fasten and Join Plastics. Materials Engineering 75: 38-45.

Bralts, V.F., Wu, I.P. and Gitlin, H.M., 1981. Manufacturing variation and drip irrigation uniformity. Trans. Amer. Soc. Agric. Eng. 24(1): 113-119.
Bucks, D.A. and Myers, L.E., 1973. Trickle irrigation: Application uniformity from simple emitters. Trans. Amer. Soc. Agric. Eng. 16(6): 1108-1111.
Daugherty, R.L. and Franzini, J.B., 1977. Fluid Mechanics with Engineering Applications. McGraw-Hill, New York, 564 pp.
DuBois, P., 1978. Plastics in Agriculture. Applied Science Publishers, London, 176 pp.
Engineer's Guide to Plastics., 1971. Materials Engineering 75: 38-45.
Guide to Plastics., 1980. McGraw Hill, New York, 287 pp.
Harper, C.A., 1975. Handbook of Plastics and Elastomers. McGraw Hill, New York, 1008 pp.
Howell, T.A. and Barinas, F., 1980. Pressure losses across trickle irrigation fittings and emitters. Trans. Amer. Soc. Agric. Eng. 23(4): 928-933.
Hughes, T.C. and Jeppson, R.W., 1978. Hydraulic friction loss in small diameter plastic pipelines. Water Resour. Bull. 14: 1159-1166.
Karmeli, D. and Keller, J., 1975. Trickle Irrigation Design. Rain Bird Sprinkler Manufacturing Corp., Glendora, California, 133 pp.
Modern Plastics Encyclopedia., 1982. McGraw Hill, New York, Vol. 59, No. 10A, 840 pp.
Parchomchuk, P., 1976. Temperature effect on emitter discharge rates. Trans. Amer. Soc. Agric. Eng. 19(3): 690-692.
Shipston, C., 1976. Familiarization with polyethylene. Drip/Trickle Irrig. 1: 6-10.
Solomon, K., 1977. Evaluation criteria for trickle irrigation emission devices. Proc. 4th Ann. Int. Drip Irrig. Assoc. Meet., Fresno, California, pp. 65-76.
Solomon, K., 1979. Manufacturing variation of trickle emitters. Trans. Amer. Soc. Agric. Eng. 22(5): 1034-1038, 1043.

CHAPTER 2

DESIGN PRINCIPLES

2.2 SYSTEM DESIGN

I. P. WU, H. M. GITLIN, K. H. SOLOMON, AND C. A. SARUWATARI

Trickle irrigation uses a pressurized pipe network similar to sprinkler irrigation, but with generally lower operating pressures and rates of application. The trickle irrigation pipe network consists of laterals, submain and main lines. The lateral line is usually a plastic tubing combined with emitters or simply a small thin-wall tubing with outlets or orifices. The laterals are designed to distribute irrigation water throughout the field with an acceptable degree of uniformity. The submain delivers water to the laterals and also needs adequate designing to achieve uniform flow to the lateral lines. The main line serves as a conveyance system for delivering the total amount of water at the required water pressure for the irrigation system.

2.2.1 *Hydraulics*

The ideal trickle irrigation system is one in which all emitters (or orifices) deliver the same volume of water in a given irrigation time so that each plant would receive the same quantity of water in an irrigation period. From a practical point of view, it is impossible to achieve this idealized performance requirement because the emitter (or orifice) flow will be affected by variations in water pressure and manufacturing characteristics. The emitter flow variation caused by water pressure variation in a trickle irrigation system can be controlled by hydraulic design and is called hydraulic variation. The emitter flow variation caused by manufacturing inconsistencies is designated as manufacturer's variation.

2.2.1.a Hydraulics of emitters

Trickle emitters vary from elaborate variable flow rate types to simple orifices which are made by punching, drilling or burning holes in the pipe. In general, the flow rate through the emitter is controlled by the hydraulic pressure at the emitter and the flow path dimensions of the emitter. There are three major groups of emitter types: (a) the orifice or nozzle emitters, (b) the long flow path emitters, and (c) special type emitters such as pressure compensated, vortex and porous-pipe.

The orifice and nozzle types usually have fixed emitter geometry so the flow area is constant. The flow and hydraulic pressure relations are shown to be

$$q_e = kH^{0.5} \qquad (2.2.1)$$

in which q_e = emitter flow rate (L/hr), k = a constant, and H = pressure head (m).

The long flow path type can be represented by flow in a small microtube. If the area of the flow path is fixed, the emitter flow function can be given as a simple power function

$$q_e = kH^x \qquad (2.2.2)$$

in which x = 1 for laminar flow, x = 0.57 when flow is considered as turbulent flow in the smooth pipe, and x = 0.50 when the flow is considered as full turbulent in the small tube. Hydraulic studies of microtube emitters show the following empirical relations (Khatri, Wu and Gitlin, 1979):

<u>Laminar flow</u>

$$q_e = 1.272\ d^{2.70}\ \left(\frac{H}{L}\right)^{0.80} \qquad (2.2.3)$$

<u>Turbulent flow</u>

$$q_e = 1.776\ d^{2.73}\ \left(\frac{H}{L}\right)^{0.56} \qquad (2.2.4)$$

where q_e = the flow rate in L/hr, d = the diameter in mm, H = pressure head in m, and L = microtube length in m. The long path can be built into an emitter such as the continuous spiral groove type design. Since the length and diameter are fixed, the emitter flow and pressure relationship can be expressed by equation 2.2.2. The long path in the emitter can also be designed with different shapes and passages to form the 'labyrinth' type emitter. This emitter usually has an x value of 0.5 or a little larger and is considered a turbulent flow emitter.

Pressure compensating emitters are designed so that the flow area (orifice or nozzle) changes with respect to pressure, decreasing when the pressure increases. The flow area and the hydraulic pressure are related by

$$A = bH^{-y} \qquad (2.2.5)$$

in which A is the orifice area, b and y are two constants in the power function, and using equations 2.2.1 and 2.2.5 the resultant emitter flow function becomes

$$q_e = kH^{0.5-y} \qquad (2.2.6)$$

This shows that the x value in the power function, equation 2.2.2, can be less than 0.5. If the y value is 0.5, the x value will be zero and thus means that the emitter is fully pressure compensating to give a constant flow rate even when the hydraulic pressure changes.

Vortex emitters are orifices which have circular cells. The water enters the cell tangentially through an orifice which causes a jet to form a vortical flow against the cell wall. The resulting pressure loss is usually greater than that of a simple orifice of the same dimension, so that for a given operating pressure and emitter flow rate, the orifice could be larger for a vortex emitter compared to a simple orifice.

Porous-pipe emitters are pipes with many small pores or perforations in the pipe wall that allow water to exit from the pipe. The flow rates from porous pipes depend on the geometry of the material and the applied pressure. Prediction of emitter flow rates based on theory is difficult and empirical methods must be used.

2.2.1.b Hydraulics of lateral line and submain

Flow in the lateral line and submain is hydraulically steady, spatially varied pipe flow with lateral outflows. The flow within a trickle irrigation lateral, submain or main line decreases in the downstream direction. The lateral and submain can be considered as having similar hydraulic characteristics and are designed to maintain a small pressure variation along the line.

2.2.1.b.1 Friction equations for trickle irrigation lines

The trickle irrigation lines made of plastic are usually considered as smooth pipes. Both the Darcy-Weisbach equation for pipe flow and the Williams-Hazen empirical equation can be used to determine friction drop along the lateral line and submain.

In general, the friction drop equation for pipe flow has a simple form

$$\Delta H = aQ^{m}\Delta L \qquad (2.2.7)$$

where ΔH = total energy drop (m), a = constant for a given pipe size and a type of flow (or a special type of empirical equation used), Q = flow rate within the pipe section (L/sec), ΔL is a section length (m), m = 1 for laminar flow, m = 1.75 for turbulent flow in smooth pipe, m = 1.852 for turbulent flow using the Williams and Hazen formula and m = 2 for a fully turbulent flow where the friction coefficient is constant.

The Williams-Hazen equation for smooth pipe (using c = 150, c is the friction factor in the original Williams-Hazen equation) is as follows:

$$\Delta H = 1.135 \times 10^{6} \frac{Q^{1.852}}{D^{4.871}} \Delta L \qquad (2.2.8)$$

in which ΔH and ΔL are expressed in m, Q is the total discharge or flow rate (L/sec) and D is inside diameter of the pipe (mm). Since the discharge in the lateral line or submain decreases with respect to the length, the total friction drop can be determined by using the total discharge and total length of the line as follows:

$$\Delta H = a \frac{Q^m}{m+1} L \tag{2.2.9}$$

where Q is the total discharge at the inlet and L is the total length. Using the Williams-Hazen equation, the friction drop at the end of the lateral or submain can be determined by

$$\Delta H = 3.98 \times 10^5 \frac{Q^{1.852}}{D^{4.871}} L \tag{2.2.10}$$

where ΔH is the total friction drop at the end of the line (m), L is the total length (m), Q is the total discharge at the inlet (L/sec), and D is the diameter (mm). Equation 2.2.10 is only applicable to lines with more than twenty outlets.

Equations 2.2.9 and 2.2.10 can determine only the flow friction in the laterals and submain, the losses through emitter connection and pipe fittings are not included. These losses in the ordinary hydraulic design of pipe flow are usually considered as minor losses. When the outlets are spaced closely as in the lateral line or submain, these minor losses may result in a considerable energy loss along the line. The minor losses are usually expressed as an equivalent length of pipe and given by the equation

$$L_e = 3.43\ H_e\ D^{4.871}\ Q^{-1.852} \tag{2.2.11}$$

where L_e = equivalent length of pipe in mm, H_e = emitter connection or pipe fitting friction loss in m, Q = flow rate in L/hr and D = pipe diameter in mm. The equivalent length for all connections and fittings can be calculated from equation 2.2.11 and combined in the lateral and submain design. The friction losses of connections and fittings H_e are determined by laboratory experiments.

2.2.1.b.2 Energy gradient line for trickle irrigation laterals (or submains)

The total specific energy at any section of a trickle line can be expressed by the energy equation

$$\overline{H} = z + H + \frac{v^2}{2g} \tag{2.2.12}$$

where $\overline{H}$ is the total energy, z is the potential head or elevation, H is the pressure head and $v^2/2g$ is the velocity head all expressed in m. As the flow rate in the line decreases with respect to the length because of emitter discharges from laterals and submain outflows into the laterals, the energy gradient line will not be a straight line but

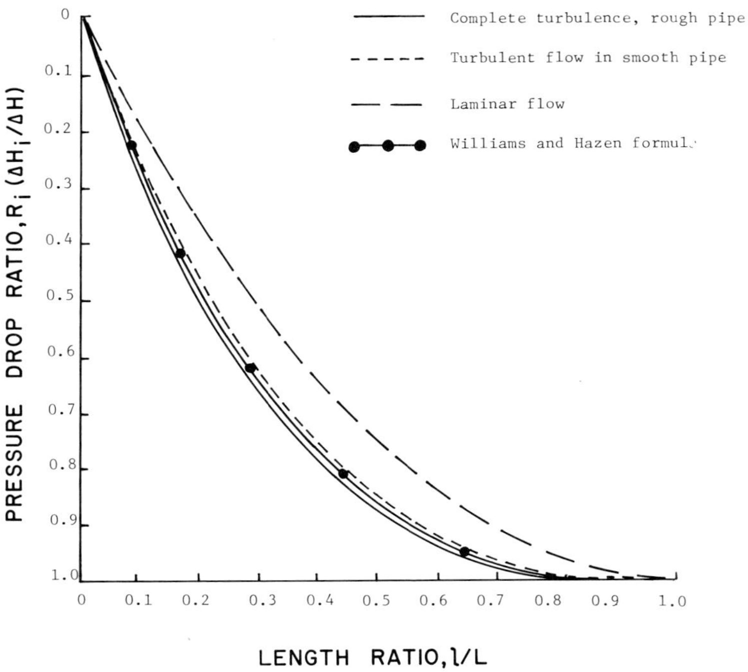

Fig. 2.2.1 Dimensionless curves showing the friction drop pattern caused by laminar flow, flow in smooth pipe, and complete turbulent flow in the lateral line.

an exponential type curve. The shape of the energy gradient line for level irrigation lines can be expressed by a dimensionless energy gradient line

$$R_i = 1 - (1 - i)^{m+1} \tag{2.2.13}$$

in which R_i is $\Delta H_i/\Delta H$ and is called the energy drop ratio, m is the exponent of the flow rate in the friction equation, ΔH_i is the total pressure drop expressed in meters at a length ratio i ($i = \ell/L$), ΔH is the total energy drop at the end of the line, L is the total length of the line and ℓ is a given length measured from the head end of the line. When the Williams-Hazen formula is used for the pipe flow, the dimensionless energy gradient line can be expressed as

$$R_i = 1 - (1 - i)^{2.852} \tag{2.2.14}$$

The dimensionless energy gradient lines for different flow conditions are shown in figure 2.2.1.

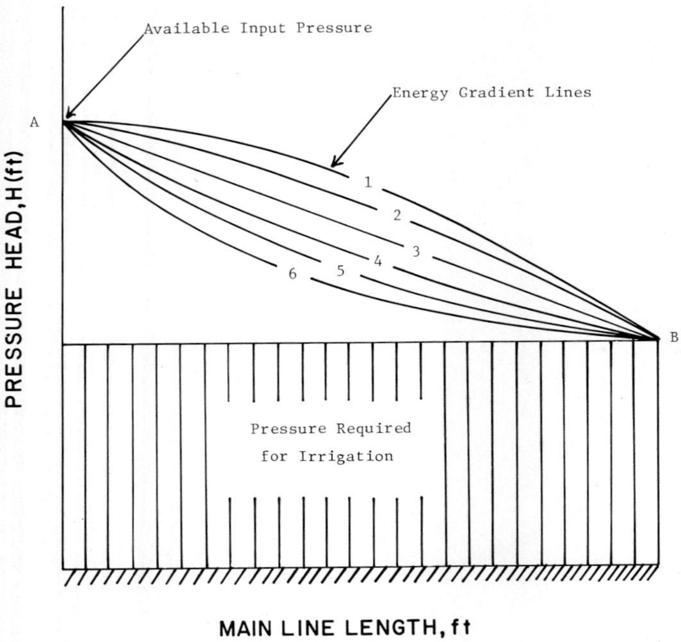

Fig. 2.2.2 Main line profile and energy gradient lines.

The dimensionless energy gradient line will serve to determine the energy gradient curve when the total energy drop ΔH is known.

2.2.1.c Hydraulics of main lines

The main line design is based on input energy (from a reservoir or a pumping station), main line slope, required operating pressure for irrigation, and needed energy gradient which will give a total energy higher than that required at any submain for irrigation. The design parameters of the main line are: first, allowable energy drop for each main line section; and second, main line sizes, selected from the allowable friction drop, ΔH, by using the pipe flow equation 2.2.8. The design procedure can be very simple if the main line supplies water to only a single field (one submain). This main line design can be done by considering a pipe flow condition in which the pipe size can be determined by the allowable energy drop ΔH, total required discharge Q, and the main line length L. When a main line system is supplying water to a series of fields, the main line flow capacity (discharge in the pipe) changes with respect to length. There will be different amounts of discharge in different main line sections. This design requires the estimation of the energy gradient curve so that the energy drop for each section can be determined.

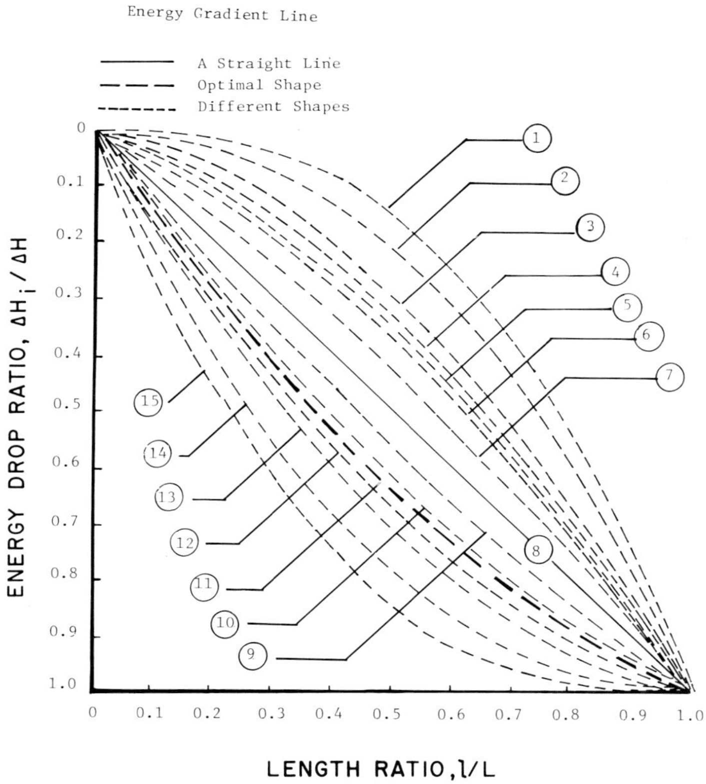

Fig. 2.2.3 Dimensionless energy gradient lines for main line design.

2.2.1.c.1 Energy gradient line

The main line design is a series of pipe flow designs. Once the field layout is set, the required discharge rate in each section can be determined. In a common practice, the Williams-Hazen formula is used to determine pipe sizes. This formula for plastic pipe given in equation 2.2.8 can be rearranged as

$$\frac{\Delta H}{\Delta L} = 1.135 \times 10^{6} \frac{Q^{1.852}}{D^{4.871}} \qquad (2.2.15)$$

in which $\Delta H/\Delta L$ is the energy slope. This shows that the pipe size, D, can be determined for a given discharge, Q, when the energy slope is known. The energy slope or the slope of the energy gradient line should be selected so that the energy gradient line is above the required water pressure along the main line as shown in figure 2.2.2.

The water pressure in the main line will be always equal to or greater than the required water pressure for trickle irrigation operation along the main line.

2.2.1.c.2 The optimal shape of energy gradient line

The energy gradient line can be a straight line or any one of a set of curves, as shown in figure 2.2.2. As long as the total energy is greater than the required operating water pressure, the design is hydraulically sound. Figure 2.2.2 shows the main line profile drawn according to a nearly level topographic condition and a required water pressure along the line. If an available inlet pressure at point A is determined, and point B indicates the pressure required at the downstream end, a straight line and curves will connect A and B. The straight energy gradient AB is one solution, and all the curves connecting A and B are the other possible solutions. Each solution will result in a set of main line design parameters. There is an optimum shape of the energy gradient line which will produce the minimum cost (pump and fuel costs are not included in the analysis).

The optimum shape of the energy gradient line is a curve just slightly below the straight line shown in figure 2.2.3. Fifteen energy gradient curves are plotted dimensionlessly: No. 8 is a straight line, and No. 11 is the optimal energy gradient.

2.2.1.c.3 A straight energy gradient line

Examination of the optimal energy gradient line indicates that the difference in cost between the optimal shape and the straight line is only about 2%. This provides a very fast and convenient method of design. When the main line profile, discharge, inlet pressure and required operating pressure are known, the straight energy gradient line can be determined. The design can be done using equation 2.2.14, considering the constant energy slope $\Delta H/\Delta L$ determined from the straight energy gradient line.

2.2.2 *Distribution uniformity*

The uniformity of emitter flow depends on the water pressure variation along the lateral lines and submain. The water pressure variation is obtained by evaluating the energy relations of the line. Friction drop will cause energy loss and the line slope will cause either a loss (upslope) or gain (downslope) of potential energy.

2.2.2.a Pressure variation along a trickle irrigation line

The total energy at any section of a trickle irrigation line can be expressed by the energy equation, equation 2.2.12. The change in energy with respect to the length of line can be expressed as

$$\frac{d\overline{H}}{dL} = \frac{dz}{dL} + \frac{dH}{dL} + \frac{d(v^2/2g)}{dL} \qquad (2.2.16)$$

Considering that the outflow from emitters is low in a trickle irrigation line, the change of velocity head with respect to the length is small and can be neglected. Therefore, the energy equation reduces to

$$\frac{d\overline{H}}{dL} = \frac{dz}{dL} + \frac{dH}{dL} \tag{2.2.17}$$

where $d\overline{H}/dL$ is the slope of the energy line or the energy slope, and

$$\frac{d\overline{H}}{dL} = -S_f \tag{2.2.18}$$

The minus sign indicates friction loss with respect to the line length.

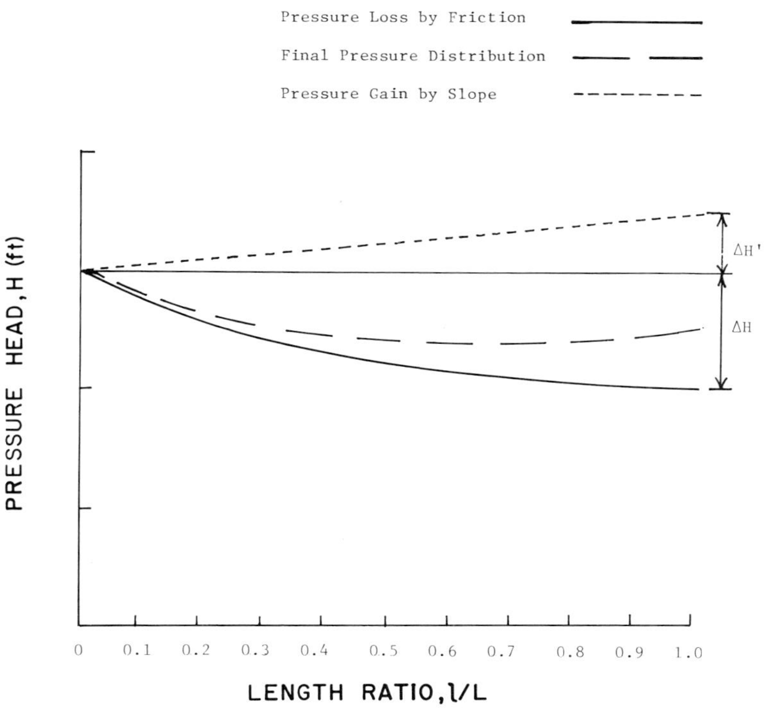

Fig. 2.2.4 The pressure distribution along a drip irrigation line (downslope).

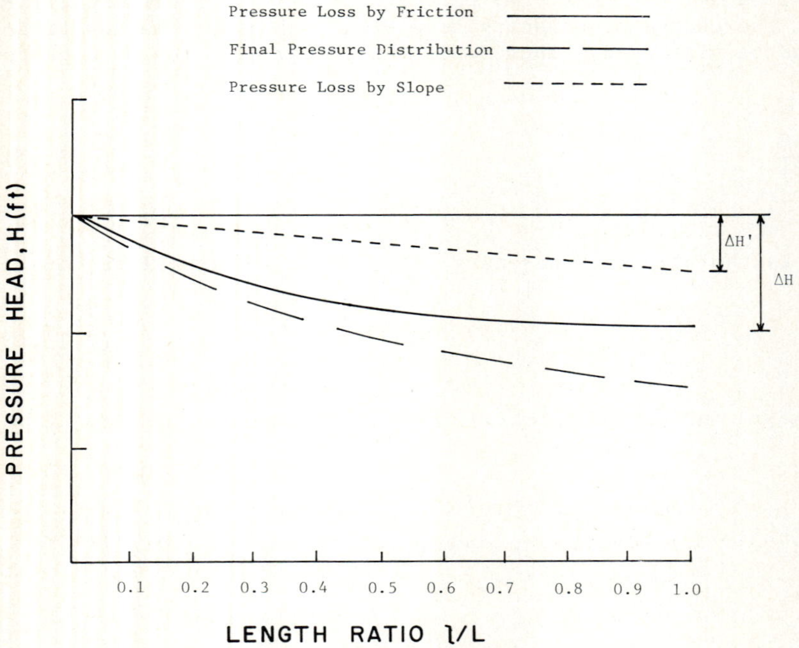

Fig. 2.2.5 The pressure distribution along a drip irrigation line (upslope).

The ratio dz/dL represents the slope of the line, as

$$\frac{dz}{dL} = -S_o \quad \text{(downslope)} \tag{2.2.19}$$

and

$$\frac{dz}{dL} = S_o \quad \text{(upslope)} \tag{2.2.20}$$

The pressure variation along a trickle irrigation lateral when laid downslope is

$$\frac{dH}{dL} = S_o - S_f \tag{2.2.21}$$

The pressure variation for a trickle irrigation lateral when laid up slope is

$$\frac{dH}{dL} = -S_o - S_f \qquad (2.2.22)$$

Equations 2.2.21 and 2.2.22 show that the change of pressure with respect to the length of a trickle irrigation line is a linear combination of the line slope and energy slope. Using the dimensionless energy gradient line, the friction drop at any given length of the line can be determined when a total friction drop (ΔH) is known. If the length of the line and slope are known, the pressure head gain or drop can be determined. If an input pressure is given, the pressure distribution along a drip irrigation lateral can be determined as shown in figures 2.2.4 and 2.2.5 and can be expressed mathematically as

$$H_i = H - \Delta H_i \pm \Delta H_i' \qquad (2.2.23)$$

in which H_i is the pressure expressed as hydrostatic head, in meters, at a given length ratio i, H is the input pressure, ΔH_i is the total friction drop at a given length ratio i, and the $\Delta H_i'$ is the energy gain or loss ('+' sign for downslope, '−' sign for upslope), at a given length ratio i. Equation 2.2.23 can be expressed by using the energy drop ratio R_i from the dimensionless energy gradient line and an enery gain (or loss) ratio by slopes, R_i',

$$H_i = H - R_i \Delta H \pm R_i' \Delta H' \qquad (2.2.24)$$

in which ΔH is the total energy drop by friction and $\Delta H'$ is the total energy gain (or loss) by slopes, $R_i = \Delta H_i / \Delta H$ and $R_i' = \Delta H_i' / \Delta H'$. The energy relation shown in equation 2.2.24 can be used for both uniform and nonuniform slopes. For uniform slopes, R_i' is the same as the length ratio i; the pressure along the trickle irrigation line will be

$$H_i = H - R_i \Delta H + i \Delta H' \qquad (2.2.25)$$

In the case where the trickle irrigation line is laid on non-uniform slopes, and assuming the total length is divided into n sections where the slope for each section is determined by S_1, S_2, ...S_j,....S_n, the pressure along the trickle irrigation line for non-uniform slope can be expressed as

$$H_i = H - R_i \Delta H + \frac{L}{n} \sum_{1}^{n} S_j \qquad (j = 1, 2, ...n) \qquad (2.2.26)$$

in which S_j = slope of the j section along the line using '+' sign for downslope (energy gain) and '−' for upslope (energy loss).

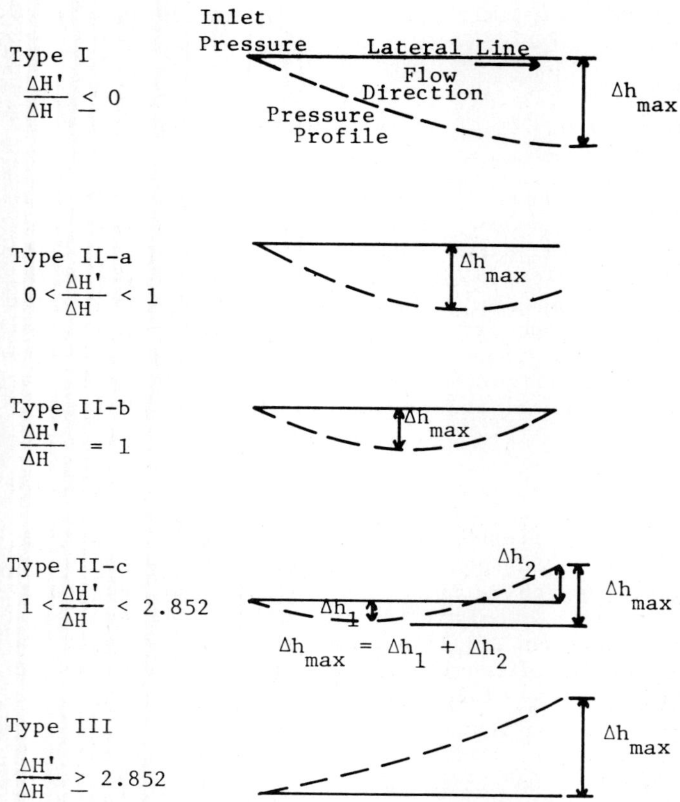

Fig. 2.2.6 Pressure profiles along a lateral line for different $\Delta H'/\Delta H$ values.

2.2.2.b Pressure profiles

The pressure profiles for uniform slope situations can be classified according to a dimensionless ratio, $\Delta H'/\Delta H$, in which $\Delta H'$ is the energy gain or loss by slope at the end of the line and ΔH is the total energy drop by friction at the end of line. Figure 2.2.6 shows the five possible pressure profiles. These are as follows:

Pressure profile type I - The pressure decreases with respect to the lateral line length. This occurs when the lateral line is laid on zero or uphill slopes. In this condition the dimensionless ratio $\Delta H'/\Delta H$ is less than or equal to 0.

Pressure profile type II - The pressure decreases with respect to the lateral line length, reaches a minimum point and then increases with respect to the lateral line length. This profile can be also classified into three types according to the slope situation:

(1) Type II-a: This occurs under the slope situation where $\Delta H'/\Delta H$ is larger than zero but less than 1. The pressure at the end of the line is less than the operating pressure.
(2) Type II-b: This occurs under the slope situation where $\Delta H'/\Delta H$ is equal to the operating pressure.
(3) Type II-c: This occurs under the slope situation where $\Delta H'/\Delta H$ is larger than 1 but less than 2.852. For this condition the pressure at the end of line is larger than the operating pressure. The constant 2.852 is determined as the ratio of the total friction drop at the end of a pipe to the total friction drop, ΔH, at the end of a lateral line, assuming that both have the same total inlet discharge, diameter, and length.

Pressure profile type III - The pressure increases with respect to the lateral line length. This is caused by a steep downslope situation where $\Delta H'/\Delta H$ is equal to or larger than 2.852 (the energy gain is larger than the friction drop for all sections along the lateral line).

The pressure profile for the non-uniform slope situation depends on the non-uniform slope pattern of each individual line and has to be determined from equation 2.2.26.

2.2.2.c Emitter flow variation along a trickle irrigation line and the uniformity coefficient of emitter flow

As indicated by equation 2.2.1, the emitter flow is controlled by the hydrostatic pressure at the emitter. This means there will be an emitter flow variation caused by the pressure profile along the irrigation line. For an orifice type of emitter, the emitter flow can be shown as a square root function of the pressure

$$q_i = c\,(H - R_i\,\Delta H + R_i'\,\Delta H')^{1/2} \qquad (2.2.27)$$

in which q_i = emitter flow at a given length ratio i. For other types of emitters, the proper value of x should be used as shown in equation 2.2.2.

The emitter flow profile for a trickle irrigation line on uniform slope will have the shape similar to that of one of the pressure profiles as shown in figure 2.2.6. The degree of emitter flow variation is expressed by the Uniformity Coefficient as defined by Christiansen (1942) for sprinkler irrigation. The uniformity coefficient for emitter flow can be expressed as

$$U_s = 1 - \frac{\overline{\Delta q}}{\overline{q}} \qquad (2.2.28)$$

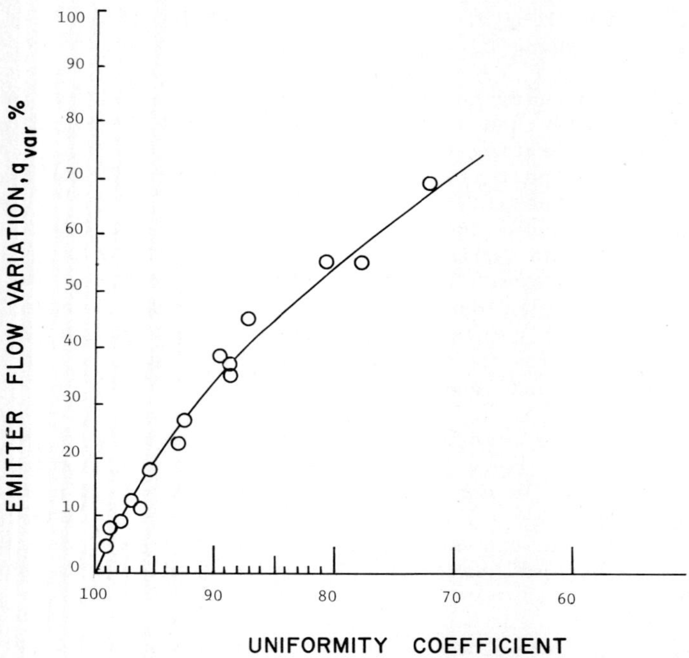

Fig. 2.2.7 Relationship between emitter flow variation and uniformity coefficient.

in which U_s is the uniformity coefficient, $\overline{q}$ is the mean emitter flow and $\overline{\Delta q}$ is the mean of the absolute deviations from the mean emitter flow. The uniformity coefficient is a quantitative expression of the emitter flow variation. Another way of computing the emitter flow variation is by comparing the maximum with the minimum emitter flow. The most commonly used definition is represented by

$$q_{var} = \frac{q_{max} - q_{min}}{q_{max}} \tag{2.2.29}$$

in which q_{var} is the emitter flow variation, q_{max} is the maximum emitter flow, and q_{min} is the minimum flow along the line.

2.2.2.d. Design criteria

As shown in equations 2.2.25, 2.2.26, and 2.2.27, the pressure and emitter flow variations along a lateral line can be calculated. A

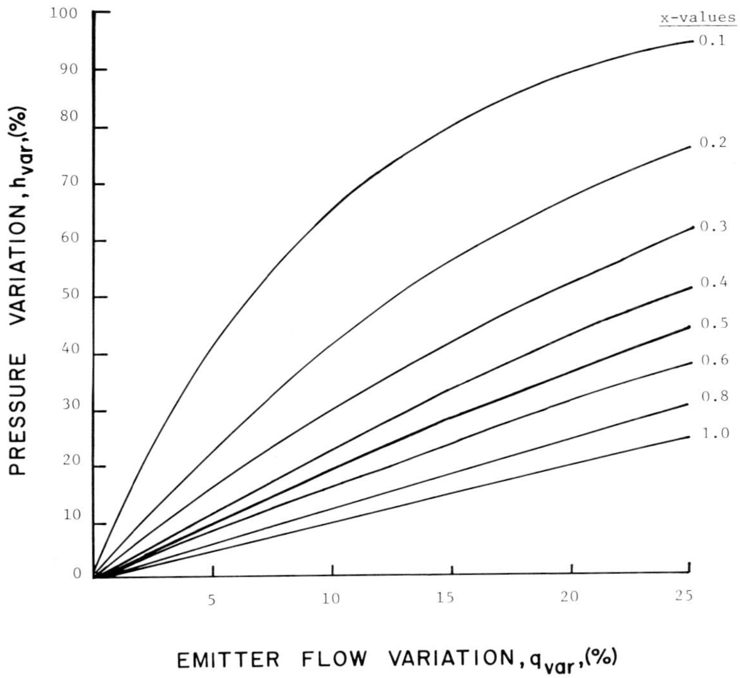

Fig. 2.2.8 Relationship between emitter flow variation and pressure variation for different x-values.

lateral line with a given length, operating pressure and slope situation will hydraulically create an emitter flow variation along the line. A design criterion is a quantitative value of emitter flow variation used as the basis to accept, reject or revise a design. The uniformity coefficient of emitter flow, in essence, determined by all the emitter flows, should be considered as the best quantitative value for the design criterion. However, the calculation is tedious and impractical. The uniformity coefficient of emitter flow can be calculated by using only limited samples of emitter flow along a line. The quantitative expression of emitter flow variation shown in equation 2.2.29 is simple to obtain because it requires only the maximum and minimum emitter flows. Since the emitter flow profiles (see figure 2.2.6. for pressure profiles) are smooth curves, a relationship between the emitter flow variation, q_{var}, and uniformity coefficient, U_s, can be obtained as shown in figure 2.2.7. A uniformity coefficient of about 98% equals an emitter flow variation of 10%, and a uniformity coefficient of about 95% equals an emitter flow variation of 20%.

The pressure and emitter flow variations are related by the x-value shown in the emitter flow function (equation 2.2.2). The relation can be expressed as

$$q_{var} = 1 - [1 - H_{var}]^{x} \qquad (2.2.30)$$

and

$$H_{var} = \frac{H_{max} - H_{min}}{H_{max}} \qquad (2.2.31)$$

in which H_{var} = pressure variation, H_{max} = the maximum pressure in the line, and H_{min} = the minimum pressure in the line. The relationship between emitter flow variation and pressure variation for different x-values is plotted in figure 2.2.8. When the x-value is 0.5, which is true for most of the orifice type of emitters, a pressure variation of 20% is equivalent to a 10% emitter flow variation, and a pressure variation of 10% is equivalent to a 5% emitter flow variation.

2.2.3 *Hydraulic design*

The main goal of hydraulic design is to maintain an acceptable uniformity of emitter flows throughout the trickle irrigation system. The design procedure follows the order from downstream to upstream starting with the emitter, the lateral line, the submain and finally the main line which is similar to the design procedure for a sprinkler irrigation system.

2.2.3.a Emitter or lateral line selection

The emitters must supply enough water to the plant root zone to meet the plant water requirements. Normally, one-third to as much as three-fourths of the plant rooting volume should be supplied with adequate water. The larger the percentage of the wetted rooting volume becomes relative to the total volume, the safer the design becomes. This allows for temporary system breakdown, salinity problems, and conversion from surface irrigation to trickle irrigation on mature orchards. However, if the wetted percentage becomes too large, many of the advantages of trickle irrigation are lost. The wetted soil volume depends on the emitter flow rate, irrigation duration, emitter spacing and soil type.

Normally, the emitters are located near the plant or the areas of high root concentration. Installing emitters close to the tree trunks, however, should be avoided. Several possible emitter placement geometries can be used to irrigate the desired percentage of the root zone. Emitters can be on single laterals with equal spacing, double laterals, laterals with loops, or other configurations. When designing irrigation systems for new orchards, the increasing water demand as the plants mature should be kept in mind.

Trickle irrigation emitters can be designed as a point source or line source to supply water into the plant root zone depending on the type of crops. Row crops such as sugarcane or vegetables with high density planting usually require the line source. A lateral line with in-line emitters or thin wall plastic hose with orifices are used to supply the line source. Cropping systems with low density tree plantings are usually designed for point source.

The required emitter (or orifice) flow can be calculated based on the water requirement, number of emitters, irrigation application efficiency of the trickle irrigation system and the irrigation duration and can be expressed as

$$q_r = \frac{q_t I}{T E_a e} \tag{2.2.32}$$

where q_r is the required emitter flow rate in liters per hour (L/hr), q_t is the water requirement per plant per day in liters per day (L/day), I is the irrigation interval in days, T is the irrigation time per set in hours, E_a is the trickle irrigation application efficiency and e is the number of emitters per plant. For line source emitters, e can be determined as the ratio of the total number of emitters (or orifices) to the total number of plants along the line. (In addition, see chapter 4.1 on Irrigation Scheduling).

2.2.3.b Lateral line design

The emitter flow variation along a lateral line is affected largely by the manufacturer's and hydraulic variations. Information regarding the manufacturer's variation of emitters and lateral lines can be obtained from laboratory tests and manufacturer's reports. In general, the manufacturer's and hydraulic variations are treated separately; that is, the emitter or lateral line with an acceptable manufacturer's variation is initially selected and the lateral line design is mainly based on hydraulic variation only. The criterion for manufacturer's variation is usually set at less than 10% (See chapter 2.1 on Emitter Construction). As stated in the section on lateral line and submain hydraulics (2.2.1.b), the pressure variation is determined by combining friction drop and slope gain (or loss) along the line. Since the energy gradient line for a fixed emitter with constant spacing is a curve, and energy gain or loss for uniform conditions is a straight line, there will certainly be pressure variations which will cause emitter flow variation along a lateral line. The major criterion in lateral line design is to achieve an acceptable emitter flow variation or uniformity coefficient.

When the emitter (or lateral line) and emitter spacing are selected based on crop requirement and soil conditions, the total discharge at the inlet section for a given length under an operating pressure can be determined (assuming all emitters are operated at the same inlet pressure). The pressure variation can be calculated for a selected size of lateral by equations 2.2.25 and 2.2.26 for the uniform and non-uniform slope situations, respectively. The emitter flow variation can be determined from equation 2.2.30 when the pressure variation, H_{var}, is known.

There are two ways for deriving the proper lateral design. One is to select the lateral line size for a given length of run which can deliver the required amount of water to the plants within a desired range of uniformity. The other is to determine the maximum lateral line length for the given flow conditions and ground slope when the lateral line size is limited to a specific size. The lateral line sizes range from 12 to 21 mm (0.5 to 0.825 inch) inside diameter of polyethylene (PE) or polyvinyl chloride (PVC) material.

Commercial PE or PVC tubes are made of fixed nominal sizes. The commonly used size is 16 mm (1/2 inch nominal size with an inside diameter (I.D.) of 0.625 inch). Lateral line design utilizes available lateral line sizes and designs a length to maintain an acceptable emitter flow uniformity. The design can be made by either hydraulic calculations or using design charts.

2.2.3.b.1 Hydraulic calculation

Since the total discharge of a lateral line depends on the total length of the lateral, design starts by preselecting a length. The emitter flow variation along the line is determined and compared to the design criterion. The design procedure is as follows:

Step 1. Assume a lateral length for the selected lateral (with a known size).

Step 2. Determine the total inlet discharge based on the operating pressure.

Step 3. Calculate the total friction drop ΔH at the end of the lateral using equation 2.2.10.

Step 4. Calculate the pressure profile along the line using equation 2.2.24.

Step 5. Determine the pressure variation H_{var} using equation 2.2.31.

Step 6. Determine the emitter flow variation q_{var} using equation 2.2.30.

Step 7. Compare the emitter flow variation q_{var} with the design criterion of emitter flow variation. The design is accepted when the calculated q_{var} is less than or equal to the design criterion; otherwise, the design is rejected.

Step 8. If the design is rejected, use a shorter length and repeat the design procedure until a proper length is found to meet the design criterion.

Step 9. If the design is accepted, but the q_{var} is significantly smaller than the design criterion, use a longer length and repeat the design procedure. This part, however, is optional.

Example 2.2.1

Problem: Determine the pressure variation and emitter flow variation of a 16 mm (I.D.) lateral line with a length of 100 meters. Emitter flow is 4 liters per hour at an operating pressure head of 10 m and spaced every 0.66 m along the lateral. The lateral line is 1% uniform downslope. The x-value in the q-H relationship of the emitter is assumed to be 0.5 (equation 2.2.2).

Solution: Lateral line length $L = 100$ m
Lateral line size $D = 16$ mm (inside diameter)
Total inlet discharge $Q = 150 \times 4 = 600$ L/hr $= 0.167$ L/sec

Total friction drop at the end of the lateral ΔH, using equation 2.2.10.

$\Delta H = 1.97$ m

Pressure head along the lateral line is calculated using equation 2.2.24.

i=0	H = 10 m
i=0.1	H = 10 - 1.97 x 0.26 + 0.1 = 9.59 m
i=0.2	H = 10 - 1.97 x 0.47 + 0.2 = 9.27 m
i=0.3	H = 10 - 1.97 x 0.64 + 0.3 = 9.04 m
i=0.4	H = 10 - 1.97 x 0.77 + 0.4 = 8.88 m
i=0.5	H = 10 - 1.97 x 0.86 + 0.5 = 8.81 m
i=0.6	H = 10 - 1.97 x 0.93 + 0.6 = 8.77 m
i=0.7	H = 10 - 1.97 x 0.97 + 0.7 = 8.79 m
i=0.8	H = 10 - 1.97 x 0.99 + 0.8 = 8.85 m
i=0.9	H = 10 - 1.97 x 0.999 + 0.9 = 8.93 m
i=1.0	H = 10 - 1.97 x 1.0 + 1.0 = 9.03 m

Pressure variation H_{var} is calculated by equation 2.2.31

$$H_{var} = \frac{10 - 8.77}{10} = 0.123$$

Emitter flow variation q_{var} is calculated by equation 2.2.30

$$q_{var} = 1 - (1 - 0.123)^{0.5} = 0.064$$

2.2.3.b.2 Design charts

The hydraulic calculations presented in the previous section are simple, but tedious, especially when the process has to be repeated to meet the design criterion. A computer can be used to determine emitter flow variation for each of the many possible combinations of lateral length, total discharge, operating pressure, lateral line size and uniform slope conditions. Furthermore, design charts can be prepared from computer simulation.

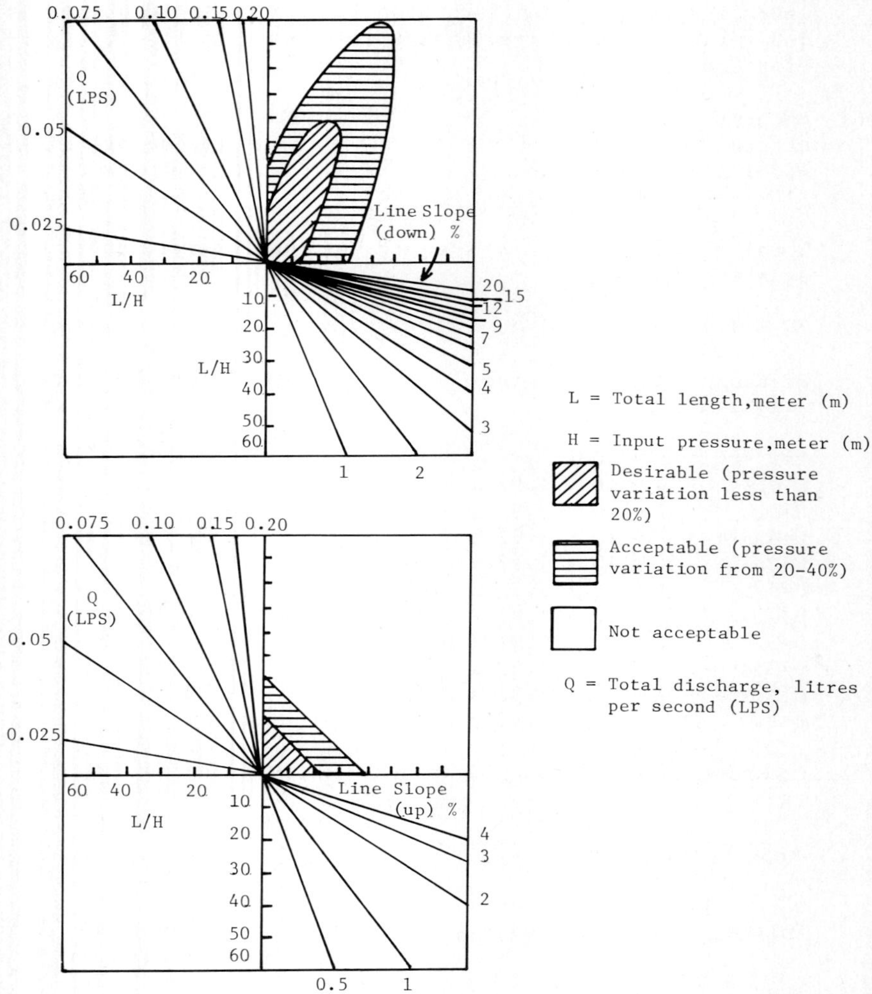

Fig. 2.2.9 Trickle irrigation design calculator (12 mm).

Two sets of design charts developed for lateral line design on uniform slopes are shown in figures 2.2.9, 2.2.10, 2.2.11, 2.2.12, and 2.2.13. The first set is specifically made for two lateral lines sizes, 12 mm and 16 mm (figures 2.2.9 and 2.2.10). The second set is a general design chart which can be used for any lateral line size (figures 2.2.11, 2.2.12, and 2.2.13).

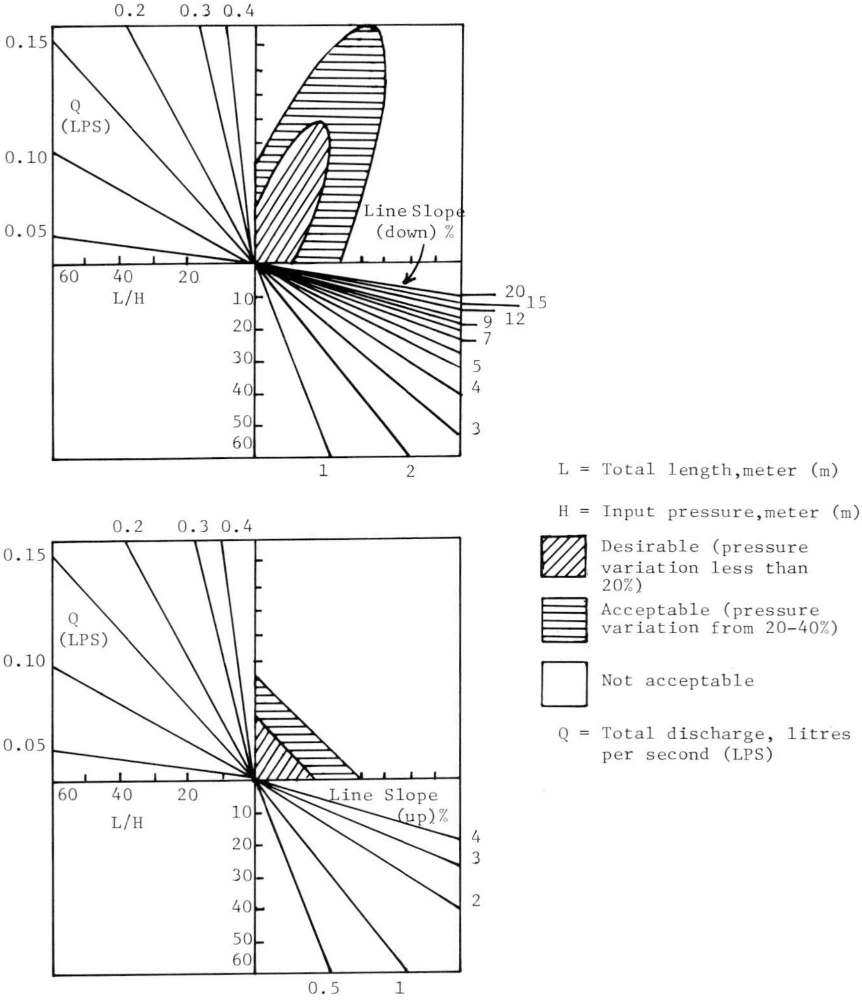

Fig. 2.2.10 Trickle irrigation design calculator (16 mm).

The 12 mm and 16 mm lateral line sizes are commonly used design charts in the field practice. The design procedures for such specific design are as follows:

Step 1. Establish the lateral length (L) to operating pressure head (H) ratio L/H and determine the total discharge (Q) in liters per second (L/sec).

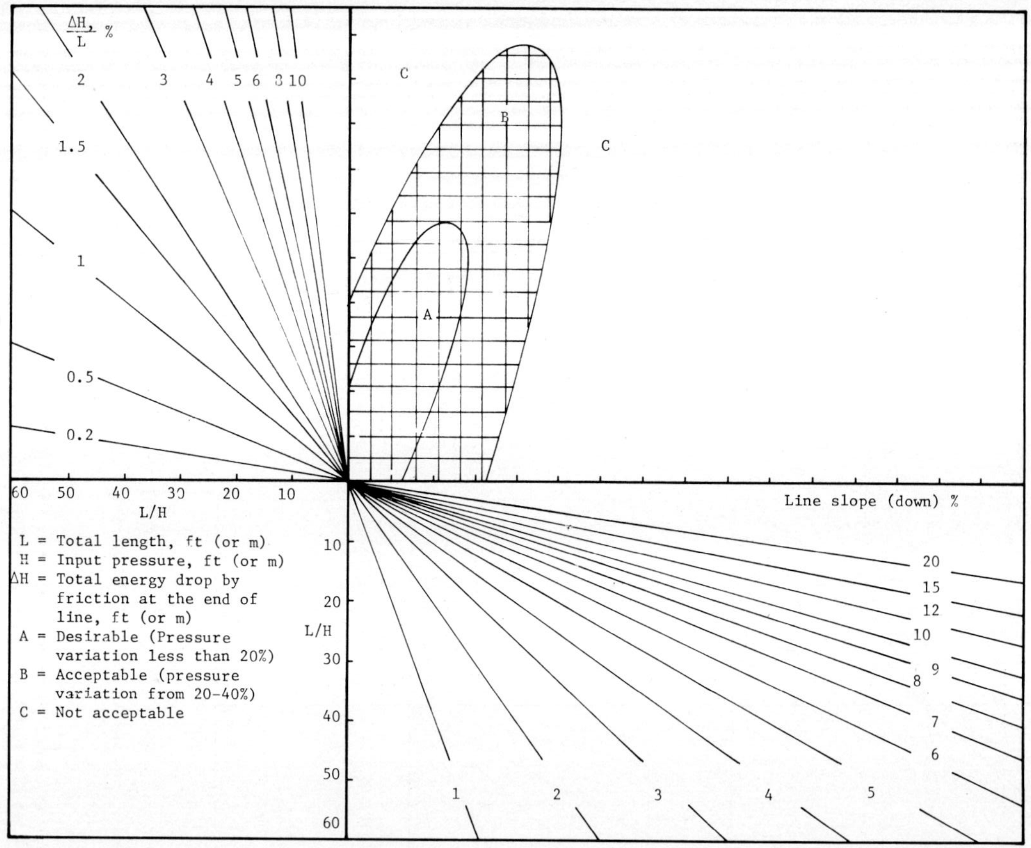

Fig. 2.2.11 Dimensionless general design chart (downslope).

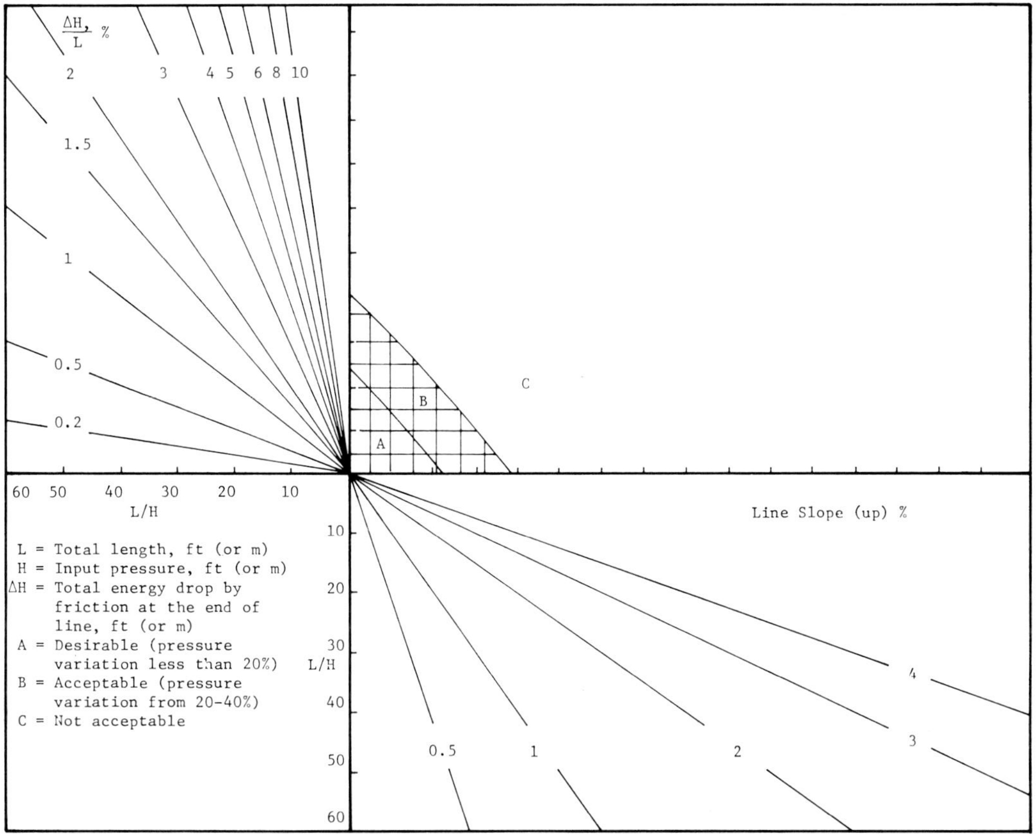

Fig. 2.2.12 Dimensionless general design chart (upslope).

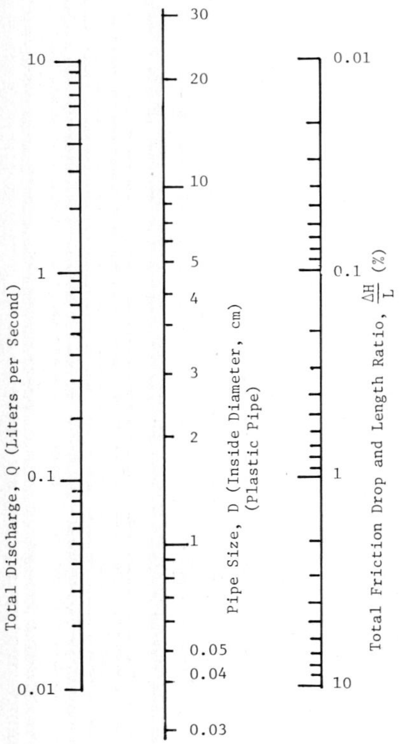

Fig. 2.2.13 Nomograph for trickle irrigation laterals and submain design in metric units.

Step 2. Move vertically from L/H (in Quadrant III) to the given total discharge L/sec line in Quadrant II, then establish a horizontal line into Quadrant I in the appropriate chart.

Step 3. Move horizontally from L/H (Quadrant III) to the percent slope line in Quadrant IV, then establish a vertical line into Quadrant I.

Step 4. The intersection point of these two lines in Quadrant I determines the acceptability of the design.

a. Desirable (pressure variation less than 20% or emitter flow variation less than 10%).
b. Acceptable (pressure variation 20-40% or emitter flow variation about 10-20%).
c. Not recommended (pressure variation greater than 40% or emitter flow variation larger than 20%).

Example 2.2.2

Problem: Use the information given in example 2.2.1 to check the acceptability of the design.

Given:

Lateral line length	L = 100 m
Lateral line size	D = 16 mm
Operating pressure head	H = 10 m
Total discharge	Q = 0.167 L/sec
Line slope	S = 1% (down)

Solution:

a. Calculating L/H = 10
b. Using figure 2.2.10 and following steps 1, 2, 3 and 4, the intersection point is found in the desirable zone in which the pressure variation is within 20% and emitter flow variation is within 10%.
c. The design is, therefore, accepted because the design criterion of q_{var} is 10%.

The general design charts for lateral line design as shown in figures 2.2.11 and 2.2.12 are plotted dimensionlessly for uniform downslope and upslope, respectively, and can be used for all pipe sizes. Note that the dimensionless term, the total friction drop to length ratio, ΔH/L, is plotted in Quadrant II of the design charts. A nomograph for determining ΔH/L from total discharge, Q, and pipe size, D, is given in figure 2.2.13 and is a graphical solution of the friction drop equation 2.2.10. This set of design charts (figures 2.2.11, 2.2.12, and 2.2.13) can be used to check the acceptability of a design when the lateral line size is given, or to select the proper size of a lateral line to meet the design criterion. The design procedure are:

(1) To check acceptability of design when the lateral size is given.

Step 1. Establish a trial L/H and total discharge Q (L/sec).
Step 2. From the nomograph (figure 2.2.13) use the total discharge and lateral line size to determine ΔH/L.
Step 3. Move vertically from L/H (Quadrant III) to the determined ΔH/L in Quadrant II of the appropriate figure (figure 2.2.11 or 2.2.12), then establish a horizontal line into Quadrant I.
Step 4. Move horizontally from L/H to the % slope line in Quadrant IV, then establish a vertical line into Quadrant I.
Step 5. The intersection point of these two lines in Quadrant I determines the acceptability of the design:

a. Desirable (emitter flow variation less than 10%).
b. Acceptable (emitter flow variation from 10-20%).
c. Not recommended (emitter flow variation greater than 20%).

(2) To select proper lateral size.

Step 1. Determine L/H and total discharge Q (L/sec).
Step 2. Move horizontally from L/H to the percent of slope line in Quadrant IV of the appropriate figure (figure 2.2.11 or 2.2.12). From that point establish a vertical line into Quadrant I.
Step 3. Establish a point along this line in Quadrant I at the upper boundary of the desirable region A or acceptable region B depending on the design criterion. From that point establish a horizontal line into Quadrant II.
Step 4. Establish a vertical line in Quadrant II from the L/H value so that it intersects the horizontal line of Step 3 at a point.
Step 5. Determine the $\Delta H/L$ value in Quadrant II at this point.
Step 6. From the nomograph (figure 2.2.13) using total discharge and the $\Delta H/L$ value, establish the minimum lateral size according to the selected design criterion.

Exampble 2.2.3

Problem: A lateral line length in a vegetable field is 100 m and the slope is 1% down slope. Emitters spaced 0.3 m apart are installed in the lateral line. The emitter flow is 4 liters per hour at an operating pressure head of 10 m. Design the lateral line size.

Given: Lateral length = 100 m
Operating pressure H = 10 m
Number of emitters = 300
Total discharge Q = 1200 L/hr
= 0.334 L/sec

Solution:

Step 1. Determine L/H = 100/10 = 10.
Step 2. From figure 2.2.11 move horizontally from L/H = 10 to 1% down slope line in Quadrant III. From that point establish a vertical line into Quadrant I.
Step 3. Establish a point along this line in Quadrant I at the upper boundary of 'desirable' region A. From that point establish a horizontal line into Quadrant II.
Step 4. Establish a vertical line in Quadrant II from the L/H = 10 so that it intersects the horizontal line of step 3 above at a point.
Step 5. At this point, determine $\Delta H/L$ = 3.5% in Quadrant II.
Step 6. From nomograph (figure 2.2.13) using the total discharge Q = 0.334 liters per second and $\Delta H/L$ = 3.5%, the minimum lateral line size is determined as 19 mm (or use the available size larger than 19 mm).

The lateral line slope depends on the topographic conditions and the field layout of the irrigation system. In most cases, the field can be graded to form a uniform slope or can be assumed to have a uniform slope. For very uneven terrain situations in which a non-uniform slope has to be used, design can be made based on hydraulic calculations using equations 2.2.23 and 2.2.26. Graphical solutions are available using polyplots and design charts for non-uniform slope situations.

2.2.3.c Submain design

The design of submains is based on both capacity and uniformity. Capacity means the submain size should be large enough to deliver the required amount of water to irrigate the field. Uniformity means the submain should be designed to maintain an allowable pressure variation so the flow into all lateral lines will have little variation.

The submain, hydraulically, is the same as a lateral line having a steady, spatially-varied flow with lateral out flows. Design of a submain is based on the study of hydraulics and energy relations. Most of the developed design charts for laterals can also be used for submains.

2.2.3.c.1 A general submain design chart for a single size on uniform slopes.

This set of design charts for submain design is the same as presented in section 2.2.3.2 and shown in figures 2.2.11, 2.2.12, and 2.2.13. The same design procedure can be used. The only difference is the calculation of total discharge. In case of the submain design, total discharge is the summation of all the lateral line discharges.

2.2.3.c.2 A simplified submain design chart for single size on uniform downslopes (or zero slope)

Since the length of a submain is relatively short (approximately 25 to 75 meters), the design can be made by considering that the total friction drop is equal to the total energy gain by slope, $\Delta H = \Delta H'$. This design can, in general, achieve a high uniformity of lateral discharge along the submain. When the ΔH is made equal to $\Delta H'$, equation 2.2.10 can be written as

$$\Delta H' = 3.98 \times 10^5 \frac{Q^{1.852}}{D^{4.871}} L \qquad (2.2.33)$$

or

$$S_o = 3.98 \times 10^5 \frac{Q^{1.852}}{D^{4.871}} \qquad (2.2.34)$$

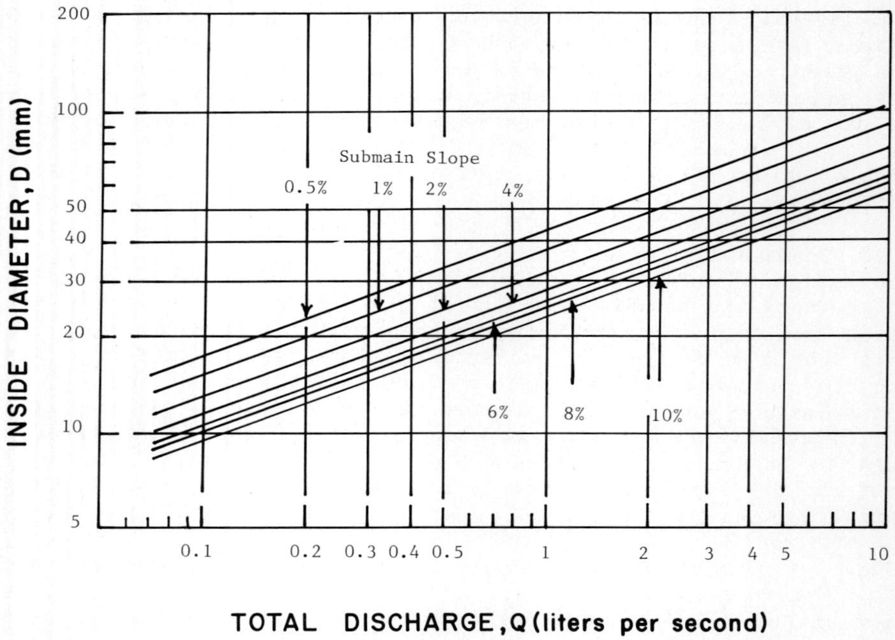

Fig. 2.2.14 Submain design chart — slope equal or larger than 0.5%.

in which S_o is simply the submain slope. When equation 2.2.34 is used, the maximum pressure difference will be near the middle section of the submain with a magnitude equal to 0.36 $\Delta H'$. This value can be used to determine the maximum pressure variation of a submain design. Equation 2.2.34 is plotted in figure 2.2.14. The simple submain design chart (figure 2.2.14) is designed for slopes equal to or larger than 0.5%. When the slope is less than 0.5%, it is considered as level or zero and equation 2.2.34 cannot be used. Under this condition it is assumed that the pressure variation along a submain is affected by friction drop only and the maximum energy drop (or pressure drop) is at the end of the submain. Therefore, equation 2.2.10 can be written as

$$\Delta H/H = 3.98 \times 10^5 \frac{Q^{1.852}}{D^{4.871}} \frac{L}{H} \tag{2.2.35}$$

By setting $\Delta H/H$ = 10%, which is the allowable pressure variation, equation 2.2.35 becomes

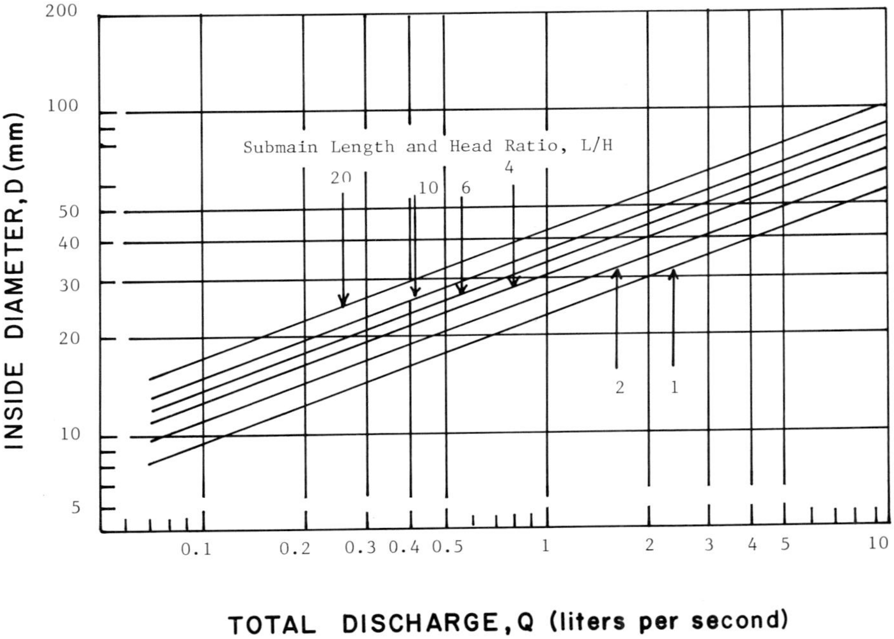

Fig. 2.2.15 Submain design chart — slope less than 0.5% and allowable pressure variation 10%.

$$0.10 = 3.98 \times 10^5 \frac{Q^{1.852}}{D^{4.871}} \frac{L}{H} \tag{2.2.36}$$

Equation 2.2.36 can be plotted to give a design chart as illustrated in figure 2.2.15 and can be used for designing submain based on the total discharge and the length and pressure head ratio L/H. The design procedure for using the simplified design charts, figures 2.2.14 and 2.2.15 is as follows:

Step 1. Determine the total discharge, Q, for the submain.

Step 2. Determine the length and pressure ratio L/H.

Step 3. Determine the submain slope. If the submain slope is less than 0.5%, use figure 2.2.15 to design the submain size.

Step 4. If the submain slope is equal to or larger than 0.5%, the submain size can be determined from figure 2.2.14. The maximum pressure variation is 0.36 $\Delta H'$.

Example 2.2.4

Problem: Using the lateral line as given in example 2.2.2, design a submain when the submain length is 40 meters and the lateral line spacing is 2 meters. Design the submain size when the submain slope is

(a) zero, and (b) 5% uniform downslope.

Given: Lateral line length = 100 m
Discharge per lateral line = 0.167 L/sec
Lateral line size = 16 mm
Number of lateral lines = 20
Total discharge of submain inlet, Q = 20 x 0.167 = 3.34 L/sec
Submain length, L = 40 m
Operating pressure head, H = 10 m

Solution: a. S = 0
L/H = 4

Using the simplified submain design chart, figure 2.2.15, the submain size is designed as 50 mm.

b. S = 5% uniform downslope

Using the simplified submain design chart, figure 2.2.14, the submain size is designed as 40 mm.

The general and simplified design charts for submain as used in Design Example 2.2.4 are for rectangular shaped fields where lateral line lengths are the same in the field. For irregular shaped field, a shape coefficient can be applied to determine an adjusted total discharge in order that the simplified charts, figures 2.2.14 and 2.2.15, can still be used. Under certain field conditions where the submain length is relatively long (100 m or greater), the submain design can be modified using multiple pipe sizes.

2.2.3.d Main line design

The main line is designed to supply the required discharge at a specified operating pressure into each field for trickle irrigation. It should be designed in such a way that the total energy at any outlet along the main line is equal to or higher than the energy required for operating the entire irrigation system. Thus, the design approach is mainly to determine the allowable energy drop for all main line sections. The developed straight energy gradient line concept provides a nearly optimal solution (initial pipe cost only, operation cost is not included). An example of main line system design is as follows:

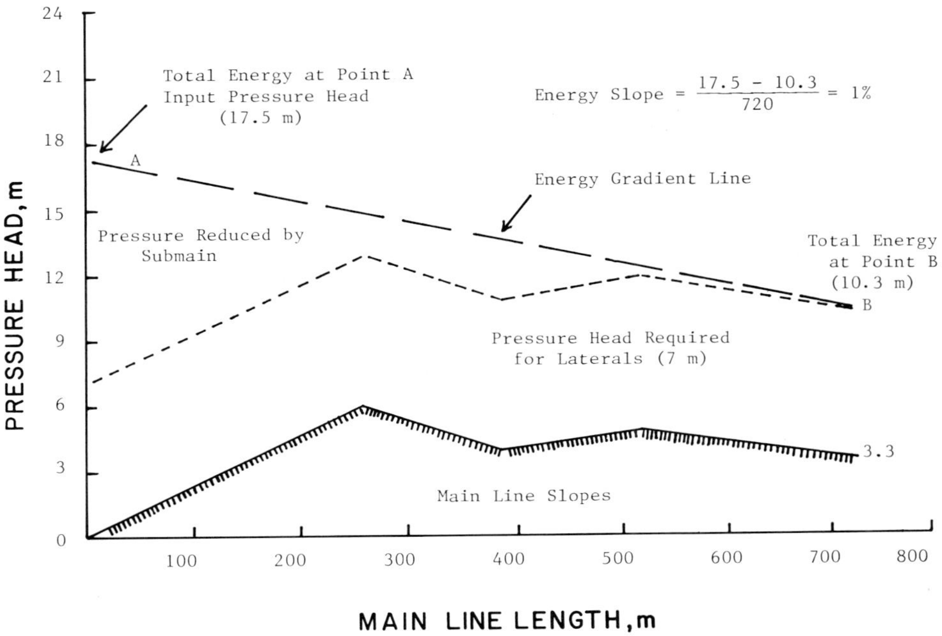

Fig. 2.2.16 Main line layout and energy gradient line.

Example 2.2.5

Problem: A trickle irrigation system design is needed for a 20 hectare field. The main plot is rectangular and divided into 0.4 hectare subplots which are controlled by submains connected from the main line. The main line is laid in the center of the field with 25 subplots on each side of the main line. Each subplot is about 130 meters long and 30 meters wide, and each main line section is 30 meters long. The design capacity is 2 liters per second for each subplot. There is a total of 24 outlets on each side of the main lines to supply 2 L/sec to each subplot. If the main line slopes, the required operating pressure (7 m) at the lateral lines and available input pressure (17.5 m) at the beginning of the main line are given in figure 2.2.16, design the main line system.

Given: Main line length, L = 720 m
Main line profile, as shown in figure 2.2.16
Operating pressure for trickle irrigation laterals = 7 m
Available input pressure at the inlet = 17.5 m (Point A)
Required total energy at the last field = 10.3 m (Point B)

Solution:

Step 1. Plot main line profile as shown in figure 2.2.16.
Step 2. Plot the required pressure head, 7 m, along the main line profile as shown in figure 2.2.16.
Step 3. Determine the energy slope. From figure 2.2.16, the slope of the straight energy gradient line is determined as 1%.
Step 4. Design the main line size. The main line size is determined from the nomograph, figure 2.2.17, using 1% energy slope. The results are shown in table 2.2.1.

TABLE 2.2.1

Main line sizes determined from nomograph (figure 2.2.17).

Main line section	Discharge L/sec	Main line size (inside diameter, cm)
0*	100	
1	96	25
2	92	25
3	88	25
4	84	20
5	80	20
6	76	20
7	72	20
8	68	20
9	64	20
10	60	20
11	56	20
12	52	20
13	48	20
14	44	20
15	40	15
16	36	15
17	32	15
18	28	15
19	24	15
20	20	12.5
21	16	12.5
22	12	10
23	8	10
24	4	7.5

* There is an outlet at the entrance of section 1 for irrigating the subplots on both sides of section 1.

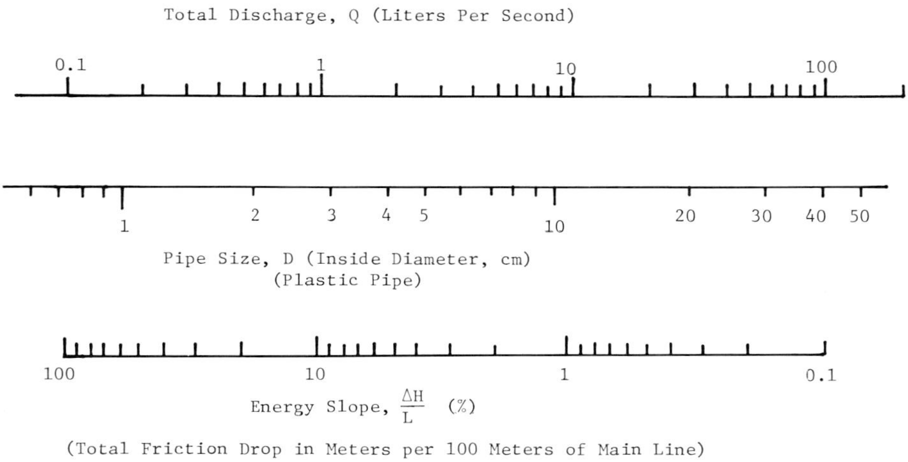

Fig. 2.2.17 Nomograph for trickle irrigation main line design in metric units.

2.2.4 *Irrigation efficiency*

The design of a trickle irrigation pipe network may be based on the hydraulics of pipe flow. There are two irrigation efficiency terms that are affected by the system design: distribution and application. Distribution efficiency determines how uniformly irrigation water can be distributed through a trickle irrigation system into the field. An application efficiency shows how well irrigation water is applied, i.e., what percentage of water applied is stored in the root zone as required and available for plant use. The distribution efficiency can be determined from the emitter flow variation along a lateral line (or submain) of a field trickle irrigation system and can be expressed by the equation

$$E_d = 100\left(1 - \frac{\overline{\Delta q}}{\bar{q}}\right) \qquad (2.2.37)$$

in which E_d is the distribution efficiency and is the same as the uniformity coefficient, $\bar{q}$ is the mean emitter flow rate and $\overline{\Delta q}$ is the average absolute deviation of all emitter flow from the average emitter flow. The application efficiency is defined as the ratio of water required in the root zone to the total amount of water applied. If the root zone can be fully irrigated by the minimum flow rate and the total irrigation time, the application efficiency is

$$E_a = \frac{eq_{min}T}{V} \tag{2.2.38}$$

in which E_a is the application efficiency, e is the total number of emitters, q_{min} is the minimum emitter flow rate, T is the total irrigation time, and V is the total amount of water applied. Since the mean emitter flow is

$$\bar{q} = \frac{V}{eT} \tag{2.2.39}$$

the application efficiency can also be expressed as

$$E_a = \frac{q_{min}}{\bar{q}} \tag{2.2.40}$$

2.2.4.a Determination of application efficiency

Both equations 2.2.38 and 2.2.40 can be used to determine trickle irrigation application efficiency. This is based on the condition that the minimum emitter flow rate and irrigation time will meet the water requirement, and all other emitters will irrigate more than that required. This is also based on the assumption that over irrigation is simply wasting water and will not affect crop production. When any emitter flow q_i between the maximum and minimum flow rates and irrigation time T is used to meet the water requirement, there will be certain areas of over irrigation and certain areas of under irrigation known as deficit as shown in figure 2.2.18. When the maximum emitter flow, q_{max}, and irrigation time T are used to meet the water requirement, there will be no over irrigation, and all emitters except q_{max} will cause a deficit situation in the field. Under this situation, all water applied is stored in the root zone and available for crop use so that the application efficiency is 100%.

Different emitter flow profiles were used to determine the relationship between trickle irrigation application efficiency and the percent of deficit for a design criteria of 10% and 20% flow variation, q_{var}. The results are plotted in figure 2.2.19 which can be used to determine application efficiency when the emitter flow variation q_{var} and the allowable deficit are known.

2.2.4.b Trickle irrigation scheduling

Irrigation scheduling determines when the system must be turned on and how much water must be applied. The quantity of water delivered is determined by adjusting the irrigation period using automatic or manual controls. The irrigation set time is related to the designed

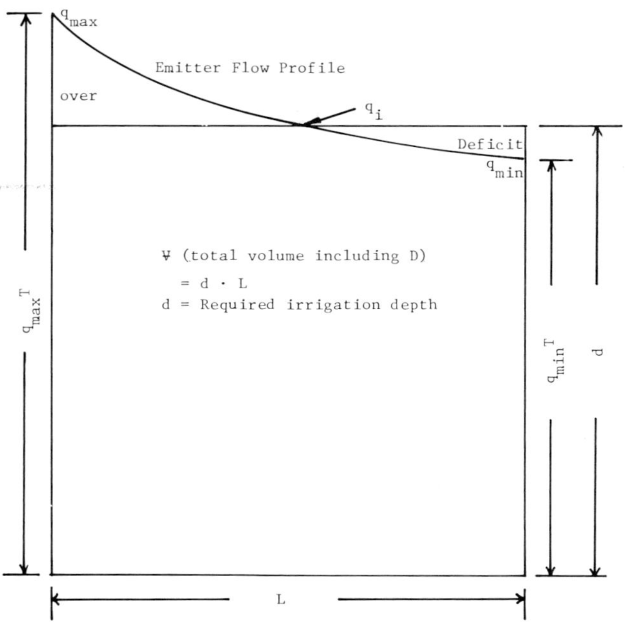

Fig. 2.2.18 A typical emitter flow profile and amount of irrigation along a lateral line showing over irrigation and deficit.

trickle irrigation system, the application efficiency, the total water requirement and any allowable deficit situation and can be expressed by the equation

$$T = \frac{W\ (1 - P_D)}{3600\ Q_a\ E_a} \qquad (2.2.41)$$

where T is the irrigation set time in hours, W is the required amount of water to supply the root zone in liters (L), P_D is the percent of deficit, Q_a is the actual total discharge delivered by the system (L/sec) and E_a is the application efficiency. The required amount of water W is determined from the daily consumptive use or evapotranspiration requirement of the crop and irrigation interval. Detailed discussions on water requirement and evapotranspiration measurements are presented in chapter 4.1 on Irrigation Scheduling. The allowable deficit is estimated from the relationship between crop yield and amount of water application. This is an economic consideration and usually the allowable deficit is assumed not to cause a significant reduction in crop yield. The actual total discharge, Q_a, can be

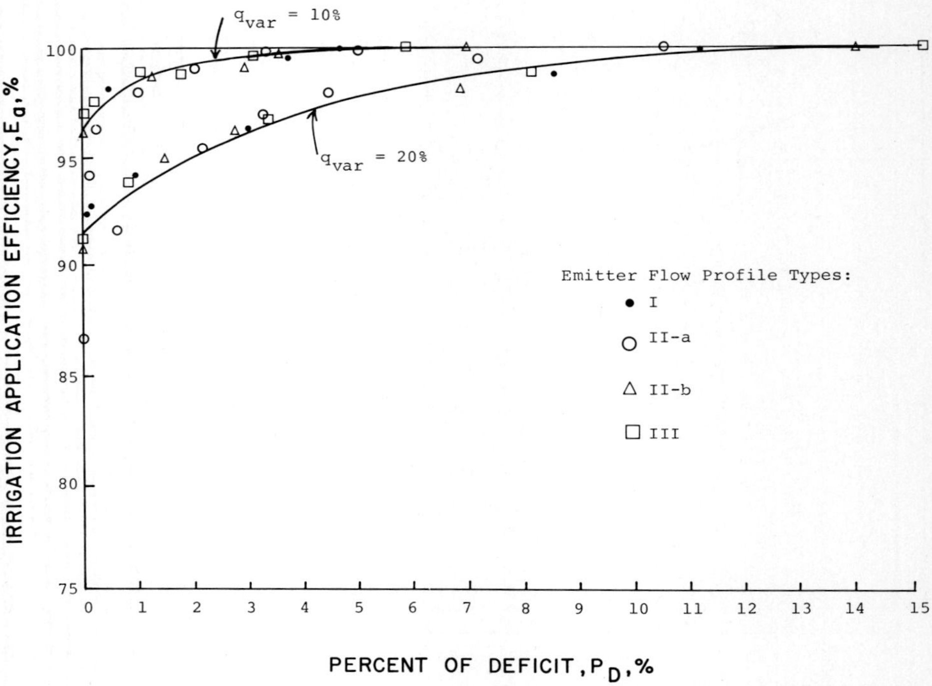

Fig. 2.2.19 The relationship between irrigation application efficiency and deficit for emitter flow variations (by hydraulics), q_{var} of 19% and 20%.

measured with a totalizing flow meter installed in the trickle irrigation system or estimated from the designed total discharge and emitter flow variation q_{var}. The irrigation application efficiency can be determined using figure 2.2.19 from emitter flow variation, q_{var}, and the percentage of deficit.

Example 2.2.6

Problem: A trickle irrigation system is installed for a one hectare vegetable plot. The trickle irrigation system has a total design discharge of 5 liters per second under an operating pressure head of 10 m. The emitter flow variation is 20% and the actual total discharge is 4.5 L/sec. If the water requirement is 5 mm, determine the irrigation time for

(a) zero deficit condition, and (b) 5% deficit condition.

Solution: 1. The total amount of water required

$$W = 0.5 \text{ cm} \times 10{,}000 \text{ cm} \times 10{,}000 \text{ cm} = 5 \times 10^4 \text{ L}$$

2. The actual total discharge

$$Q_a = 4.5 \text{ L/sec}$$

3. The irrigation application efficiency can be determined from from figure 2.2.19.

$P_D = 0$ $E_a = 91.5\%$
$P_D = 5\%$ $E_a = 97.8\%$

4. The irrigation time is calculated using equation 2.2.41.

a. $P_D = 0$

$$T = \frac{5 \times 10^4}{3600 \times 4.5 \times 0.978} = 3.37 \text{ hr}$$

b. $P_D = 5\%$

$$T = \frac{5 \times 10^4 \times 0.95}{3600 \times 4.5 \times 0.978} = 3.00 \text{ hr}$$

2.2.5 *System design limitations*

The trickle irrigation system design presented in this chapter (section 2.2) is based on the hydraulics of pipe flow. The emitter flow variation q_{var} obtained from emitter flow profiles shown in figure 2.2.6 represents the hydraulic variation only. Manufacturer's coefficient of variation is not included or it is assumed to be small and neglected. When the manufacturer's variation is included, the total emitter flow variation should be determined as a combination of hydraulic variation and manufacturer's variation. The distribution of emitter flow will have a tendency to be more normally distributed than that shown in figure 2.2.6. The determination of application efficiency and the irrigation scheduling will be different from that presented in sections 2.2.4.a and 2.2.4.b.

PROBLEMS

1. Discuss some of the merits of the Darcy-Weisbach equation versus the Williams-Hazen equation for the determination of the friction drop along the lateral line submain line.

2. Describe the types of equations used to determine pressure variation along a trickle irrigation lateral line for a uniform, upslope, and downslope condition.

3. Discuss the three types of pressure profiles.

4. Discuss some of the factors that need to be considered in emitter and lateral line selection.

5. Discuss the irrigation efficiency terms that are directly related to the system design.

6. Discuss the irrigation schedule and water conservation under deficit irrigation condition.

7. Determine the pressure variation and emitter flow variation of a 16 mm (I.D.) lateral line with a length of 100 meters. Emitter flow is 4 liters per hour at an operating pressure head of 10 m and spaced every 0.66 meter along the lateral. The lateral slope is 1% uniform upslope. The x-value in the q-H relationship of the emitter is assumed to be 0.5 [equation 2.4.2].

8. Use the information given in problem 7 to check the acceptability of the design. The design criterion of emitter flow variation, q_{var} is 10%.

REFERENCES AND SUPPLEMENTAL READING

Bralts, V.F., Wu, I.P. and Gitlin, H.M., 1981. Manufacturing variation and drip irrigation uniformity. Trans. Amer. Soc. Agric. Eng. 24(1): 113-119.

Bralts, V.F., Wu, I.P. and Gitlin, H.M., 1981. Drip irrigation uniformity considering emitter plugging. Trans. Amer. Soc. Agric. Eng. 24(5): 1234-1240.

Bralts, V.F., Wu, I.P. and Gitlin, H.M., 1982. Emitter plugging and drip irrigation lateral line hydraulics. Trans. Amer. Soc. Agric. Eng. 25(5): 274-1281.

Bucks, D.A. and Myers, L.E., 1973. Trickle irrigation--Application uniformity from simple emitters. Trans. Amer. Soc. Agric. Eng. 16(6): 1108-1116.

Christiansen, J.E., 1942. Hydraulics of sprinkling systems for irrigation. Trans. Amer. Soc. Civ. Eng. 107: 221-239.

Gillespie, V.A., Phillips, A.L. and Wu, I.P., 1979. Drip irrigation design equations. J. Irrig. Drain. Div., Amer. Soc. Civil Eng. 105(IR3): 247-257.

Hart, W.E. and Reynolds, W.N., 1965. Analytical design of sprinkler systems. Trans. Amer. Soc. Agric. Eng. 8(1): 83-89.

Howell, T.A. and Barinas, F.A., 1980. Pressure losses across trickle irrigation fittings and emitters. Trans. Amer. Soc. Agric. Eng. 23(4): 928-933.

Howell, T.A. and Hiler, E.A., 1974. Designing trickle irrigation laterals for uniformity. J. Irrig. Drain. Div., Amer. Soc. Civil Eng. 100(IR4): 443-454.

Howell, T.A. and Hiler, E.A., 1974. Trickle irrigation lateral design. Trans. Amer. Soc. Agric. Eng. 17(4): 902-908.

Hughes, T.C. and Jeppson, R.W., 1978. Hydraulic friction loss in small diameter plastic pipelines. Water Resour. Bull. 14: 1159-1166.

Karmeli, D., 1977. Classification and flow regime analysis of drippers. J. Agric. Eng. Res. 22: 165-173.

Karmeli, D. and Keller, J., 1974. Evaluation of a trickle irrigation system. Proc. 2nd Int. Drip Irrig. Congr., San Diego, California, pp. 287-292.

Karmeli, D. and Keller, J., 1975. Trickle Irrigation Design. Rain Bird Sprinkler Manufacturing Corporation, Glendora, California, 133 pp.

Keller, J. and Karmeli, D., 1974. Trickle irrigation design parameters. Trans. Amer. Soc. Agric. Eng. 17(4) :678-684.

Kenworthy, A. L., 1972. Trickle Irrigation--The Concept and Guideline for Use. Michigan Agric. Exp. Sta. Res. Report 165, 19 pp.

Khatri, K.C., Wu, I.P., Gitlin, H.M. and Phillips, A.L., 1979. Hydraulics of microtube emitters. J. Irrig. Drain. Div., Amer. Soc. Civil Eng. 105(IR2): 163-173.

Myers, L E. and Bucks, D.A., 1972. Uniform irrigation with low pressure trickle system. J. Irrig. Drain. Div., Amer. Soc. Civil Eng. 98(IR3): 341-346.

Paraqueima, J.R. 1977., Study of Some Frictional Characteristics of Small Diameter Tubing for Trickle Irrigation Laterals. Unpubl. M.S. Thesis, Utah State University, Logan.

Parchomchuk, P., 1976. Temperature effects on emitter discharge rates. Trans. Amer. Soc. Agric. Eng. 19(4): 690-692.

Pair, C.H., Hinz, W.W., Sneed, R.E., Frost, K.R., and Schiltz, T.J. (Eds.), 1983. Irrigation (formerly Sprinkler Irrigation). Fifth Edition, Irrigation Association, Silver Spring, Maryland, 686 pp.

Perold. P.R., 1977. Design of irrigation pipe laterals with multiple outlets. J. Irrig. Drain. Div., Amer. Soc. Civil Eng. 103(IR2): 179-193.

Ramirez-Guzman, H. and Manges, H., 1971. Uniform flow from orifice in irrigation pipe. Trans. Amer. Soc. Agric. Eng. 14(1): 179- 193.

Stevenson, D.S., 1973. Guide to Design and Operation of Trickle Irrigation Systems. Canada Agric. Pub. Res. Sta., Summerland, B.C., 17 pp.

Solomon, K., 1976. Emission uniformity--An introduction. Proc. 4th Ann. Int. Drip Irrig. Assoc. Meet., Fresno, California, pp. 56-64.

Solomon, K., 1979. Manufacturing variation of emitters in trickle irrigation systems. Trans. Amer. Soc. Agric. Eng. 22(5): 1034-1038,1043.

Solomon, K., 1977. Performance comparison of different emitter types. Proc. 7th Int. Agric. Plastics Congr., San Diego, California, pp. 97-102.

Solomon, K. and Keller, J., 1978. Trickle irrigation uniformity efficiency. J. Irrig. Drain. Div., Amer. Soc. Civil Eng. 104(IR3): 293-36.

Van't Woudt, B.D., 1964. Uniform discharge from multiorificed pipes. Trans. Amer. Soc. Agric. Eng. 7(2): 352-355.
Watters, G.Z. and Keller, J., 1978. Trickle irrigation tubing hydraulics. ASAE Paper 78-2015, Amer. Soc. Agric. Eng., St. Joseph, Michigan, 18 pp.
Wilke, O.C. and Wendt, C.W., 1971. Performance of a twin-wall trickle irrigation hose. ASAE Paper 71-204, Amer. Soc. Agric. Eng., St. Joseph, Michigan, 8 pp.
Williams, G.S. and Hazen, A., 1960. Hydraulic Tables. Third Edition, John Wiley and Sons, New York, pp. 25-58.
Wu, I.P., 1975. Design of drip irrigation main lines. J. Irrig. Drain. Div., Amer. Soc. Civil Eng. 101(IR4): 265-278.
Wu, I.P. and Gitlin, H.M., 1975. Energy gradient line for drip irrigation laterals. J. Irrig. Drain. Drain. Div., Amer. Soc. Civil Eng. 101(IR4): 323-326.
Wu, I.P. and Gitlin, H.M., 1977. Design of drip irrigation submains. J. Irrig. Drain. Div., Amer. Soc. Civil Eng. 101(IR2): 231-243.
Wu, I.P. and Gitlin, H.M., 1979. Drip irrigation design on non-uniform slopes. J. Irrig. Drain. Div., Amer. Soc. Civil Eng. 105(IR3): 289-303.
Wu, I.P., Howell, T.A. and Hiler, E.A., 1979. Hydraulic Design of Drip Irrigation Systems. Hawaii Agric. Exp. Sta. Tech. Bull. 105, University of Hawaii, Honolulu, 80 pp.

Chapter 2

DESIGN PRINCIPLES

2.3 SOIL WATER DISTRIBUTION

A. W. WARRICK

Soil water distribution is determined by the soil properties and the way water is added and withdrawn from the profile. Factors which generally differentiate the soil water regime for trickle from other irrigation systems are (1) the flow regime is 2 or 3-dimensional rather than vertical only; (2) the water is added at a high frequency; and (3) soil water is maintained within a relatively narrow range. The multidimensional nature of flow from point or line sources leads to more complex mathematics if the system is to be modeled. The high frequency and narrow moisture range tends to negate somewhat the concept of field capacity — the volume of water stored remains a key factor, but water is added more often and presumably is very well controlled. Plant rooting patterns under limited volume and nonstressed conditions also leads to some new concepts relative to conventional irrigation. For example, when expressing water for evapotranspiration, should it be per unit of total land area, per unit of a strip which is farmed or per unit area of canopy? In addition, the desirability of distributing water to encourage root development over an extensive area is debatable and perhaps site specific.

Example 2.3.1 Water application time for point emitters

Problem: Assume emitters are spaced on a 2 x 5 m grid. If each emitter produces 6 liters/hr, how long does it take to add 10 mm of water calculated for the total area of a 10 hectare field? Repeat assuming 10 mm of water over a 1 m diameter wetted circle for each emitter.

Answer: 16.7 and 1.31 hr

Example 2.3.2 Water volume determination for line emitters

Problem: Consider parallel line emitters on a 2 m regular spacing. What is q_L the flow discharge rate per meter of line if 20 mm of water are delivered in 4 hours? (Note: On an individual line the orifices or emitters are assumed to be very close.)

Answer: q_L = 10 liters $hr^{-1}m^{-1}$

2.3.1 *Soil properties*

2.3.1.a Soil water potential

Soil water is described either according to the amount present or by energy level. Measurements and parameters usually reflect which description is most useful. For example, in describing the plant stress, or moisture movement, the potential is preferred.

The two most common modes of expressing amount present are water on a mass basis Θ_m or on a volume basis Θ_v. The definitions are

$$\Theta_m = \text{(Mass of water in sample)/(Dry mass of soil)} \quad (2.3.1)$$

and

$$\Theta_v = \text{(Volume of water in sample)/(Apparent volume of sample)} \quad (2.3.2)$$

The relationship between the two is

$$\Theta_v = (\rho_b/\rho_w)\Theta_m \quad (2.3.3)$$

where ρ_b is the bulk density of the soil (Mg m^{-3}) and ρ_w the density of water. In general, "water content" without any qualification refers to Θ_m, but Θ_v is more useful than Θ_m when discussing water storage in the soil.

The energy level is expressed as the soil water potential ϕ_T as

$$\phi_T = z + h + \pi \quad (2.3.4)$$

with z elevation (m), h pressure head (m), and π the osmotic potential (m). The potential is the amount of work per unit weight to transfer a small quantity of water from a reference state to the point and conditions to be measured. Factors which are neglected in equation 2.3.4 include air pressure, temperature and overburden effects. The pressure head h is positive for saturated conditions and negative for unsaturated conditions. It is equal to the soil water pressure minus the soil air pressure. For unsaturated conditions, h is also commonly called the matric potential and -h is referred to as the suction or tension — note that suction is a positive quantity.

There are alternative ways of expressing potential. Equation 2.3.4 gives an energy/(unit weight) rather than energy/(unit mass) or energy/(unit volume) which are also commonly used. Some equivalences, in terms of atmospheres, are listed in the appendix section. The "head" is given as the associated pressure divided by $\rho_w g$ in consistent units (g is the gravitational acceleration constant). "Atmospheres" and "bars" are roughly interchangeable and both are equivalent to approximately 10 m of water head.

For saturated conditions, the water pressure is normally greater than the adjacent atmosphere. For these conditions, h may be measured with a piezometer tube and is equal to the water level height within.

For low suctions (fairly wet conditions), the unsaturated h can be evaluated with a tensiometer as presented in example 2.3.3.

Example 2.3.3 A tensiometer problem

Problem: Consider the mercury-water tensiometer in figure 2.3.1.a. The density of Hg is 13.1 Mg m^{-3}.

a. What is the matric potential at the cup expressed as cm of water? As kPa?
b. What is the pneumatic (air) component of pressure potential at the cup also expressed as cm of water?
c. Suppose the Hg-H_2O manometer is replaced by a vacuum-gage tensiometer (see figure 2.3.1.b) with the gage 15 cm above the soil surface. What will the gage read if 0 is no pressure and 100 is -1 bar?

Answer: a. -263 cm, -26.1 kPa
b. 1020 cm (approx.)
c. 36

Tensiometers may be used as dry as about -8 m of water, but difficulties occur when the lowest pressure in the instrument approaches a total vacuum. On the other hand, tensiometers are sensitive in the wet range and must be carefully placed at a precise location. Models are available which can be used for automatically activating an irrigation system. Also a commercial instrument has been introduced providing digital output from a portable, rapid-response pressure transducer. For dryer conditions, alternative instrumentation is needed. Methods include psychrometers for h less (i.e. drier) than -2 m or so.

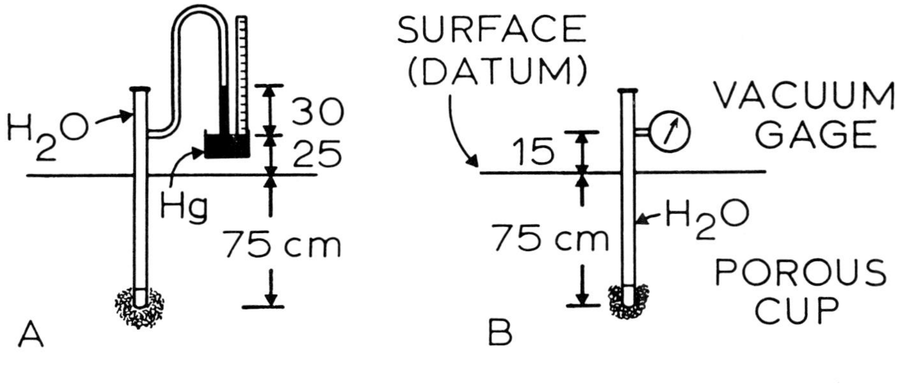

Fig. 2.3.1 Mercury-water (A) and vacuum-gage (B) tensiometers.

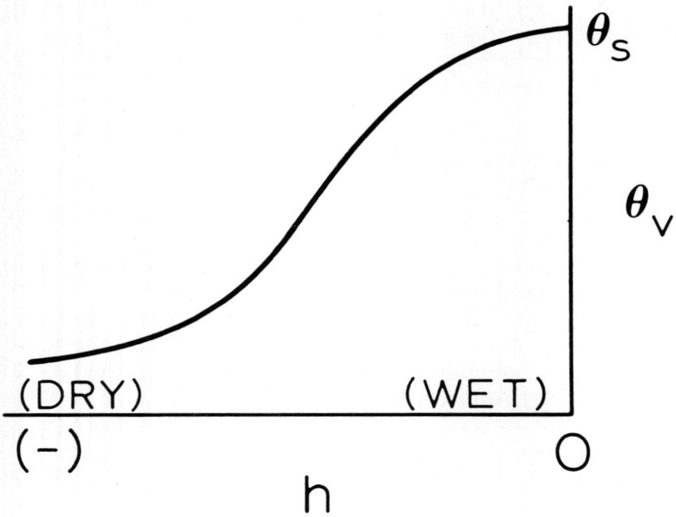

Fig. 2.3.2 A soil water characteristic curve (h vs. Θ).

Soil psychrometers measure the relative humidity of the soil air which is normally very near saturation. Once the relative humidity is known, $h + \pi$ is easily evaluated. Also, the matric potential may be evaluated over a fairly wide range by measuring dissipation of a heat pulse in a porous material in equilibrium with the soil water. Such devices are commercially available. Still another device is the ordinary moisture resistance block which is useful for h of -10 m and dryer. The principle is that the electrical conductivity is dependent on the amount of water in the block which in turn is in equilibrium with the soil water.

The relationship of h to Θ_v (or Θ_m) is the soil water characteristic curve. This is also called the moisture release curve. A typical relationship is in figure 2.3.2. As h decreases, Θ_v becomes less. For a coarse-textured (sandy) soil, the change in Θ_v is more abrupt; whereas for a fine soil, the higher water contents are maintained at higher suctions. Such relationships are determined by measuring Θ_v and h simultaneously, most commonly in the laboratory. For the wet range, a hanging water column is useful; for the drier range, a pressure plate is used. For h between -2 and 0 m, an "undisturbed" sample is desirable as soil structure influences the results. For conditions drier than -2 m, macro-pore structure becomes less of a factor and a disturbed sample is more acceptable. Values for h vs Θ_v can be influenced by the history of the system. This effect, known as "hysteresis," complicates matters in that the relationship is not unique — Θ_v tends to be greater for a given h during drainage than while wetting.

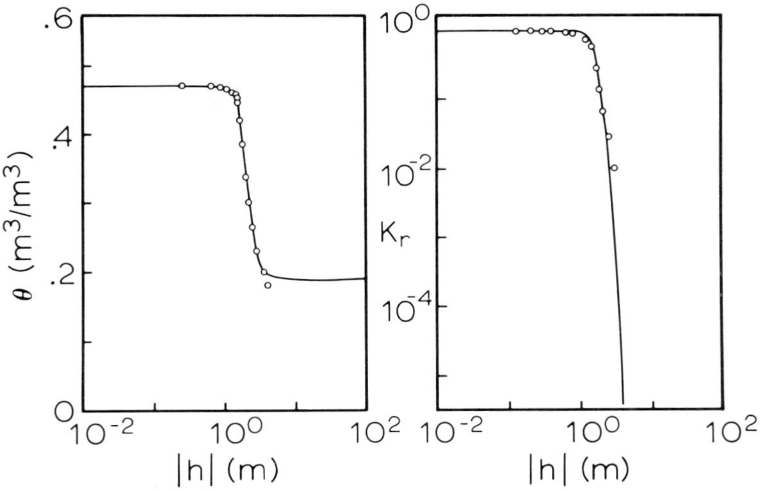

Fig. 2.3.3 Soil water characteristic and hydraulic conductivity for G.E. 3 silt loam (after van Genuchten, 1980).

Many different algebraic and statistical fits have been used to approximate soil water characteristics. One of these due to van Genuchten (1980) is illustrated in example 2.3.4. This is a first step to soil water modeling.

Example 2.3.4 A soil water characteristic curve (Θ_v vs. h)

Problem: Calculate and plot Θ_v vs. h for G.E.3 silt loam using equation 2.3.5 with $\Theta_s = 0.469$, $\Theta_r = 0.190$, $\alpha = 0.079$ cm^{-1} and $n = 10.4$. Compare results to figure 2.3.3a. (Θ_s is saturated water content; Θ_r the "residual" water content).

$$\Theta = \frac{\Theta_v - \Theta_r}{\Theta_s - \Theta_r} = [1 - (\alpha h)^{1-1/n}]^{-n} \qquad (2.3.5)$$

2.3.1.b Soil water dynamics

The soil water potential can be used to describe dynamic conditions, even though the strict definition is for equilibrium. Flow occurs from regions of higher to lower potential and is most often assumed to follow Darcy's Law:

$$\underline{v} = -K(h) \text{ grad } H \tag{2.3.6}$$

where $H = h + z$ is the hydraulic head.

The unsaturated hydraulic conductivity $K(h)$ is a function of water status and "grad" the vector gradient. In cartesian coordinates, equation 2.3.6 may be written equivalently as

$$v_x = -K(h) \; \partial H/\partial x \tag{2.3.7a}$$

$$v_y = -K(h) \; \partial H/\partial y \tag{2.3.7b}$$

$$v_z = -K(h) \; \partial H/\partial z \tag{2.3.7c}$$

with the assumption of isotropic conditions and v_x, v_y, and v_z the components of $\underline{v}$. Thus, $\underline{v}$ is a vector quantity with both a magnitude and a direction.

Darcy's law for unsaturated flow is a generalization of that for saturated flow, the major difference being that K is a function of the pressure potential or the water content. The unsaturated relationship was formulated based on the idea that as the soil dries out, movement of the water (liquid) through the soil matrix becomes much more tortuous and hence v is reduced for a given potential gradient. The measurement of K is much more difficult for unsaturated than for saturated conditions. (Methods are discussed in the general references at the end of the chapter.) Several algebraic relationships have been used to approximate K, for example

$$K(h) = K_o \exp(\alpha h) \tag{2.3.8}$$

$$K(h) = a/[b + |h|^n] \tag{2.3.9}$$

The values of α, K_o, a, b and n are chosen by best fitting experimentally determined K values. Since $h = 0$ would correspond to saturated conditions, K_o or a/b may be taken as the saturated K. However, in most cases the constants are fit over a range of h values and K_o and a/b should be considered as empirical factors. Table 2.3.1 contains a list of K_o and α values corresponding to equation 2.3.8. This form has been used advantageously in mathematical modeling, more because of convenience rather than because of any general superiority over other forms.

TABLE 2.3.1

Hydraulic conductivity parameters for different soils for the form $K_o \exp(\alpha h)$. (For original references and additional values see Bresler 1978; Warrick et al., 1981; Amoozegar-Fard et al., 1984.)

Soil	α (cm^{-1})	K_o (cm/sec)
Clay loam	0.1258	1.12×10^{-3}
Columbia sandy loam	0.100	1.39×10^{-3}
Dackley sand	0.513	1.00×10^{-4}
Gila fine sandy loam	4.43×10^{-2}	2.43×10^{-4}
Guelph loam	3.4×10^{-2}	3.67×10^{-4}
Ida silt loam	2.6×10^{-2}	2.92×10^{-5}
Indio loam	1.69	3.9×10^{-4}
Latene clay loam	3.86×10^{-2}	5.21×10^{-5}
Panoche loam	4.16×10^{-2}	1.1×10^{-3}
Pima clay loam	1.4×10^{-2}	1.15×10^{-4}
Plainfield sand	0.126	3.44×10^{-3}
Sandy loam	0.1112	1.00×10^{-3}
Silt loam	1.39×10^{-2}	5.74×10^{-5}
Yolo clay	3.67×10^{-2}	9.33×10^{-6}
Yolo fine sandy loam	2.5×10^{-2}	4.07×10^{-5}

Approximate values of a conductivity may be calculated from a soil water characteristic curves (h vs. Θ_v). For these relationships, the porous medium is modeled as a network of capillary tubes. One such general relationship is

$$K/K_s = \Theta^{0.5}[(\int_o^{\Theta} h^{-1} d\Theta)/\int_o^1 h^{-1} d\Theta)]^2 \qquad (2.3.10)$$

with K_s the saturated hydraulic conductivity.

For h vs. Θ_v of the form of equation 2.3.5 becomes

$$K/K_s = \Theta^{0.5}[1-(1 - \Theta^{n/(n-1)1-1/n}]^2 \qquad (2.3.11)$$

For discussion of these and others, see particularly Mualem (1976) and van Genuchten (1980).

Example 2.3.5 Calculation of K vs. Θ_v

Problem: a. Calculate K vs. Θ_v for the example 2.3.4 using equation 2.3.11.

b. Plot K vs. h also and compare to figure 2.3.3b.

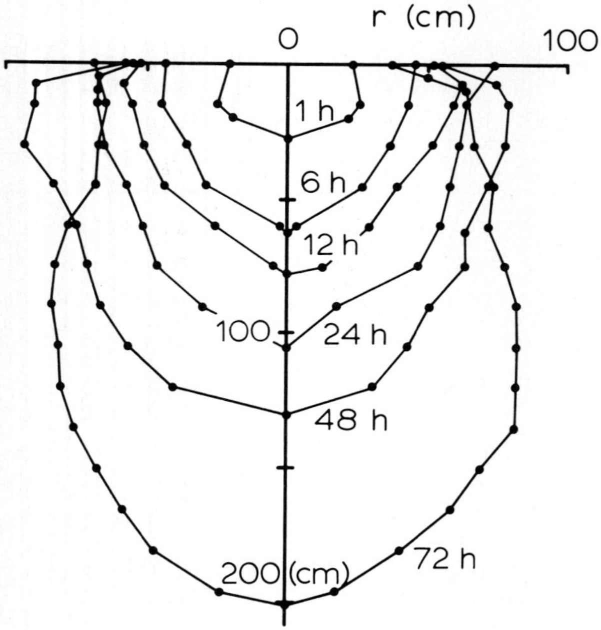

Fig. 2.3.4 Wetted profiles for 4 liters/hr tests with Superstition sand (after Roth, 1982).

Example 2.3.6 Unsaturated K for the exponential with h form.

Problem: Plot log K vs. h for Pima clay loam using α and K_o from table 2.3.1. If K and h were tabulated values only, how would you find K_o and α ?

2.3.2 *Soil water modeling*

Models take many forms — including physically-scaled, analog and mathematical models. Laboratory hydraulic models have been used to depict conditions for movement from trickle sources. In the following sections, however, the emphasis will be on mathematical models — both simple formulas based on volume balance and complex solutions to Richards' equation depicting water movement.

2.3.2.a Simplified models based on volume balance

A simple relationship follows for infiltration from a point (single) emitter by assuming (1) the wetted volume is everywhere a

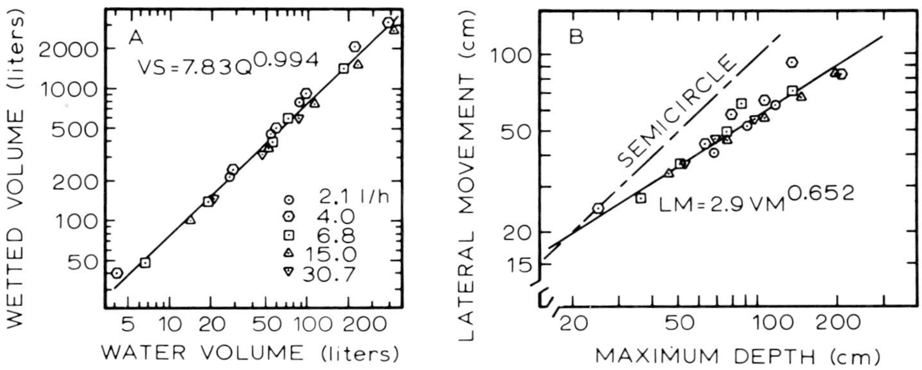

Fig. 2.3.5 Regression results for wetted soil volume VS and total water Q (A) and for maximum lateral movement LM and depth VM for Superstition sand (B) (after Roth, 1982).

constant value, say Θ_o, (2) the wetting front advances as a hemisphere and (3) the initial soil water is Θ_i. With these assumptions, conservation of mass requires that

$$q_e = (2/3)\pi(\Theta_o - \Theta_i)r_f^3 \text{ (hemisphere)} \tag{2.3.12}$$

with q_e the emitter discharge and r_f the wetting front.

The preceding relationship has been generalized based on field tests on the Superstition fine sand (Roth, 1982) . Various flow rates and amounts were used on 22 proximate locations. Wetted soil volumes were estimated by excavating quickly with a "back hoe," and measuring the wetted front position. Sample wetted volumes for the 4 liter/hr test are shown in figure 2.3.4. Note for the larger times the fronts are much deeper, but that the horizontal movement is not much more than for earlier times. The results for wetted volumes are summarized by figure 2.3.5.a. All of the results follow the power relationship with VS the wetted volume of soil (liters) and Q the total water added (liters)

$$VS = (7.83)Q^{0.994} \tag{2.3.13}$$

The fact that the exponent is very close to 1 and that the coefficient of determination is high (0.995) indicated that for these conditions the wetted soil volume is nearly independent of the discharge rate.

Shown in figure 2.3.5.b is a plot of maximum lateral movement (LM) vs. maximum vertical movement (VM) approximated by

$$LM = (2.90)\ VM^{0.652} \tag{2.3.14}$$

where LM and VM are expressed in cm. The VM is greater as expected due to gravity. For comparative purposes, the 1:1 line using the hemispherical approximation is also shown.

The ratio

$$c = LM/(VS)^{1/3} \qquad (2.3.15)$$

was examined for the Superstition sand as well as for laboratory results on the Gadsen clay. The c's satisfied approximately a linear relationship

$$c = 0.93 - 0.15\ (VM/LM) \qquad (2.3.16)$$

Of course, for the hemispherical model, c would be $r/[(2/3)\pi\ r^3]^{1/3} = 0.78$ and VM/LM would be 1 which is reasonably close to the data points.

Equations 2.3.13 to 2.3.16 may be combined to find LM given the total discharge Q for the Superstition sand. This results in the implicit form

$$LM = 18.5\ Q^{0.331}\ (1 - 0.0292\ LM^{0.534}) \qquad (2.3.17)$$

where LM is in cm and Q is in liters. (Note: The coefficients, 7.83 of equation 2.3.13, 2.90 of 2.3.14, and 18.5 and 0.0292 of 2.3.17 are all unique to the units indicated.) Equation 2.3.17 may be solved for LM several ways. The foolproof method is to specify LM, solve for Q and make a plot of Q vs. LM or log Q vs. LM. A second method is by successive approximations by estimating LM, substituting the estimate into the right-hand side and solving for an improved LM. The improved LM is then used a new estimate and the process repeated. This is demonstrated in the next example problem.

Example 2.3.7 Estimating wetting fronts.

Problem: Assume 3 liters/hr are added from a single emitter for 12 hours on Superstition sand.

a. If the soil water content increases on the average from $\theta = 0.05$ to 0.15, what is the wetted soil volume using the hemispherical model?

b. What is the wetted radius using the hemispherical model?

c. What is the wetted soil volume using the regression relationship 2.3.17?

d. What is the maximum lateral movement LM and maximum vertical movement VM?
(Hint: Use the answer from "b" as an initial estimate and obtain successive approximations to LM by equation 2.3.17.)

Answers: a. 360 liters

b. 55.6 cm

c. 276 liters

d. LM = 46.8 cm, VM = 71.2 cm

2.3.2.b Models based on Richards' equation

A general continuity relationship valid over any volume of soil is

I Rate of change of water stored	=	II Net rate of flow into the volume	–	III Net plant water uptake rate

In particular, if the volume is a small element ΔV then I, II and III will be given by

I. $(\partial\Theta_v/\partial t)\Delta V$

II. $[-(\partial v_x/\partial x) - (\partial v_y/\partial y) - \partial v_z/\partial z]\Delta V$

III. $S\Delta V$

where v_x, v_y and v_z are the components of the velocity (equation 2.3.7) and S is defined as the volume of water extracted per unit volume of soil per unit time. By use of Darcy's law and the above continuity relationship, the result is Richard's equation

$$\partial\Theta_v/\partial t = \partial(K\partial H/\partial x)/\partial x + \partial(K\partial H/\partial y)/\partial y + \partial(K\partial H/\partial z)/\partial z - S \quad (2.3.18)$$

or in vector notation

$$\partial\Theta_v/\partial t = \nabla\cdot(K\nabla H) - S \quad (2.3.19)$$

with "∇" the vector gradient operator.

Richards' equation is difficult to solve for several reasons:

(1) there are 2 dependent variables Θ_v and H.

(2) the relationship is nonlinear as K is a function of water status.

(3) the uptake function S is nebulous.

The first difficulty can be circumvented assuming H = h + z only and taking $h = h(\Theta_v)$, i.e. h as a unique function of Θ_v. The left-hand side of equation 2.3.18 is written alternatively as

$$\partial\Theta_v/\partial t = (d\Theta_v/dh)\partial h/\partial t = C(\partial h/\partial t) \quad (2.3.20)$$

where

$$C = d\Theta_v/dh \quad (2.3.21)$$

is the "specific water capacity" (units are L^{-1}). With these substitutions, equation 2.3.19 becomes

$$C(\partial h/\partial t) = \nabla\cdot(K\nabla h) + \partial K/\partial z - S \quad (2.3.22)$$

Similarly, h could be taken as a function of Θ resulting in

$$\partial\Theta_v/\partial t = \nabla\cdot(D\nabla\Theta_v) + \partial K/\partial z - S \quad (2.3.23)$$

with the soil water diffusivity defined as

$$D = K/C \quad (2.3.24)$$

and having units L^2/T.

Except for a few special cases, Richards' equation can be solved only by numerical techniques -- e.g. finite differencing and finite elements. Examples of numerical techniques applied to trickle problems with the general references listed at the end of the chapter. Rather than consider the more general conditions, we will concentrate here on a specialized case for which Richards' equation becomes linear and for which analytical solutions exist.

Richards' equation greatly simplifies for steady-state conditions when $K = K_o\exp(\alpha h)$. Taking the z-axis positive for the downward direction, the steady-state form without uptake is

$$\nabla^2\phi - \alpha\partial\phi/\partial z = 0 \quad (2.3.25a)$$

where

$$\phi = (K_o/\alpha)\exp(\alpha h) \quad (2.3.25b)$$

Equation 2.3.25a is linear (i.e., linear combinations of solutions are also solutions) and is amenable to standard analytical techniques. Geometries and solutions to equation 2.3.25a relevant to trickle systems can be made.

Solutions for point sources

Some possible emitter geometries are shown in figure 2.3.6A-F. (When the horizontal axis is labeled as x or r, the figure is valid for both 2 and 3-dimensional cases.) In 2.3.6A, a source of water provides a constant inflow q and the boundaries extend to infinity. For the point source $\phi = \phi_{3B}$ is

$$\phi_{3B} = (\alpha q/8\pi\rho)\exp(Z - \rho) \quad (2.3.26)$$

with

$$Z = \alpha z/2$$

$$R = \alpha r/2$$

$$\rho^2 = Z^2 + R^2 \quad (2.3.27)$$

This is one of the simplest analytical expressions relevant to trickle irrigation. It is useful for calculating pressure head distributions near an emitter, especially for large times, for approximating steady conditions or for evaluating conditions far away from the surface. In order to calculate a pressure head, values of α, K_o and q must be known or assumed. First ϕ_{3B} is found by equation 2.3.26 and then h by equation 2.3.25b. For all of the point and line source solutions, ϕ becomes undefined as the singularity point is approached. The region for which ϕ is large and $h > 0$ should be disregarded or an alternative solution sought.

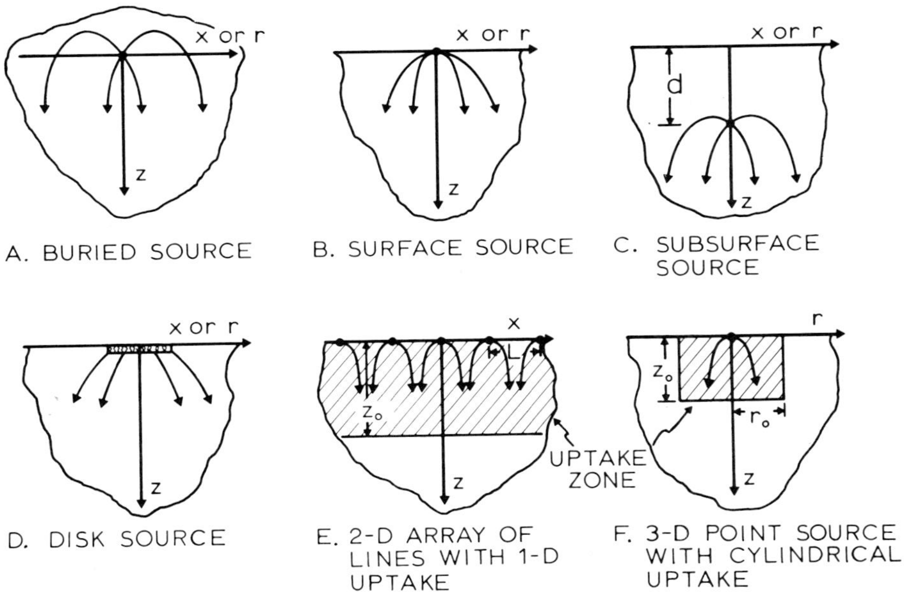

Fig. 2.3.6 Trickle flow geometries.

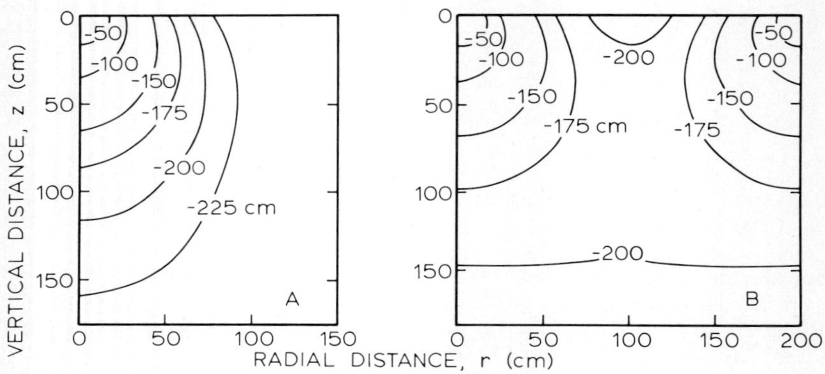

Fig. 2.3.7 Soil water pressure distribution for a single (A) and double source (B), (see examples 2.3.8 and 2.3.9).

A more realistic model for a point emitter is given in figure 2.3.6B where the flow region is semi-infinite. If no flow occurs through the surface away from the source, i.e.

$$v_z = 0, \quad z = 0 \text{ and } r > 0 \tag{2.3.28}$$

then the solution is

$$\phi_{3S} = 2\{\phi_{3B} - \exp(2Z)\int_Z^\infty \exp(-2Z')[\phi_{3B}]_{Z=Z'}dZ'\} \tag{2.3.29}$$

The integration result is listed in table 2.3.2.

Example 2.3.8 Steady-state point source

Problem: Assume $K = 9.9 \exp(0.014\ h)$ cm/day (with h as cm) and $q = 48$ liters/day. Calculate h for several points for $r < 100$ and $z < 100$ cm using equation 2.3.26 for an unbounded domain. Compare results to figure 2.3.7A for which the surface is assumed a barrier. For which solution is h less? Where in the profile do the results tend to agree more favorably?

Example 2.3.9 Double point sources

Problem: Repeat the above example, but assume you have 2 point sources separated by 100 cm. How many planes of symmetry exist for a double source? Compare results to figure 2.3.7B.

TABLE 2.3.2

Solutions to equation 2.3.25a for point and line sources. To convert ϕ to h use equation 2.3.25a.

Point sources	
1. Buried (Figure 2.3.6A)	$\phi_{3B} = (\alpha q/8\pi\rho)\exp(Z - \rho)$ $Z = \alpha z/2 \quad R = \alpha r/2$ $\rho^2 = R^2 + Z^2$
2. Surface (Figure 2.3.16B)	$\phi = (\alpha q/4\pi)[(1/\rho)\exp(Z - \rho) - \exp(2Z)E_1(Z+\rho)]$ $E_1(u) = \int_u^\infty t^{-1}\exp(-t)dt$ ("exponential integral")
3. Array	$\phi = \phi_1 + \phi_2 + \ldots\ldots$ where ϕ_1, ϕ_2 is for an individual source
Line sources	
1. Buried (Figure 2.3.6A)	$\phi_{2B} = (q/2\pi)\exp(Z)K_o[(X^2 + Z^2)^{0.5}]$ (K_o modified Bessel function of the second kind)
2. Surface (Figure 2.3.6B)	As in equation 2.3.29
3. Subsurface (Figure 2.3.6C)	$\phi = \phi_{2B}(X,Z-D) + \exp(-2D)\phi_{2B}(X,Z+D) + \int_{Z+D}^{\infty} \exp[2(Z-Z')]\phi_{2B}(X,Z')dZ'$ $(D = \alpha d/2)$
4. Array of lines (Figure 2.3.6E)	$\phi = \phi_1 + \phi_2 + \ldots$ where ϕ_1, ϕ_2, ... for individual sources.

Solutions for discs and cylinders

For most soils and flow rates, some ponding will occur at the soil surface. This is particularly significant for high flow rates and soils of low permeability. The radius of the saturated zone may be shown to be approximately

$$r_o = [(4/\alpha^2\pi^2) + (q/\pi K_o)]^{0.5} - 2/\alpha\pi \qquad (2.3.30)$$

(see Bresler, 1978, esp. p. 12). Note that r_o will be large as q increases or K_o (the saturated conductivity) decreases. Relevant solutions exist and are referenced at the end of the chapter. The solution for a cylindrical sink (figure 2.3.6F) is also available. This is of use for modeling plant water uptake, particularly for trees and shrubs for which the root system is symmetric. Such an example will be demonstrated shortly in section 2.3.3.

Solutions for line sources

If the emitters are sufficiently close, then the flow system can be analyzed as for a line source. The simplest case is as in figure 2.3.6A. For a line source in infinite space, $\phi = \phi_{2B}$ is

$$\phi_{2B} = (q_L/2\pi)\exp(Z)\mathbf{K}_o[(X^2 + Z^2)^{0.5}] \qquad (2.3.31)$$

with $\mathbf{K}_o$ a modified Bessel function of the second kind, and q_L the line strength (units L^2/T) and $X = \alpha x/2$. The solution for a line source at the surface is by equation 2.3.29 where ϕ_{2B} is substituted for ϕ_{3B}. The integral needs to be evaluated numerically, but is generally well behaved. The solution for a subsurface line (figure 2.3.6C) can be modeled either by ϕ_{2B} or by equation 2.3.31.

Example 2.3.10 Calculation of h near a single line source

Problem: Assume K_o and α as in example 2.3.6. Take q_L = 240 cm^2/day and assume a single line source. Calculate h for several z and r values by equation 2.3.31.

2.3.3 *Emitter spacing and discharge*

The design and operation of trickle systems must integrate the plant, soil and system parameters. For modeling, gross simplifications are necessary. As a consequence, numerical results must be complemented by common sense and experience.

For our purposes, a steady-state analysis is assumed sufficient. Obviously, plant water uptake will vary with plant growth stage, evapotranspirative demand and available water. A steady-state analysis will be most nearly valid for relatively constant environmental conditions, a plant canopy that is stable and for highly frequent applications of water. With this in mind, we relate the following factors for the irrigation, soil and plant systems:

(1) Irrigation
average discharge q (or q_L)
spacing
depth (for buried lines)
(2) Soil
hydraulic properties K_o and α
soil moisture characteristic Θ vs. h
a reference point value $h = h_M$
(3) Plant
average ET rate
rooting depth

Of these factors, some are invariant, some are consequences of the others and some are defined by the system. For example, the soil properties may be taken invariant (although secondary effects of compaction etc. may dominate a specific situation). The rooting depth and average ET are physiological in nature, dependent on climate and interrelated with the other factors as well.

Following are examples utilizing solutions of the steady-state Richards' equation in the last section. The "answers" obtained are exact for the assumed relationships but should not be viewed as precise answers for a real situation. Nevertheless, the relative effects of spacings, discharge, root zone depth, etc. should be mimicked correctly. Additional nomographs are in Amoozegar-Fard et al. (1984).

Example 2.3.11 Calculation of h near a single point source

Problem: a. Assume a steady q of 24 liters/day and take K_o for the Yolo fine sandy loam of table 2.3.1. By consulting table 2.3.3, find the resulting pressure head at r - 100 cm and z = 100 cm.

In consistent units note

$$q = (2.4)(10)^4 \text{ cm}^3/\text{day}$$

$$K_o = (4)(10)^{-5}(8.64)(10)^4 = 34.6 \text{ cm/day}$$

$$\alpha = 0.025 \text{ cm}^{-1}$$

From table 2.3.3 with $\alpha r/2 = 1.25$ and $z = r$

$$(8\pi K_o/\alpha^2 q)\exp(\alpha h) \approx 0.339$$

Therefore, solving for h

$$h = (1/\alpha)\ln(\frac{0.339\ \alpha^2 q}{8\pi K_o})$$

$$= (1/0.025)\ln(0.0586) = -113 \text{ cm}$$

b. Suppose q is doubled to an average of 48 liters/day. What is the corresponding h, also at r = 100, z = 100 cm?

$$h = (1/0.025)\ln(0.117) = -85.8 \text{ cm}$$

(Note the soil is wetter than for part "a" as it should be.)

c. Suppose 2 sources are used and are separated by 200 cm. The discharge for each is 24 liters/day. What is the value of h midway between (100 cm from each) and at an 100 cm depth?

For midway between double sources, double the tabulated value:

$$h = (1/\alpha)\ln(\frac{0.678\ \alpha^2 q}{8\pi K_o}) = -85.8 \text{ cm}$$

which is the same answer as for part "b". Similarly, results can be obtained for the center point of n-sources located on the vertices of a n-sided regular polygon -- such as for an equilateral triangle or square.

Example 2.3.12 Determination of the pressure head near a point source assuming a cylindrical uptake.

Problem: Let us assume a uniform uptake, 15 liters/day, within a cylinder 85 cm in diameter and 100 cm in depth. The soil is a Pima clay with $\alpha = 0.014\ \text{cm}^{-1}$ and $K_o = 9.9$ cm/day. The point source is located at the soil surface 60 cm away from the central axis of the cylinder (i.e. p = 60 cm with q = 25 liters/day. The reference point (point of measurement) is at $z_o = 100$, $r_o = 85$ cm and the horizontal distance (r^*) between the point source and reference point is 70 cm.

1. Examine the appropriate nomograph (figure 2.3.8) where Φ is the dimensionless matric flux potential $8\pi\Phi/\alpha q$.

2. For the cylindrical uptake, find the value of $\alpha z_o = 1.4$ (Point I) on the lower part of the nomograph.

3. Draw a horizontal line to intersect the curve $\alpha r_o = 1.19 \approx 1.2$ (Point II).

4. Construct a vertical line to intersect the upper axis Φ_{sink} (Point III).

5. For the point source, find the value of $\alpha z_o = 1.4$ for the reference (see Point IV) on the upper right section of the nomograph.

6. Construct a vertical line to intersect curve $\alpha r^* = 1$, (Point V) and draw a horizontal line to intersect Φ source at VI.

7. Find the value of $u = q/q_{uptake} = 0.6$ on the scale (Point VII), extend a straight line through points III and VII to intersect $u\Phi$ at VIII.

8. Draw a vertical line from VIII and a horizontal line from VI to intersect at Point IX.

9. Follow the guide curves and find $\Phi_{total} = 0.55$ (Point X). (This is $\alpha q\phi/8\pi$)

10. Calculate $8\pi K_o/\alpha^2 q = 51$ (Point XII).

11. Connect points X and XII and read the value of $\alpha h = -4.5$ at XI.
Calculate $h = -320$ cm.

Example 2.3.13 Determination of spacing L for an array of subsurface lines

Problem: Suppose we want to calculate a spacing L for an array of line sources located at d = 20 cm depth for a Columbia sandy loam soil. We choose the reference point at a 10 cm depth giving $\alpha z_M = 1$. Given are the values of discharge, $q = 50$ $cm^3/cm/day$; the allowable potential at the depth $z_M = 10$ cm directly below the midpoint between the lines, $h_M = -70$ cm; and the soil hydraulic properties α and K_o equal to 0.1 cm^{-1} and 120 cm/day, (table 2.3.1), respectively.

1. Find the value of $\alpha h_M = -7.0$, Point I on figure 2.3.1.

2. Find $\alpha q/K_o = 0.042$, Point II.

3. Draw a line through αh_M and $\alpha q/K_o$ (Points I and II) to intersect with scale A, Point III.

4. Proceed horizontally to curve $\alpha d = 2$, Point IV.

5. Proceed vertically downward and read $\alpha L = 7.5$, Point V, then calculate $L = 75$ cm.

Example 2.3.14 Determination of discharge q for an array of line sources with plant uptake

Problem: Consider a Pima clay soil with $\alpha = 0.014$ cm^{-1} and $K_o = 9.9$ cm/day where an array of lines is located on the soil surface (i.e. $d = 0$) which are at a spacing $L = 200$ cm. The rooting depth $z_o = 100$ cm with the uptake amount as $u = 0.75$ cm/day. The matric potential at the reference point located at depth $z_M = 35$ cm directly below the midpoint between the lines is $h_M = -350$ cm.

1. Find $\alpha z_o = 1.4$, (Point I on figure 2.3.10).

2. Proceed horizontally to curve $\alpha z_M = 0.5$ (Point II).

3. Proceed vertically upward to curve $(K_o/u)\exp\alpha h_M = 0.098$ (Point III).

4. Proceed horizontally to intersect scale C, Point IV.

5. Find $\alpha L = 2.8$ (Point VIII).

6. Proceed vertically upward to curve $\alpha z_M = 0.5$ (Point VII).

7. Proceed horizontally to intersect scale A (Point VI).

8. Connect the two points on scales C and A (Points IV and VI).

9. Read the value of $\alpha q_L/u = 3.2$ (Point V), then calculate $q_L = 171$ cm^2/day.

Note that a discharge of 171 cm^2/day is equivalent to a depth of $q_L/L = 0.86$ cm/day over the entire area. Choosing the spacing between lines as 100 cm results in a calculated q_L of 75 cm^2/day.

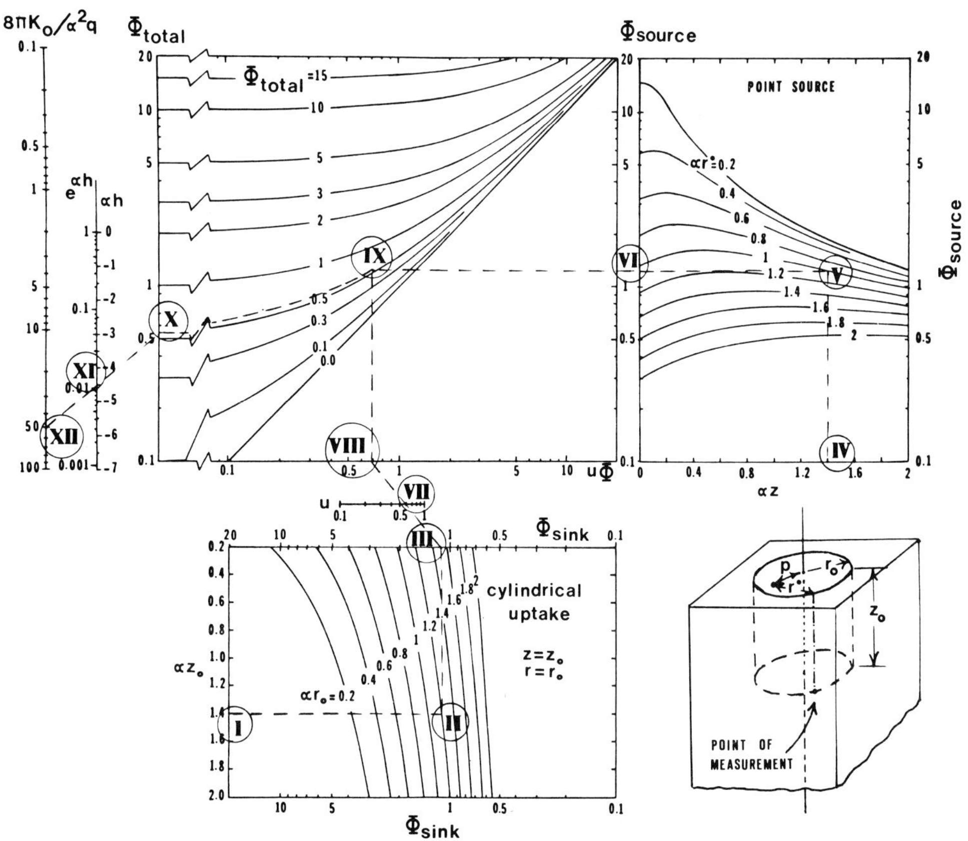

Fig. 2.3.8 Nomograph for point emitter on the surface and uptake from a cylindrical soil volume. The total water uptake per unit time is u, the cylinder is of radius r_o and depth z_o. The $\Phi = 8\pi\phi/\alpha q$ are dimensionless matric flux potentials (after Amoozegar-Fard et al., 1984).

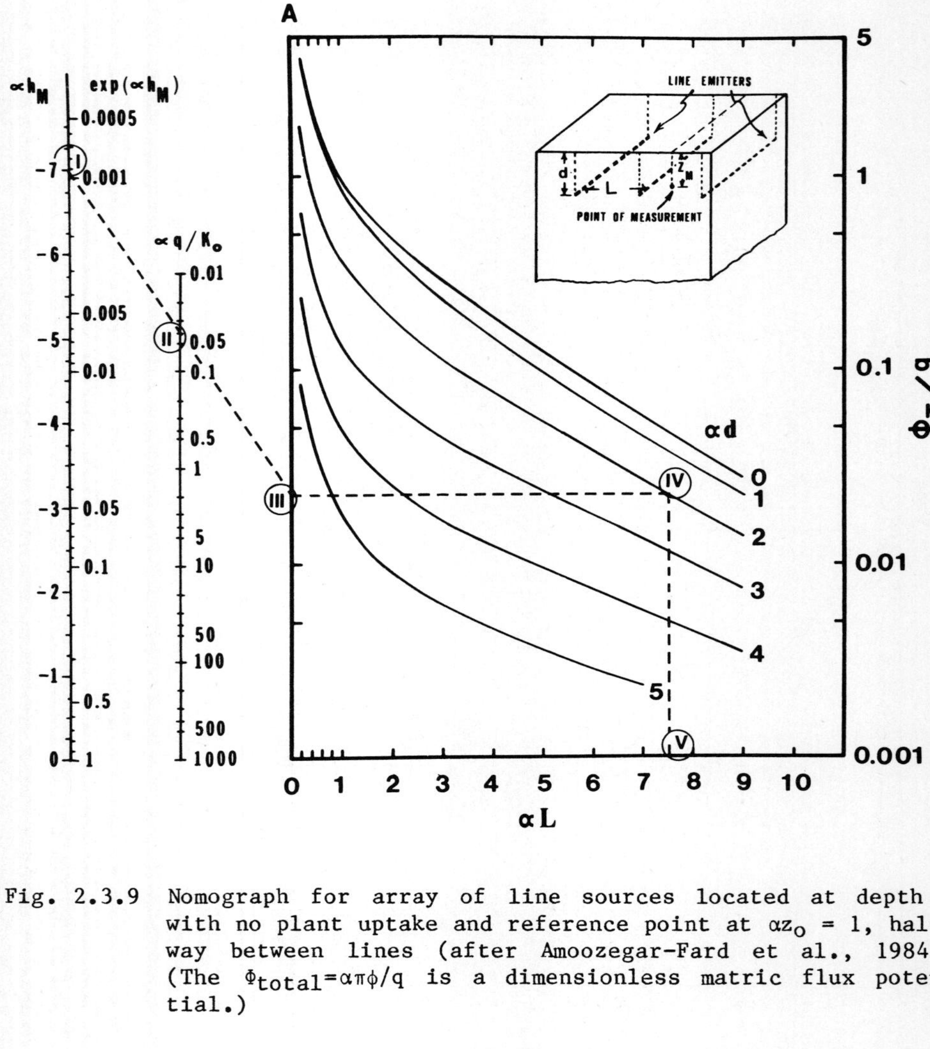

Fig. 2.3.9 Nomograph for array of line sources located at depth d with no plant uptake and reference point at $\alpha z_o = 1$, halfway between lines (after Amoozegar-Fard et al., 1984). (The $\Phi_{total}=\alpha\pi\phi/q$ is a dimensionless matric flux potential.)

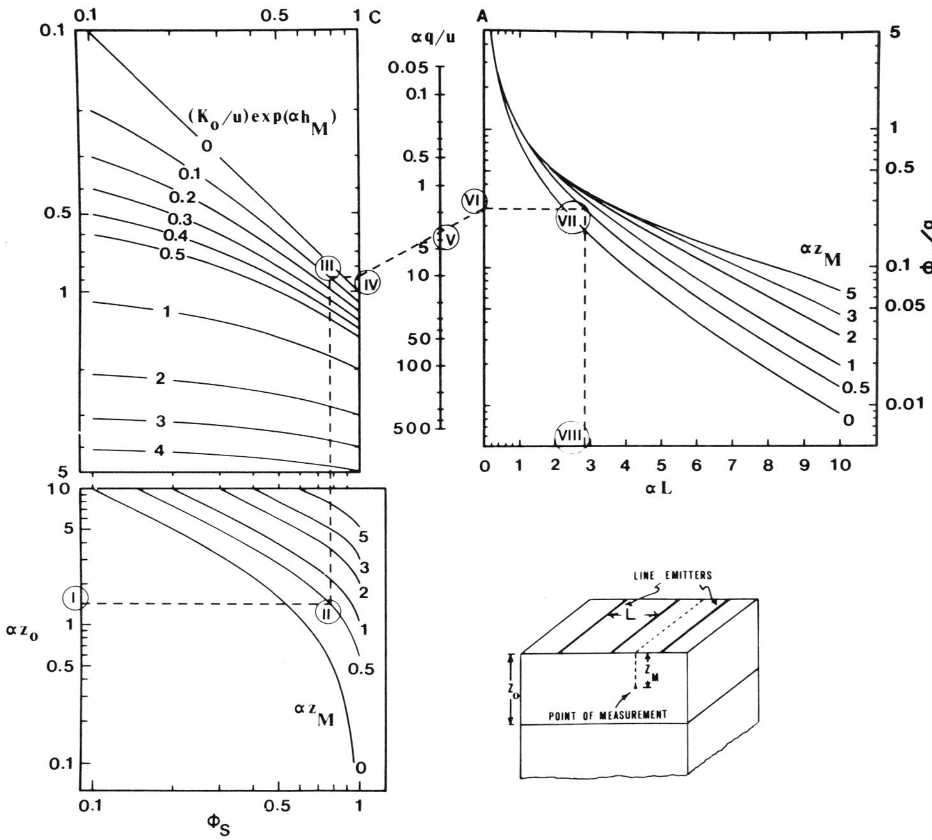

Fig. 2.3.10 Nomograph for array of line sources located at the soil surface with plant uptake to depth to depth z_o and reference point at depth z_M halfway between lines (after Amoozegar-Fard et al., 1984). (The uptake per unit total area is $\Phi_{total}=\alpha\pi\phi/q$, a dimensionless flux potential.)

TABLE 2.3.3

Generalized distance vs. pressure head relationships for a point source. (For double sources take r as the distance midway between emitters and double the tabulated h-expression.)

$\alpha r/2$	$(8\pi K_o/\alpha^2 q)\exp(\alpha h)$		
	z = 0	z = r/2	z = r
0.1	14.4	13.7	11.0
0.2	5.74	5.83	4.77
0.5	1.31	1.62	1.45
0.75	0.579	0.844	0.811
1.0	0.297	0.507	0.522
2.0	3.75×10^{-2}	0.117	0.155
5.0	3.99×10^{-4}	6.16×10^{-3}	1.61×10^{-2}
7.5	1.58×10^{-5}	8.29×10^{-4}	3.74×10^{-3}
10.0	7.66×10^{-7}	1.29×10^{-4}	9.81×10^{-4}

REFERENCES AND SUPPLEMENTAL READING

Amoozegar-Fard, A., Warrick, A.W. and Lomen, D.O., 1984. Design nomographs for trickle and subsurface irrigation. J. Irrig. and Drain. Div., Amer. Soc. Civil Eng. 110(IR2): 107-120.

Bresler, E., 1978. Analysis of trickle irrigation with application to design problems. Irrig. Sci. 1: 3-17.

Bucks, D., Nakayama, F.S. and Warrick, A.W., 1982. Principles, practices and potentialities of trickle (drip) irrigation. In Hillel, D., (Ed.) Advances in Irrigation 1: 219-295. Academic Press, New York.

Mualem, Y., 1976. A new model for predicting the hydraulic conductivity of unsaturated porous media. Water Resources Res. 12: 513-522.

Roth, R.L., 1982. Moisture Movement from a Point Source. Unpub. Ph.D. Thesis, The University of Arizona (available as DA 83 15305, University Microfilms, Ann Arbor, Michigan).

Van Genuchten, M.Th., 1980. A closed-form equation for predicting the hydraulic conductivity of unsaturated soils. Soil Sci. Soc. Amer. J. 44: 892-898.

Warrick, A.W., Lomen, D.O. and Tonellato, P.J., 1981. Soil moisture flow for point and line sources -- computer programs for linearized solutions. Agric. Eng. and Soil Sci., The University of Arizona, Tucson, Arizona, 78 pp.

Chapter 2

DESIGN PRINCIPLES

2.4 SALT DISTRIBUTION

R. S. BOWMAN AND F. S. NAKAYAMA

This chapter considers the role of salt distribution in soil-water-plant relationships, with special emphasis on trickle irrigation. Section 2.4.1 discusses general salinity effects on plants and the influence of irrigation methods on soil salt distribution. For the casual reader, this section will provide sufficient background for understanding differences in salt distribution under trickle as compared to other irrigation methods. Sections 2.4.2, 2.4.3, and 2.4.4 provide more detail on the behavior of salts in the soil. Section 2.4.2 discusses the important topic of salt leaching, and gives examples of the prediction of leaching requirements from irrigation water quality and salt tolerance data. Section 2.4.3 looks in detail at soil chemical reactions which influence salt distribution and soil physical properties, which in turn affect plant growth. Section 2.4.4 reviews current efforts to model salt movement in soils, again placing special emphasis on salt distribution under trickle irrigation.

2.4.1 *Theory and concepts*

More than 200 million hectares in the world are currently under irrigation (Finkel, 1982). Of this irrigated land, at least one-third is estimated to be salt-affected, i.e., salts are present at levels sufficient to interfere with optimum crop production (Skogerboe and Law, 1971). Salts in the soil water limit plant growth by making this water less available for uptake by plant roots. Salt movement is intimately tied to water movement, and therefore salinity management is largely a function of water management in any irrigation system. Trickle irrigation is a unique water application technique which introduces special problems and opportunities in controlling soil salt distribution.

2.4.1.a Water-salt interactions

Water moves from the soil into plant roots due to differences in potential between the bulk soil water and the root cell cytoplasm. Neglecting gravitational influences, the total soil water potential, Φ, in the vicinity of the plant root is determined by the sum of the matric potential, h, and the osmotic potential, π:

$$\Phi = h + \pi \qquad (2.4.1)$$

Since the potential of free water at the surface of a pool of pure water at 25°C is defined as zero, water in the presence of matric or osmotic forces which restrict the freedom of movement of water molecules has a potential of less than zero.

The matric potential results from adhesive forces between the water and soil particles and is negative under unsaturated conditions. Thus, h represents the energy required to pull or suck this adhering water away from the soil, and the absolute value of the matric potential (-h) is sometimes referred to as the soil water suction. h is a function of both soil water content and soil composition. For a given volumetric soil water content, Θ, h will be greater (less negative) in a coarse-textured soil with high sand-size fraction than in a fine-textured soil with high clay-size fraction. In any given soil, h decreases as Θ decreases. Thus, as a soil dries out between rainfall events or irrigations, h becomes more negative and, all other factors being equal, plants have more difficulty extracting water from the soil.

The osmotic potential, π, is a function of the salt concentration in the soil water. As the salt concentration increases, π becomes more negative. The causes of the osmotic effect can be envisaged in several ways. Since individual salt particles attract and bind water molecules, energy is required to separate the water from the salt. Root cell membranes allow water to pass through freely, but exclude or limit the passage of most dissolved solutes; therefore, the potential for water movement into the root is lowered as the salt content of the soil solution rises. Another way of viewing the osmotic effect is to imagine water molecules from the soil solution colliding with root cell membranes, with some of the molecules passing through the membrane into the cytoplasm. As the salt concentration increases, more salt particles (and hence, fewer water molecules) impact upon the cell membrane, and less water passes through the membrane per unit area per unit time compared to salt-free water.

A theoretical equation which relates π to salt content for dilute salt solutions is (Dickerson, 1969):

$$\pi = -RTC_s \tag{2.4.2}$$

where R is the gas constant, T is the absolute temperature, and C_s is the molar concentration of individual salt particles. At 25°C, RT has a value of 2.47 MPa L mol^{-1} (one MPa equals ten bars, or approximately ten atmospheres pressure).

Example 2.4.1

Problem: Estimate the osmotic potential of an irrigation water which has a salt concentration of 550 mg/L, with the major cations and anions being calcium and chloride, respectively.

Solution: We need first to calculate the molarity of $CaCl_2$ in the water

$$\text{Molarity } CaCl_2 = \frac{550 \text{ mg } CaCl_2}{L} \times \frac{1g \ CaCl_2}{1000 \text{ mg } CaCl_2}$$

$$\times \frac{1 \text{ mole } CaCl_2}{111 \text{ g } CaCl_2}$$

$$= 5.0 \times 10^{-3} \text{ mol/L}$$

Each mole of $CaCl_2$ dissolves in water to yield one mole of Ca^{2+} ions, and two moles of Cl^- ions. Thus

$$\text{Molarity of salt particles} = 5.0 \times 10^{-3} \frac{\text{mole } Ca^{2+}}{L}$$

$$+ 2(5.0 \times 10^{-3} \frac{\text{mole } Cl^-}{L})$$

$$= 1.5 \times 10^{-2} \text{ mol/L}$$

Now we use equation (2.4.2) to calculate π

$$\pi = -(2.47 \text{ MPa L/mol})$$

$$(1.5 \times 10^{-2} \text{ mol/L})$$

$$= -3.7 \times 10^{-2} \text{ MPa}$$

$$= -37 \text{ kPa}$$

Thus, the osmotic potential of this irrigation water, which has a composition typical of an irrigation water of the western United States, is in the range of minus one-third bar (one bar = 100 kPa).

The above example serves to illustrate the relationship between salt concentration and the osmotic potential of the soil solution. For more concentrated solutions of mixed salts, equation 2.4.2, which was derived for ideal solutions, is not appropriate and overestimates the osmotic effect. Empirical relationships between osmotic potential and the electrical conductivity of soil solutions, which is a measure of their salt content, have been developed (United States Salinity Laboratory Staff, 1954).

For the relatively high quality irrigation water of example 2.4.1, the decrease in water potential due to salt is about equal to the matric potential at "field capacity" (considered to be in the range of -10 to -33 kPa). Compared to a soil water potential of -1.5 MPa, at which many plants show irreversible symptoms of water stress, the decreased potential due to salt in this water is not very significant. But for a crop irrigated with such a water, -37 kPa is the highest value of π, and of Φ, which is possible in the soil solution. As the soil dries due to plant water uptake and evaporation from the soil surface, the concentration of salts in the soil solution increases proportionately. A fivefold concentration of the soil solution caused by evapotranspiration would result in a significant lowering of the soil water potential due to osmotic effects. Decreased osmotic potentials following soil drying occur simultaneously with lowered matric potentials as the soil water content decreases. For saltier irrigation waters, with initial salt contents of several thousand mg/L, the osmotic contribution as the soil dries is even more significant.

As mentioned earlier, the flow of water from the soil into the plant roots depends on the water potential gradient, with water flowing from a higher potential in the soil to a more negative potential in the plant root cell cytoplasm. The water potential in turgid plant roots adequately supplied with water is in the range of -0.5 to -1 MPa. This negative potential arises mainly from low values of π in the root cellular solution due to the high concentration of solutes. As Φ decreases, and thus as the potential gradient between soil and root cell water decreases, plants respond by manufacturing solutes such as organic acids. This increases the concentration, and lowers π, of the cellular solution, and causes a steeper potential gradient between cellular solution and soil solution. Although such a response helps increase the water supply to the plant, it occurs at the expense of diverting energy and metabolites that otherwise would be used for plant growth. A decrease in π due to the salinity of the soil solution thus ultimately results in lowered crop yield, even for apparently unstressed plants.

Certain salts can also induce specific toxicity symptoms in plants. Sodium, chloride, and boron are examples of species, found in soils and irrigation waters, which can be harmful to plants when their concentrations become too great. Such toxicity effects are generally localized and often crop-specific, and will not be discussed further in this chapter.

2.4.1.b Salt distribution and irrigation method

Since the presence of salts in the soil solution affects plant growth primarily by lowering the soil water potential, the success of any irrigation regime is partially dependent on its influence on salt movement and distribution. One of the major benefits of a properly designed trickle irrigation system is the ability to minimize the salt concentration of the soil water in the vicinity of plant roots.

Salt movement, and hence salt distribution, in soils, is directly related to water movement. As water moves within the soil by mass flow or diffusion in response to water potential gradients, it carries along with it a burden of soluble salts. Except under very dry conditions

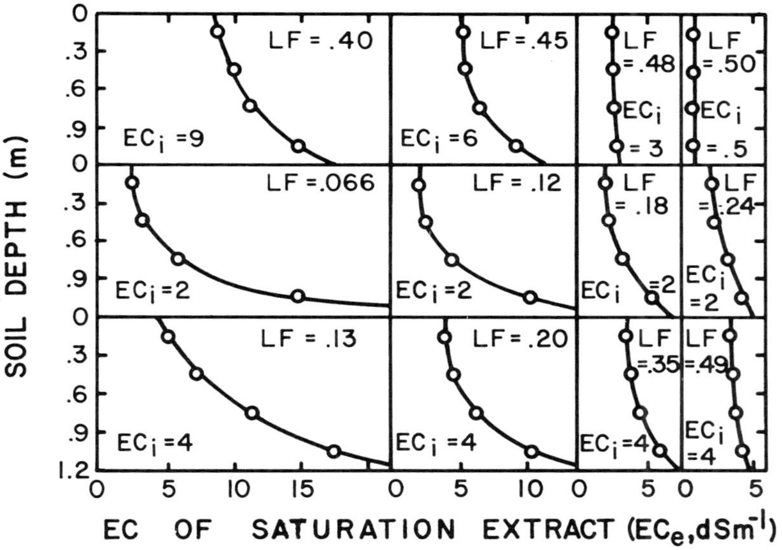

Fig. 2.4.1 Steady state salt profiles, expressed as electrical conductivity of the saturation extract (EC_e, dS/m), as influenced by the electrical conductivity of the irrigation water (EC_i, dS/m) and leaching fraction (LF), (after Bower et al., 1969).

where water flow is slight, most transport of salts occurs in response to such water movement rather than self-diffusion due to gradients in salt concentration. Only a small fraction of the soil water salt load is taken up by plant roots, and no salt is lost to the atmosphere along with water evaporated from the soil surface. As a result, salt tends to concentrate in areas of water depletion such as the plant root zone and the soil surface. Continued water extraction results in increasing salt concentrations in the soil solution, and eventually to precipitation of salts. In order to prevent salt buildup to levels harmful to plants, leaching in the form of rainfall or irrigation water is necessary to dilute or wash out excess salt.

Figure 2.4.1, taken from Bower et al. (1969) presents steady-state salinity profiles for alfalfa after long-term leaching with irrigation waters of several salinities at different leaching fractions. The leaching fraction is a measure of the degree of leaching of the soil profile, and is discussed in greater detail in section 2.4.2.b. With the exception of the high leaching fraction (ca. 0.50) treatments, the effects of root water extraction on rootzone salinity are clearly indicated. Near the soil surface, where salts are regularly washed downward by fresh irrigation water, the soil salinity, as indicated by the electrical conductivity of the saturation extract, resembles that of the irrigation water. Salinity increases with depth due to root

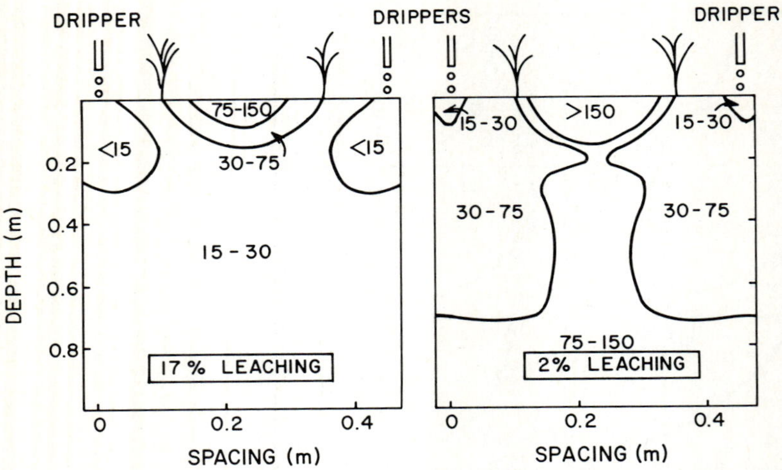

Fig. 2.4.2 Salt distribution around trickle lines with different leaching fractions. Numbers within each zone represent chloride concentration in the soil water (mol m^{-3}), (after Hoffman et al., 1980).

extraction of soil water and the associated concentration of salts in the soil solution. The profiles presented in figure 2.4.1 are representative of salt distributions expected after flood irrigation of a field for several years. Depending upon the irrigation frequency, some salt accumulation could occur near the soil surface between irrigations due to water evaporation from the soil to the atmosphere.

In contrast to the profiles developed under uniform flooding, salt distributions under furrow (line source) and trickle (point source) irrigation regimes vary areally as well as with depth. The actual distribution achieved under trickle irrigation depends upon the salinity of the irrigation water, the leaching fraction, the frequency of irrigation, the spacing of the trickle emitters, and the pattern of root water uptake. An example of the type of salt distribution encountered under a line source (in this case, actually a moving trickle source) is presented in figure 2.4.2.

Several characteristics of the salt distribution (as indicated here by chloride concentrations) presented in figure 2.4.2 are of note. Even at the lower leaching fraction, a zone of low salinity exists close to the water source. As with profiles developed under uniform flood irrigation, the salinity of the soil increases with depth in the rootzone. In addition, however, salinity near the soil surface increases with distance from the water source. This results from water movement away from the water source and the subsequent deposition of salt as the water evaporates from the soil surface. Salt continues to accumulate in this surface zone since no downward leaching with low salt water occurs in this area.

Although figure 2.4.2 presents salt distributions for a line source irrigation system, the salt distribution from a single point emitter (as in a trickle system) would be similar, with the exception that the salt profile would extend radially from the source in three dimensions.

Where emitters are spaced closely enough to permit overlap of adjacent wetted zones, the greatest salt accumulation at the surface occurs midway between the emitters. Figure 2.4.3 shows the soil water potential distribution for trickle irrigated tomatoes grown in lysimeters with a 1.1-m emitter spacing. The water potentials indicated are almost entirely due to the osmotic potentials within the soil solution in these well-watered lysimeters. As salt is transported with the moving soil water away from the emitters, it is concentrated at the point of maximum lateral water movement.

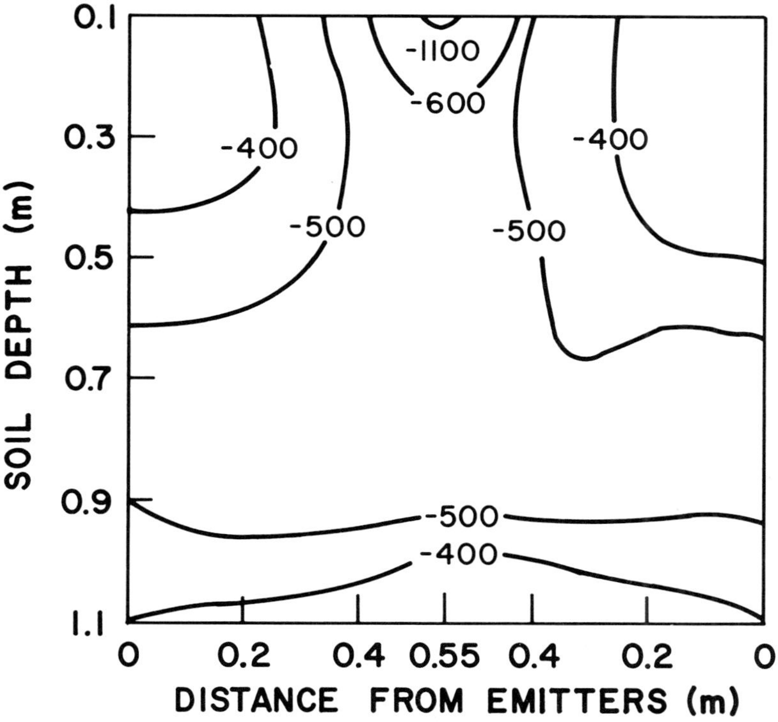

Fig. 2.4.3 Water potential pattern (kPa) due to soil salinity under trickle irrigation with overlapping wetted zones (after Tscheschke et al., 1974).

2.4.2 *Leaching*

Since excess soil salinity hinders plant growth, the design and management of any irrigation system must make provisions for maintaining root zone salinity at acceptable levels. This section describes principles of salt management which are applicable to irrigation systems in general, including trickle irrigation systems.

2.4.2.a The salt accumulation problem

Soil salinity, which affects crop productivity, can be considered both a short-term and a long-term problem. An example of a short-term problem is the situation in which the soil is already saline or sodic and seed germination and seedling establishment are of immediate concern. Salt-tolerant crops (chapter 4.4) can be used to alleviate some of the problems, but germination remains a major obstacle in such instances. Trickle irrigation can be used to ameliorate excess salts in the area close to the seed or root zone, but is not a practical solution for land reclamation where salt removal is the main objective. However, we can learn much from the experiences and practices used in other types of irrigation methods for removing and controlling salinity and sodic conditions of soils.

The long-term build up of salts has received much attention in all types of irrigation systems. Productive areas have been reduced to wastelands through poor management and the lack of understanding about salt balance and removal.

Example 2.4.2

Problem: Given irrigation waters with salt contents of 300, 600, and 1200 mg/L. Find the amounts of salt that would be applied with 0.5 meter of each irrigation water over one hectare.

Solution: The quantity of salt can be calculated by the relation:

Amount of salt = concentration x volume

Assuming that the specific gravity of the irrigation water is equal to one, 300 mg/L represents 300 gram of salt per 10^6 gram or per m^3 of water. The 0.5 m of water represents $0.5 \times 100 \times 100$ m^3 of water.

$$\text{Thus, the salt added} = (300 \text{ g m}^{-3}) \times (0.5 \times 10^4 \text{ m}^3)$$

$$= 1.5 \times 10^6 \text{ g or } 1.5 \text{ metric ton}$$

What would be the equivalent salt additon in terms of pounds per acre-foot of irrigation water?

The preceding example demonstrates the tremendous amount of dissolved salt that can be involved in irrigation. We have not specified how deep or the manner in which salts are distributed in the soil profile; this distribution pattern becomes a critical issue whenever plants are grown in such situations.

2.4.2.b Salt balance and the leaching fraction

The salt balance model for irrigation systems is based on long-term, steady state conditions. The model takes into account the salt input and removal and can be expressed by the following equation:

$$V_iC_i + V_gC_g + S_m + S_f = V_dC_d + S_p + S_c \qquad (2.4.3)$$

where V_i and C_i are the volume and salt concentration of the irrigation water, respectively; V_g and C_g the groundwater contribution from capillary rise; V_d and C_d refer to drainage water; S_m is the salt input from soil mineral weathering and salt deposits; S_f is the salt added as fertilizer or amendment; S_p represents the soluble salts in the irrigation water that precipitate in the soil; and S_c is the salt removed by the crop. When the S_m, S_p, S_f, S_c and the groundwater salt contributions are all considered negligible, equation 2.4.3 reduces to $V_iC_i = V_dC_d$. This form is often used for practical salt balance calculations.

The leaching fraction (LF) is defined as the fraction of the applied water (D_i) that actually appears as drainage water (D_d) and is represented by

$$LF = \frac{D_d}{D_i} \qquad (2.4.4)$$

In contrast, the leaching requirement (L_r) is the fraction of applied water that must pass through the root zone and become part of the drainage water to control soil salinity at any specified level. This root zone salinity that affects crop production is discussed in detail in chapter 4.4. The leaching fraction and leaching requirement are terms that are frequently used synonymously in the literature, but a distinct difference in origin and meaning exists.

The salt concentrations of the irrigation and drainage waters can give approximate estimates of LF via the relationship

$$LF = \frac{C_i}{C_d} = \frac{EC_i}{EC_d} \qquad (2.4.5)$$

where EC is the electrical conductivity of the waters, used as a measure of the salt concentrations C_i and C_d. Thus, it follows that

$$\frac{D_d}{D_i} = \frac{EC_i}{EC_d} \qquad (2.4.6)$$

When EC_d exceeds the value required for proper crop growth, the L_r is not being met. Equation 2.4.6 can be used to estimate the amount of additional irrigation water needed to increase LF until it is equal to L_r.

Example 2.4.3

Problem: Calculate the leaching requirement using irrigation waters with electrical conductivities of 1, 2, and 4 dS/m for different crops whose optimum production occurs at root zone salinities of 4, 8 and 16 dS/m.

The component D_i represents the volume of water used for crop production and includes plant transpiration and soil water evaporation (evapotranspiration or D_{et}), and the water for drainage, such that $D_i = D_d + D_{et}$. The leaching requirement can thus be expressed as $L_r = 1 - (D_{et}/D_i)$.

Other relations which can be developed in terms of the depth of water and including crop water use are

$$D_d = \frac{D_{et}}{1-L_r} \times L_r \qquad (2.4.7)$$

$$D_d = \frac{EC_i}{EC_d - EC_i} \times D_{et} \qquad (2.4.8)$$

The student should be able to derive these equations.

An irrigation efficiency term, E, is sometimes related to the total water applied and the crop water use as $E = D_{et}/D_i$ so that D_d becomes

$$D_d = D_i(1-E) \qquad (2.4.9)$$

Thus, when the irrigation efficiency is 80%, 20% of the irrigation water appears as drainage, and D_{et} would be four times D_d.

One distinct advantage of trickle over flood type irrigation under saline conditions is the ability to apply water at more frequent intervals, so that the salinity in the root zone can be maintained at an average level which is less than that existing under infrequent irrigation.

2.4.3 *Water-soil-salt interrelationship*

As water moves through the soil pores, various types of chemical reactions occur, including dissolution of clay minerals and salts, ion exchange between the solution and solid phases, and precipitation of dissolved constituents. Complex biochemical reactions brought about by microbial activity can also occur, affecting the oxidation and carbon dioxide status of the soil. The type and extent of reactions depend upon the quantity and composition of the applied water and the characteristics of the soil matrix. Such reactions can have a dramatic effect on soil salinity status. Reactions of particular concern are those which affect the soil hydraulic conductivity, which in turn influences moisture and salt distribution.

2.4.3.a Sodium status and hydraulic conductivity

The relationship of the monovalent sodium to the divalent calcium and magnesium cations in the irrigation water and soil exchange complex plays an important role in governing soil hydraulic conductivity. A special term, the sodium adsorption ratio, SAR, has been developed to define this relation as follows:

$$SAR = \frac{[Na^+]}{([Mg^{2+}] + [Ca^{2+}])^{1/2}} \qquad (2.4.10)$$

where the ionic concentrations, designated by brackets, are expressed in mmol/L.

Example 2.4.4

Problem: Given an irrigation water with 1.8 me/L Ca^{2+}, 1.0 me/L Mg^{2+} and 5 me/L Na^+. Find the SAR of the irrigation water. What happens to the SAR if salt concentrations increase by a factor of two following evaporation? By a factor of four?

Solution: First, the milliequivalents per liter must be converted to millimoles per liter. For divalent Ca and Mg, concentrations become 1.8/2 and 1.0/2, respectively. No change is needed for Na.
Therefore, $SAR = 5/(0.9 + 0.5)^{1/2} = 4.2$

High sodium concentration relative to the divalent ions calcium and magnesium increases the swelling of clay minerals and dispersion of soil aggregates, and consequently reduces the hydraulic conductivity of the soil matrix. This phenomenon is schematically illustrated in figure 2.4.4 for one type of soil material. Total electrolyte or soil solution concentration also is involved. For the same SAR, the hydraulic conductivity would be higher in the system with the higher salt concentration. Rainwater, which contains essentially no salt, can create structural and water conduction problems where the sodium

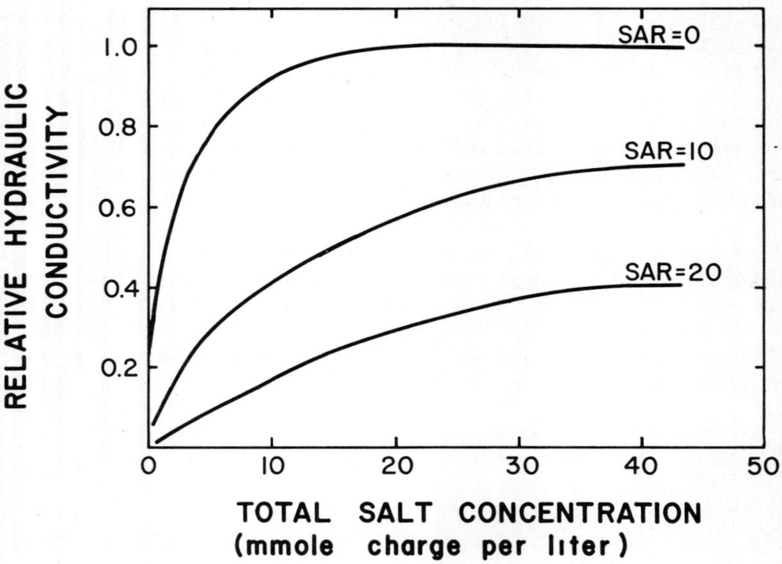

Fig. 2.4.4 Schematic diagram showing relationships among salt concentration, SAR, and hydraulic conductivity.

content or the SAR at the surface is high. When the SAR is 10 to 15 or above, soil hydraulic conductivity decreases and may cause infiltration problems.

2.4.3.b Solubility and precipitation concepts

Any change in the total electrolyte concentration or the relative concentration of an individual ion affects the SAR and ultimately the salt distribution. This can be caused by reactions such as precipitation of solute, dissolution of native minerals, fertilizer, and soil amendments, and ion exchange with the soil. The solubility product concept, which describes the equilibrium condition between the solid M_aA_m and its common ions M^{m+} and A^{a-}, is represented by an equation of the following form:

$$M_aA_m = aM^{m+} + mA^{a-} \tag{2.4.11}$$

The solubility product constant, K_s, is defined by

$$K_s = \frac{(M^{m+})^a \ (A^{a-})^m}{(M_aA_m)} \tag{2.4.12}$$

with the activities of the constituents denoted by parentheses. The activity of an ion in solution is related to, but generally less than, its concentration. Since the activity of the solid phase M_aA_m is designated as being equivalent to one, the solubility product constant is usually represented with only the ionic species.

The more common constants encountered in irrigation and soils are $K_s = (Ca^{2+})(CO_3^{2-})$ for calcium carbonate and $K_s = (Ca^{2+})(SO_4^{2-})$ for calcium sulfate or gypsum. Solubility product constants available from reference texts and handbooks are based on "pure" compounds at equilibrium. In complex soil solution systems, where supersaturation and ion-crystal interactions occur, solubility product constants are usually higher than the referenced constants (Suarez and Rhoades, 1982).

When calcium is removed from solution by precipitation, exchange, or absorption by plants, the SAR automatically increases (equation 2.4.10). This can lead to a decrease in hydraulic conductivity.

Precipitation reactions of calcium with sulfate or bicarbonate are given by $Ca^{2+} + SO_4^{2-} = CaSO_4(ppt)$, and $Ca^{2+} + 2HCO_3^- = CaCO_3(ppt) + H_2O + CO_2(gas)$. Estimating calcium carbonate precipitation is more involved than for calcium sulfate. In this instance, the calcium and bicarbonate concentrations, the pH, and the carbon dioxide partial pressure are needed together with the various constants that define the carbonate equilibrium system. A detailed discussion of this system is presented in chapter 3.2 on Water Treatment.

Pursuing the calcium sulfate example, we can envision a situation where rainwater infiltrates into a soil which contains solid-phase gypsum, $CaSO_4 \cdot 2H_2O$. Gypsum dissolves in the infiltrating water according to the relation

$$(Ca^{2+})(SO_4^{2-}) = K_s \qquad (2.4.13)$$

Example 2.4.5

Problem: (a) Calculate the equilibrium Ca^{2+} concentration when rainwater passes through a soil containing solid phase gypsum. The K_s for gypsum is $10^{-4.64}$.

(b) Estimate the saturation status of irrigation water at 25°C with 2.5×10^{-3} mol/L Ca^{2+} and 3.0×10^{-3} mol/L SO_4^{2-}.

Solution: (a) The pertinent dissolution reaction is

$$CaSO_4 \cdot 2H_2O \text{ (gypsum)} \rightleftarrows Ca^{2+} + SO_4^{2-} + 2H_2O$$

Since one calcium ion and one sulfate ion are produced for each molecule of gypsum which dissolves

$$(Ca^{2+})(SO_4{}^{2-}) = K_s = 10^{-4.64}$$

$$(Ca^{2+})^2 = 10^{-4.64}$$

$$(Ca^{2+}) = 10^{-2.32}$$

Thus, the equilibrium calcium ion activity is $10^{-2.32}$ = 0.00479. Due to ionic strength effects, the calcium ion concentration in a saturated gypsum solution is about double this value, or about 0.01 mol/L. For further discussion of activity/concentration relationships, see Bohn et al. (1979) or Lindsay (1979).

(b) The product of the calcium and sulfate concentrations is

$$(2.5 \times 10^{-3})(3.0 \times 10^{-3}) = 7.5 \times 10^{-6} = 10^{-5.12}$$

Since the concentration product is less than K_s, the solution is undersaturated and no gypsum precipitation will occur.

Predict what would happen if the solution were concentrated by a factor of three following evaporation of part of the irrigation water.

Interaction with other solution species can change the concentrations of cations in solution and thereby influence their solubility and exchange behavior. Chloride ion, for example, combines with calcium ion in solution to form the monovalent calcium chloride ion pair

$$Ca^{2+} + Cl^- \rightleftarrows CaCl^+ \tag{2.4.14a}$$

while sulfate reacts with calcium to form the uncharged calcium sulfate ion pair

$$Ca^{2+} + SO_4{}^{2-} \rightleftarrows CaSO_4{}^{\circ} \tag{2.4.14b}$$

Calcium complexed with Cl^- or $SO_4{}^{2-}$ in solution is not available for precipitation and exchange reactions, although $CaCl^+$ itself is subject to cation exchange. The $CaCl^+$ complex is relatively unimportant at typical soil chloride levels, but in some cases ion pair formation can have a significant effect on solute distribution. In a saturated gypsum solution, for example, approximately 30% of the calcium in solution is present as the uncharged $CaSO_4{}^{\circ}$ ion pair.

The long-term effects of precipitation and salt leaching can be predicted as illustrated in figure 2.4.5. As would be expected, the lower the leaching fraction the higher the amount of salt precipitation. For practical short-term conditions, salt distribution as

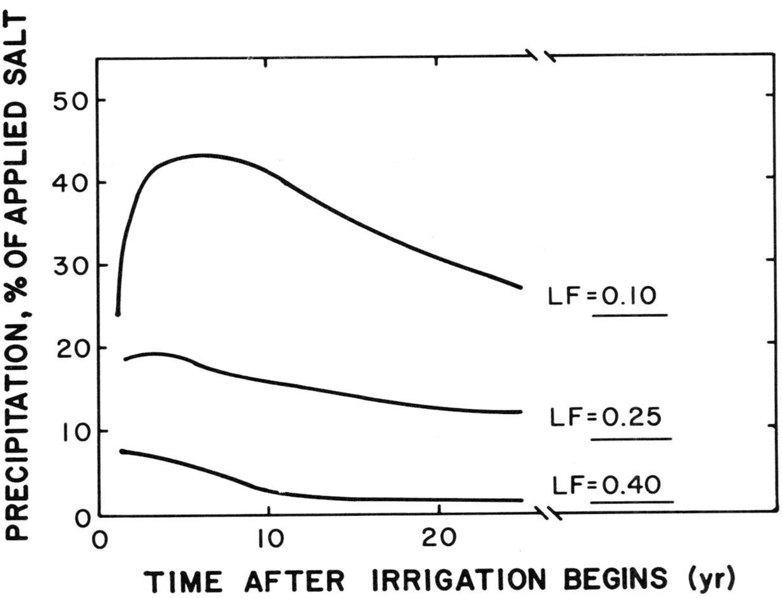

Fig. 2.4.5 Predicted extent of salt precipitation for long-term irrigation (after Jury, 1982).

related to leaching with a drip system is illustrated in figure 2.4.2. With both the low (2%) and high (17%) leaching fractions, the salts are moved away from the drip lines. The salinity patterns are somewhat different, but the important concerns for crop production are the tolerance of plant roots for the salt concentration and the yield decrement that results from such a saline environment.

Because of the importance of SAR to soil-water interrelations and the changes it undergoes as the water moves through the soil, leaching requirement - SAR relationships have been developed. A set of equations based on field observations takes the form

$$\frac{SAR_d}{SAR_i} = \frac{y(1 + 2L_r)}{(L_r)^{1/2}} \tag{2.4.15}$$

and

$$\frac{SAR_d}{SAR_i} = \frac{y(1 + 2L_r)}{(L_r)^{1/2}} x \; [1 + (8.4 - pH_c^*)] \tag{2.4.16}$$

where equation 2.4.15 refers to waters without bicarbonate and equation 2.4.16 to waters with bicarbonate that can have calcium carbonate precipitation. L_r stands for the leaching requirement, y is an empirical constant of 0.7 (Rhoades, 1974), and d and i indicate drainage water and irrigation water, respectively. The component pH_c^* is the computed Langlier pH, described in detail in chapter 3.2.

Another approach based on SAR and assuming no ion exchange or Ca interaction with the solid phase is

$$SAR_d = \frac{([Na_i]/LF)}{([Mg_i]/LF) + [Ca]_{eq})^{1/2}} \qquad (2.4.17)$$

where

$$[Ca]_{eq} = \left[\frac{K_1\ K_{IAP}\ K_{CO_2}}{K_2\ ([HCO_3^-]/[Ca^{2+}])^2\ \gamma_{Ca}\ \gamma^2_{HCO_3}}\right]^{1/3} \times (P_{CO_2})^{1/3} \qquad (2.4.18)$$

where $[Ca]_{eq}$ is the equilibrium calcium concentration, K_{IAP} is the ion activity product of $CaCO_3$, K_1 and K_2 are the first and second dissociation constants of H_2CO_3, K_{CO_2} is Henry's law constant for CO_2 solubility, γ_{Ca} and γ_{HCO_3} are the activity coefficients of Ca^{2+} and HCO_3^-, and P_{CO_2} is the partial pressure of CO_2 (Suarez, 1981).

2.4.3.c Cation exchange

In the preceding discussion, attention has been focused primarily on the liquid phase, but an important solid phase reaction, cation exchange, must also be considered. The ions on the exchange sites of the soil matrix are also participating in ion distribution. A cation exchange equilibrium equation for sodium and calcium ions can be represented by

$$2\ Na^+ + CaX = Na_2X + Ca^{2+} \qquad \text{or} \qquad (2.4.19a)$$

$$Na^+ + Ca_{1/2}X = NaX + 1/2Ca^{2+} \qquad (2.4.19b)$$

where X represents the exchange medium. Several types of exchange distribution equations can be used to describe the preceding relations, with the simpler ones taking the forms

$$\frac{[Na_2X]}{[CaX]} = K_k \frac{[Na+]^2}{[Ca^{2+}]} \qquad (2.4.20a)$$

$$\frac{[NaX]}{[Ca_{1/2}X]} = K_g \frac{[Na+]}{[Ca^{2+}]^{1/2}} \qquad (2.4.20b)$$

where the K's are the exchange coefficients. By assuming that K_g for the Mg-Na exchange is equal to K_g for the Ca-Na exchange, equation 2.4.20b becomes

$$\frac{[NaX]}{[Ca_{1/2}X] + [Mg_{1/2}X]} = K_g \frac{[Na^+]}{([Ca^{2+}] + [Mg^{2+}])^{1/2}} \qquad (2.4.21)$$

The ratio of monovalent sodium to the divalent cations on the soil exchange sites is the exchangeable sodium ratio, ESR, and thus is similar to SAR of the solution phase (compare equation 2.4.21 to equation 2.4.10).

Example: 2.4.6

Problem: Using equation 2.4.21, find the exchangeable sodium ratio for the irrigation water given in example 2.4.4, assuming $K_g = 0.015$.

2.4.4 *Modeling of salt movement*

Modeling is an attempt to translate assumptions concerning the behavior of a system into mathematical form. Models are useful for predictive purposes, and for testing how well a given system is truly understood. In this latter regard, models are invaluable in identifying gaps in knowledge and in pointing towards areas needing further research.

The modeling of salt movement and distribution in soils is directly related to the modeling of water movement. The physical, chemical, and biological processes which control soil-salt interactions are not directly affected by the water application method. The ultimate distribution of salt is, however, highly dependent upon how water enters into and is redistributed within the soil. Therefore, any model of solute transport must be tied to a water flow model. Various approaches to modeling water movement from a point (trickle) source have been discussed in chapter 2.3.

This section begins with a description of the relatively simple case of modeling salt transport under idealized, one dimensional flow conditions. This simulates the situation found under areally uniform water application, such as natural rainfall or flood irrigation. An example of the extension of this basic transport model to a three dimensional, trickle irrigation water regime is then presented. Finally, models of soil-salt interactions, and their incorporation into solute transport models, are discussed.

2.4.4.a Basic transport equations

The transport of salts in a soil is the result of the combined effects of mechanical dispersion and mass flow. Dispersion results

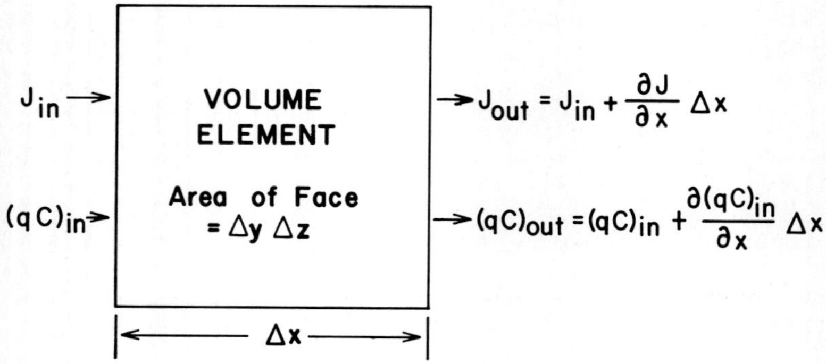

Fig. 2.4.6 Schematic diagram of water and salt flows through a volume element of soil.

from the mixing which occurs as water and salt move through the soil. Mass flow represents the bulk movement of salt as it is carried along with the flowing soil water. Looking first at the dispersion effect, we can describe the flux of salt, J, in one dimension, as

$$J = -D \frac{dC}{dx} \tag{2.4.22}$$

where C is the salt concentration, x is distance, and D is the dispersion coefficient (also known as the apparent diffusion coefficent). D is a function of the water velocity and the diffusion coefficient for the salt within the soil matrix. For a detailed discussion of the physical factors which contribute to dispersion, see Wagenet (1983) and the references cited therein. D has units of L^2T^{-1} and thus J has units $ML^{-2}T^{-1}$, where M is a unit of mass, L is a unit of length, and T is a unit of time.

Referring to figure 2.4.6, we can visualize the change in salt flux in passing through a volume element, Δx, with a cross-sectional area $\Delta y \Delta z$, as

$$J_{out} = J_{in} + \frac{\partial J}{\partial x} \Delta x \tag{2.4.23}$$

where J_{in} and J_{out} represent the fluxes of salt in and out of the volume element, respectively.

The net rate of change of the mass of salt within the volume element is thus

$$(J_{in} - J_{out}) \Delta y \Delta z = [J_{in} - (J_{in} + \frac{\partial J}{\partial x} \Delta x)] \; \Delta y \Delta z \tag{2.4.24}$$

$$= - \frac{\partial J}{\partial x} \Delta x \Delta y \Delta z \qquad (2.4.25)$$

We can also express the rate of change of mass within the volume element in terms of the salt concentration in the soil water. The volume of water within the volume element $\Delta x \Delta y \Delta z$ is given by

$$\Theta \ \Delta x \Delta y \Delta z \qquad (2.4.26)$$

where Θ is the volumetric water content. Thus, the total amount of salt in the volume element can be expressed as

$$C\Theta \ \Delta x \Delta y \Delta z \qquad (2.4.27)$$

and the net rate of change of mass during flow through the volume element is equal to

$$\frac{\partial (C\Theta)}{\partial t} \Delta x \Delta y \Delta z \qquad (2.4.28)$$

Since equations 2.4.25 and 2.4.28 both represent the rate of change of the mass of salt within the volume element, we have

$$\frac{\partial (C\Theta)}{\partial t} \Delta x \Delta y \Delta z = - \frac{\partial J}{\partial x} \Delta x \Delta y \Delta z \qquad (2.4.29)$$

$$\frac{\partial (C\Theta)}{\partial t} = - \frac{\partial J}{\partial x} \qquad (2.4.30)$$

Using the relationship for J in equation 2.4.22 we obtain

$$\frac{\partial (C\Theta)}{\partial t} = \frac{\partial}{\partial x} [D \frac{\partial C}{\partial x}] \qquad (2.4.31)$$

Equation 2.4.31 represents the contribution of molecular diffusion and mechanical dispersion to salt movement in the soil water.

We obtain the salt movement due to mass flow of the soil water in a similar manner. Defining q as the soil water flux, the net flux of salt out of the volume element is analogous to equation 2.4.24

$$[(qC)_{in} - (qC)_{out}] \ \Delta y \Delta z = [(qC)_{in} - ((qC)_{in} + \frac{\partial (qC)}{\partial x} \Delta x)] \qquad (2.4.32)$$

$$= - \frac{\partial (qC)}{\partial x} \Delta x \Delta y \Delta z \qquad (2.4.33)$$

Adding the contribution due to mass flow, equation 2.4.33, to the contribution due to dispersion, equation 2.4.29, we obtain the total salt flux in the absence of sources or sinks within the volume element

$$\frac{\partial(C\Theta)}{\partial t}\,\Delta x\,\Delta y\,\Delta z = -\frac{\partial J}{\partial x}\,\Delta x\,\Delta y\,\Delta z - \frac{\partial(qC)}{\partial x}\,\Delta x\,\Delta y\,\Delta z \qquad (2.4.34)$$

$$\frac{\partial(C\Theta)}{\partial t} = \frac{\partial}{\partial x}\left[D\,\frac{\partial C}{\partial x}\right] - \frac{\partial(qC)}{\partial x} \qquad (2.4.35)$$

Equation 2.4.35 relates the movement of salt to the velocity and pathway of water through the soil in one dimension. Bresler (1975)

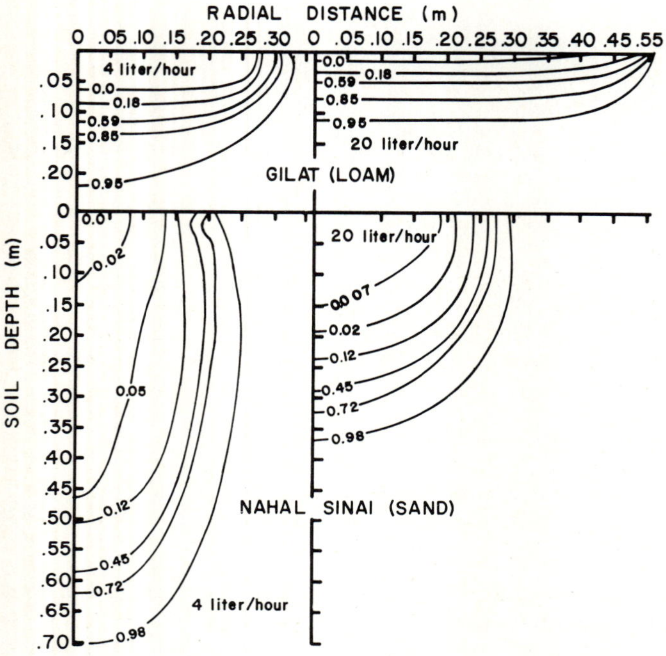

Fig. 2.4.7 Computed salt concentration fields for two different trickle discharges for two soils. The numbers labeling the curves indicate relative concentrations of salt, $(C - C_o)/C_n$, after 12 L of irrigation water infiltration. C is final soil solution concentration, C_o is irrigation water concentration, and C_n is initial soil solution concentration (after Bresler, 1975).

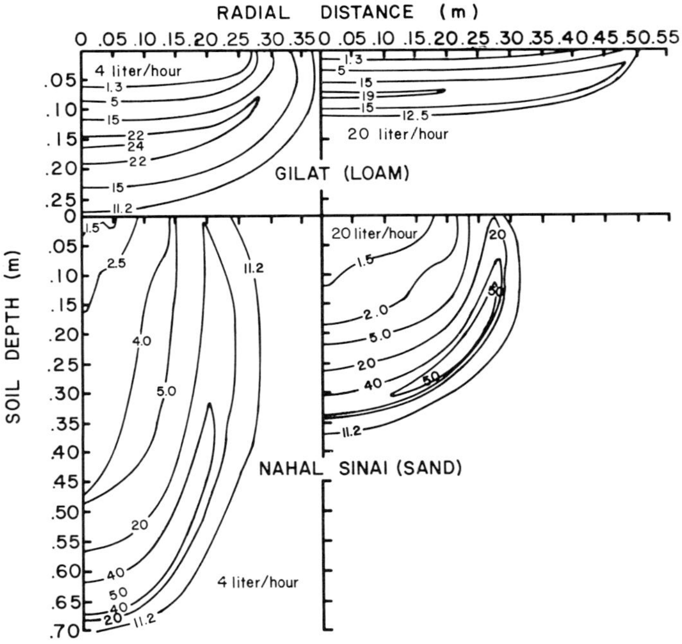

Fig. 2.4.8 Computed volumetric salt content field for two different trickle discharges for two soils. The numbers labelling the curves indicate volumetric salt content in millimoles of univalent ion per liter of bulk soil (after Bresler, 1975).

has combined a three dimensional form of equation 2.4.35 with a water transport equation to simulate numerically water and salt transport from a trickle source. Results of these studies are presented in figures 2.4.7 and 2.4.8, which show salt distribution after application of irrigation water to two different soils at two different application rates. In this example, water of low salt content from the trickle system displaces higher salinity water initially present in the soil. The effects of texture and application rate on salt distribution are clearly indicated. At slower application rates and on a coarser-textured soil, vertical salt movement is enhanced relative to horizontal movement. Near the trickle source, the soil salinity is the same as that of the irrigation water; near the wetting front, the salinity approaches that of the bulk soil. Figure 2.4.8, which presents the data in terms of volumetric salt content on a bulk soil basis, illustrates the interaction of soil water content and salt concentration in the soil solution. Although salt concentrations increase with distance from the trickle source, total salt content is greatest at a distance intermediate between the wetting front and the trickle source, due to the higher water content in the intermediate region.

The data presented in figures 2.4.7 and 2.4.8 are for water movement from a trickle source into a uniformly salinized soil, and with no water extraction by plant roots or evaporation from the soil surface. The figures serve to illustrate, however, the interaction of salt and water distribution under a trickle regime. As noted earlier, the predictive value of solute transport equations depends a great deal on the water transport models to which they are coupled.

2.4.4.b Salt movement in the presence of sources and sinks

The equations concerning salt movement which were developed in section 2.4.4.a. assumed that salt was conserved during redistribution in the soil. As has been discussed in section 2.4.3, species common in the soil solution are subject to a variety of exchange, precipitation, and dissolution reactions. These reactions change solute concentrations at different points in the soil profile, altering concentration gradients and the direction and degree of salt movement.

The degree to which dissolved salts interact with the soil medium and with each other depends upon the chemical characteristics of the individual species involved. Thus, in the modeling of solute transformations, we can no longer think in gross terms such as "salt concentration," but must instead look at the concentrations (activities) and interactions of individual ions and molecules.

The solutes which are removed from or added to solution by exchange or precipitation/dissolution reactions can be represented by a single term. Considering again the volume element of figure 2.4.6, the rate of change of mass due to such interactions can be expressed as

$$\frac{\partial S}{\partial t} \Delta x \Delta y \Delta z \qquad (2.4.36)$$

where S has units of mass of solute removed from solution per mass of soil. Note that $\partial S/\partial t$ can be either positive or negative, depending upon whether solute is entering or leaving the solid phase. Combining equations 2.4.35 and 2.4.36 and simplifying, we obtain

$$\frac{\rho}{\Theta}\frac{\partial S}{\partial t} + \frac{\partial C}{\partial t} = D' \frac{\partial^2 C}{\partial x^2} - v \frac{\partial C}{\partial x} \qquad (2.4.37)$$

The soil bulk density, ρ (ML^{-3}) and Θ are introduced in equation 2.4.37 to convert the solid phase concentration to an equivalent solution concentration basis. For purposes of simplification, we have assumed here that Θ and q are constant throughout the zone of interest. Thus, D' is the dispersion coefficient at this fixed water content, and v is the average pore-water velocity, equal to q/Θ.

Evaluation of the source/sink term, S, depends upon a knowledge of the chemical and biological interactions of the species of interest in the environment of interest. Often, a given solute will be subject to a variety of interactions simultaneously. Calcium ion for example, may enter solution via the dissolution of calcite, may be exchanged for sodium ion on soil exchange sites, and may form uncharged ion pairs in solution with sulfate ion. We can therefore envision a series of source/sink terms, S_1, S_2, S_3, ... for each species which undergoes chemical transformations within the soil.

A simple case is that in which solute is adsorbed by the exchange complex in direct proportion to its concentration in solution, according to some equilibrium constant, K. Then

$$S = KC \qquad (2.4.38)$$

and

$$\frac{\partial S}{\partial t} = K \frac{\partial C}{\partial t} \qquad (2.4.39)$$

Substituting into equation 2.4.35, we have

$$R \frac{\partial C}{\partial t} = D' \frac{\partial^2 C}{\partial x^2} - v \frac{\partial C}{\partial x} \qquad (2.4.40)$$

where $R = 1 + \rho K/\Theta$. R is referred to as the solute retardation factor. A solute which is subject to exchange reactions is thus retarded in its movement relative to solutes which do not exchange.

Due to the complexity of soil solution chemistry, and the resultant large number of factors which influence the source/sink term in equation 2.4.37, comprehensive models are required to evaluate S for real soil situations. Such models must consider exchange, precipitation/dissolution, and solution complexation behavior simultaneously. The calculations involved generally require solving a series of nonlinear algebraic equations under mass balance constraints. The chemical equilibrium programs REDEQL2 (Ingle, et al., 1978) and GEOCHEM (Sposito and Mattigod, 1979) are examples of algorithms which have been applied to soil systems.

The reactions considered thus far have been assumed to be equilibrium reactions in which the chemical transformations occur essentially instantaneously. The equilibrium assumption is generally valid for exchange reactions and for the common solution complexation reactions. Some salt transformations, notably many precipitation/dissolution reactions, are kinetically controlled. The rate of such reactions is of importance for solute transport modeling, particularly when the reactions are slow relative to soil water movement. If we assume, for example, that precipitation of a solute is governed by a first order rate reaction, we can express the quantity precipitated, S_p, as

$$\frac{\partial S_p}{\partial t} = K_p C \qquad (2.4.41)$$

where K_p is the first order rate constant. Incorporating this expression into equation 2.4.40 we derive an expression for simultaneous precipitation and retardation of a solute:

$$R \frac{\partial C}{\partial t} = D' \frac{\partial^2 C}{\partial x^2} - v \frac{\partial C}{\partial x} - K_p C \qquad (2.4.42)$$

Solutions for equations of the type represented by equation 2.4.42 for a variety of boundary conditions have been obtained (van Genuchten and Alves, 1982).

Because of the large number of processes simulated, current models must be considered approximations to real-world situations. Uncertainties with regard to thermodynamic constants, soil-solute interactions, and the nature of precipitated phases hinder precise model application. In addition, the influence of temporal and spatial variations of soil properties on salt distribution in the field is only beginning to be understood. As more knowledge is gained concerning soil chemical transformations, refinement and improvement of solute distribution models is to be expected.

REFERENCES AND SUPPLEMENTAL READING

Bernstein, L. and Francois, L.E., 1973. Comparisons of drip, furrow, and sprinkler irrigation. Soil Sci. 115: 73-86.

Bohn, H.L., McNeal, B.L. and O'Connor, G.A., 1979. Soil Chemistry. John Wiley and Sons, New York, 329 pp.

Bower, C.A., Ogata, G. and Tucker, J.M., 1969. Rootzone salt profiles and alfalfa growth as influenced by irrigation water salinity and leaching fraction. Agron. J. 61: 783-785.

Bresler, E., 1972. Control of soil salinity. In Hillel, D. (Ed.), Optimizing the Soil Physical Environment Toward Greater Crop Yields, Academic Press, New York, pp. 79-132.

Bresler, E., 1975. Two dimensional transport of solutes during non-steady infiltration from a trickle source. Soil Sci. Soc. Amer. Proc. 39: 604-613.

Bresler, E., McNeal, B.L. and Carter, D.L., 1982. Saline and Sodic Soils. Principles - Dynamics - Modeling. Springer-Verlag, New York, 236 pp.

Dickerson, R.E., 1969. Molecular Thermodynamics. Benjamin/Cummings Publishing Company, Menlo Park, California, 452 pp.

Finkel, H.J., 1982. The importance and extent of irrigation in the world. In Finkel, H.J. (Ed.), Handbook of Irrigation Technology, Vol. I, CRC Press, Boca Raton, Florida, 369 pp.

Hoffman, G.J., Ayers, R.S., Doering, E.J. and McNeal, B.L., 1980. Salinity in irrigated agriculture. In Jensen, M.E. (Ed.), Design and Operation of Farm Irrigation Systems, Monograph Amer. Soc. Agric. Eng., St. Joseph, Michigan, pp. 145-185.
Ingle, S.E., Schuldt, M.D. and Schults, D.W., 1978. A User's Guide to REDEQL. U.S. Environmental Protection Agency Report EPA-600/3-78-024, Corvallis, Oregon.
James, D.W., Hanks, R.J. and Jurinak, J.J., 1982. Modern Irrigated Soils. John Wiley and Sons, New York, 235 pp.
Jury, W.A., 1982. Use of solute transport models to estimate salt balance below irrigated cropland. In Hillel, D. (Ed.), Advances in Irrigation 1: 87-104, Academic Press, New York.
Lindsay, W.L., 1979. Chemical Equilibria in Soils. John Wiley & Sons, New York, 449 pp.
McNeal, B.L., 1974. Soil salts and their effects on water movement. In Van Schilfgaarde, J. (Ed.), Drainage for Agriculture, Amer. Soc. Agron., Madison, Wisconsin, pp. 409-431.
Rhoades, J.D., 1974. Drainage for salinity control. 1974. In Van Schilfgaarde, J. (Ed.), Drainage for Agriculture, Amer. Soc. Agron., Madison, Wisconsin, pp. 433-461.
Skogerboe, G.V. and Law, Jr., J.P., 1971. Research needs for irrigation return flow quality control. Environmental Protection Agency Water Poll. Contr. Res. Ser. 13030...11/71.
Sposito, G. and Mattigod, S.V., 1979. GEOCHEM: a computer program for the calculation of chemical equilibria in soil solutions and other natural water systems. Univ. of Calif., Riverside, California, 110 pp.
Suarez, D.L., 1981. Relation between pH_c and sodium adsorption ratio (SAR) and an alternative method of estimating SAR of soil and drainage water. Soil Sci. Soc. Amer. J. 45: 469-475.
Suarez, D. L. and Rhoades, J.D., 1982. The apparent solubility of calcium carbonate in soils. Soil Sci. Soc. Amer. J. 46: 716-722.
Tscheschke, P., Alfaro, J.F., Keller, J. and Hanks, R.J., 1974. Trickle irrigation soil water potential as influenced by management of highly saline water. Soil Sci. 117: 226-231.
United States Salinity Laboratory Staff., 1954. Diagnosis and Improvement of Saline and Alkaline Soils. U.S. Dept. of Agric. Handbook 60. Washington, DC, 160 pp.
Van Genuchten, M. Th. and Alves, W.J., 1982. Analytical solutions of the one-dimensional convective-dispersive solute transport equation. U.S. Dept. of Agric. Tech. Bull. 1661, Washington, DC, 151 pp.
Wagenet, R.J., 1983. Principles of salt movement in soils. In Nelson, D.W., Elrick, D.E. and Tanji, K.K. (Eds.), Chemical Mobility and Reactivity in Soil Systems, Amer. Soc. Agron., Spec. Publ. 11, Madison, Wisconsin, pp. 123-140.

Chapter 3

OPERATIONAL PRINCIPLES

3.1 EMITTER CLOGGING

R. G. GILBERT and H. W. FORD

3.1.1 *Evaluation*

Emitter clogging continues to be the major problem associated with trickle irrigation operation. Even though information is available on the causal factors, control measures are not always successful. Since a trickle irrigation system is expensive, longevity must be maximized to assure a favorable cost-benefit ratio. If emitters plug a short time after their installation, reclamation procedures to correct plugging increase maintenance costs and unfortunately may not be permanent. Clogging problems often discourage the operators and consequently cause the abandonment of the system and the return to a less efficient method of irrigation.

Emitter clogging is directly related to the quality of the irrigation water, i.e., suspended load, chemical composition, and microbial activity. Consequently, these factors dictate the type of water treatment necessary for the prevention of clogging. Clogging problems are often site-specific and solutions are not always available.

3.1.1.a Source of water

There is no single foolproof quantitative method for estimating clogging problems. However, by making certain water analysis, possible clogging problems can be estimated. This is especially advisable before a new trickle system is installed. Most tests can be made in the laboratory, except for some which must be made at the site because of the rapid chemical changes that can occur. Water quality can also change over the irrigation period so that samples must be taken at various times throughout the year. The water analysis data listed in table 3.1.1 can be classified into the physical, chemical, and biological factors that play major roles in the clogging process. These are further rated in terms of an arbitrary clogging hazard ranging from minor to severe; the lower the quantities of solids, salts, and bacteria in the water, the less is the clogging hazard.

Regardless of water source, trickle systems require some type of filtration to remove the bulk of the suspended materials. However, it is not practical to remove all the suspended particles. Calcium or magnesium carbonate can precipitate in filters, pipelines, or emitters when source waters have pH values above 7.5 and a high degree of hardness. Temperature, soluble organic matter, and pH are factors that influence bacterial growth and slime development. Although slime

TABLE 3.1.1

Tentative water quality criteria for indicating emitter clogging hazards (after Bucks and Nakayama, 1980).

Type of problem	Minor	Moderate	Severe
Physical			
Suspended solids[a]	50	50-100	>100
Chemical			
pH	7.0	7.0-8.0	>8.0
Dissolved solids[a]	500	500-2,000	>2,000
Manganese[a]	0.1	0.1-1.5	>1.5
Total iron[a]	0.2	0.2-1.5	>1.5
Hydrogen sulfide[a]	0.2	0.2-2.0	>2.0
Biological			
Bacterial population[b]	10,000	10,000-50,000	>50,000

[a] Maximum measured concentration from a representative number of water samples using standard procedures for analysis (mg/L).

[b] Maximum number of bacteria per milliliter can be obtained from portable field samplers and laboratory analysis. Bacterial populations do reflect increased algae and microbial nutrients.

bacteria grow best at temperatures between 20 and 30°C, they can still develop at lower temperatures. High total bacterial counts generally reflect the presence of increased nutrients that can readily support the growth of algae and bacterial populations.

Fertilizer injected into the trickle lines may also contribute to clogging. Field surveys in Florida indicate considerable variation in fertilizer solubility in different water sources (Ford, 1977a). For example, sulfide laden waters can react with dissolved iron and manganese to form precipitates that can cause emitter clogging. To ascertain a potential precipitation problem, a simple test as follows can be performed: (1) Add sufficient drops of a liquid fertilizer to a sample of the water source so that the final concentration is equivalent to the concentration of the diluted fertilizer mixture that would be flowing through the irrigation line; (2) cover and hold the mixture in the dark for 12 hrs; (3) determine whether any precipitate is formed in the bottle by directing a light beam at the bottom of the glass container. If there is no apparent precipitation, the fertilizer should be safe to use in the specified water.

3.1.1.b Surface water

Surface water sources should be tested for soluble salts, suspended solids, pH, and complexed Fe (if the water contains coloring agents). Filter clogging algae can also be identified using the illustrations reported in the American Public Health Association book, "Standard

Methods for the Examination of Water and Waste Water" (13th Ed., 1971). A general indication of suspended particulate load can be obtained by shining a beam of light through a glass bottle containing the water sample. A thick white cloud in the light beam suggests considerable suspended material, but does not indicate anything concerning the size of the particles. The amount of suspended solids in the water can best be determined by standard gravimetric laboratory procedures.

Turbidity of the water is a useful measurement when there are fine particles and other suspended materials in the water; however, this measurement must usually be carried out in the laboratory. This test can be combined with laboratory filtration that gives an indication of clogging potential and at the same time permits microscopic examination of the residue. The vacuum filter system should have a glass filter holder with a 25 mm filter made of special cellulose esters with 0.45 μm pore size. If 50 mL of the water sample can pass through the filter under suction without serious clogging and discoloration of the filter, the water can be considered free of sludges and suspended solids. The filter paper can also be stained with 2% Prussian blue in 1% of hydrochloric acid to detect the ferric or oxidized form of iron (Fe^{3+}) and with 10% carbolfuchsin to indicate the presence of organic slimes. The dried deposit on the filter paper can be examined under the microscope at 50 to 100 times magnification. To characterize the type of organism, the paper is pretreated with a special immersion oil with a refractive index of 1.5 to blankout the paper.

An additional test suitable for surface waters involves the use of flocculating agents to precipitate the fine suspended particles which can pass through most filter systems. Three hundred mL of surface water are treated with 1 mL of 24% aluminum sulfate and 1 mL of 9% sodium hydroxide. If a precipitate forms in the bottom of the bottle, suspended solids are present in the water which can contribute to clogging of drip emitters.

Total iron should be measured at the site soon after sampling because the dissolved iron can precipitate before the water sample is brought to the laboratory for analysis. Water sampled in the field can be tested for iron with portable kits containing the necessary test ingredients. Sulfide problem is rare in surface waters so that sulfide measurement is usually unnecessary.

The pH of the water should be known because it is a factor in both chemical and biocidal treatment. For example, chlorination for bacterial control is relatively ineffective above pH 7.5 so that acid additions may also be necessary to optimize the biocidal action of chlorine. The complexes formed between iron and tannins or humates are more stable at pH 6.5 than at lower pH levels.

Iron may be found in emitters when tannins are detected in quantities greater than 2 mg/L in the surface water. Tannin-like compounds, phenolics, and humic acids complex iron in water sources from lakes and streams. Tannin-like compounds can also be detected with portable test kits. Soluble organic compounds such as tannins cannot be removed by filtration. They may pass into the irrigation system and contribute to slime formation since they are nutrient sources for bacteria.

3.1.1.c Groundwater

Water from shallow wells of 6 to 15 m frequently leads to quality problems, particularly when they are open to contamination from the surface. Tests should be made for total iron, hydrogen sulfide, pH, suspended solids, and soluble salts. Actinomycetes are present in emitters utilizing shallow well water and, *Vitreoscilla*, a filamentous bacterium, is also found in profuse quantities in open shallow wells. The slime can clog emitters just by the sheer volume and gelatinous nature of the organisms. *Thiothrix* sulfur bacteria are a problem when hydrogen sulfide is present. *Pseudomonas* and *Enterobacter* type slime bacteria, which can also oxidize the complexed soluble ferrous iron are found in waters from many shallow wells.

Deep wells in Florida are usually artesian and contain 1.0 or more mg/L of hydrogen sulfide. The problem of hydrogen sulfide can also occur in other locations throughout the world, but it may not be as common. Deep wells may require tests for hydrogen sulfide, total iron, soluble salts, pH, and calcium carbonate precipitation potential. Fine sand particles can be brought up with the water during pumping.

3.1.2 *Water quality*

3.1.2.a Physical aspects: suspended solids

Physical clogging may be caused by factors such as suspended inorganic particles (sand, silt, clay or plastic), organic materials (plant fragments, animal residues, fish, snails, etc.), and microbiological debris (algae, diatoms, larva, etc.). Sand and silt may be carried into the irrigation water supply from open water canals or pumped from wells. Sand and silt introduced in the lines during installation can cause problems unless they are flushed out before the emitters are placed on the line. Physical clogging *per se* can be controlled with proper filtration and periodic flushing of laterals; however, particulate matter combined with bacterial slimes can create a type of clogging not controllable by filtration. Fine particulate matter has been collected from inside the emitters; the material has become cemented together with bacterial slimes of the genera *Pseudomonas* and *Enterobacter*. The combined mass clogs emitters even though the individual particles are small enough to pass through the emitter. Particulate-slime clogging is of major concern in certain areas. Super-chlorination at 1000 mg/L seems to be able to control the problem.

3.1.2.b Chemical aspects: dissolved solids and pH

Salinity is an important water quality factor in irrigation and does not contribute to emitter clogging unless the dissolved ions interact with each other to form precipitates or promote slime growth. Precipitation of calcium carbonate is common in arid regions with waters rich in calcium and bicarbonates. Clogging with $CaCO_3$ occurs in the narrow passages of the emitter. Waters can be analyzed in a laboratory for calcium, carbonates and bicarbonates. A simple qualitative test can be used to estimate the potential for $CaCO_3$ precipitation. A clean bottle is filled with the water sample and ammonium hydroxide is added to raise the pH to 9.2 to 9.5. After 12 hours, the sample is shaken to stir suspended solids that settled to the bottom. After shaking, the bottom of the bottle is observed in a dark room by directing a light beam at the bottom of the container. The opaque coating of white to reddish sparkling particles is $CaCO_3$. The thicker the deposit, the more severe the potential for clogging with calcium carbonate. There is an exception to the interpretation. Water containing hydrogen sulfide, and with calcium carbonate concentrations greater than 40 mg/L, will show precipitation with the ammonia test; however, in the irrigation laterals, hydrogen sulfide minimizes the precipitation of $CaCO_3$ because of its acidic property. No $CaCO_3$ precipitation has been found in trickle irrigation systems using sulfide laden water even though the calcium content may be 300 mg/L. In controlled model tests, 3 mg/L hydrogen sulfide in flowing water completely dissolved thin coatings of $CaCO_3$ (Ford, 1977a).

Iron deposits (ochre) have also created problems with trickle irrigation systems and severe clogging has been reported primarily in the United States. The soluble, reduced form of iron (ferrous ion or Fe^{2+}) is known to be present in well water from many locations so that ochre clogging may be universally present. The filamentous hydrophilic type sludge that occurs at Fe^{2+} concentrations above 0.4 mg/L is usually associated with the oxidation of Fe^{2+} and the precipitation of insoluble Fe^{3+} by iron bacteria such as ***Gallionella***, ***Leptothrix***, ***Toxothrix***, ***Crenothrix***, and ***Sphaerotilus*** plus certain nonfilamentous aerobic slime bacteria of the genus ***Pseudomonas*** and ***Enterobacter***.

3.1.2.c Biological aspects: micro- and macroorganisms

Biological clogging is most serious in trickle irrigation systems containing organic sediments plus iron or hydrogen sulfide. Clogging usually would not be a problem if the water sources are free of organic carbon which is an energy source for bacteria. There are several slime-forming organisms that contribute to clogging, particularly in the presence of Fe^{2+} and H_2S. Algae in surface waters can add organic carbon to the system. Slimes can grow on the walls of tubing. The combination of fertilizer and warming of the black tubing may further enhance growth. Most all water sources contained carbonates and bicarbonates which can serve as inorganic energy sources for certain slime forming autotrophic bacteria.

Growths of algae, actinomycetes, and fungi are present in all surface water sources. Attempts to quantify algae are of little value because of the drastic fluctuations in populations during the irrigation season. While filamentous algae can clog emitters, their most damaging feature is the formation of a gelatinous matrix in the tubing and emitters which serves as a base for bacterial slime growths. Surface waters also may contain naturally occurring complexing agents, for example tannins, phenolics, and humic acids that complex ferrous iron. Such complexing agents can be found in canal waters which can sequester up to 2.0 mg/L of Fe^{2+}. The iron bacteria in the trickle irrigation lines can precipitate the soluble complexed iron. Bacteria can also utilize and precipitate ferrous iron complexed with polyphosphates and other chelating materials that are used for iron fertilization.

3.1.3 *Causes*

3.1.3.a Physical, chemical, and biological factors

The physical, chemical and biological factors contributing to emitter clogging are summarized in table 3.1.2. These three factors are closely interrelated, and controlling one may also alleviate problems caused by the other. For example, by reducing microbial slime, the tendency of suspended particles to stick, agglomerate, and build up in the trickle lines and emitters is also reduced. In addition, small aquatic organisms such as snail eggs and larva, that are not readily observed and analyzed, can develop into large colonies in the drip lines and impose a combined physical and biological problem.

Clogging problems in the southwestern United States are caused primarily by high contents of suspended solids and water hardness, whereas problems in the southeastern United States are created by high biological activities in association with iron and sulfides. In Hawaii, major problems are associated primarily with growths of biological

TABLE 3.1.2

Principal physical, chemical and biological contributors to clogging of trickle systems (after Bucks et al., 1979).

Physical (suspended solids)	Chemical (precipitation)	Biological (bacteria and algae)
Inorganic particles: Sand Silt Clay Plastic	Calcium or magnesium carbonate Calcium sulfate	Filaments Slimes
Organic particles: Aquatic plants (phytoplankton/algae) Aquatic animals (zooplankton) Bacteria	Heavy metal hydroxides, carbonates, silicates, and sulfides Oil or other lubricants	Microbial decomposition: Iron Sulfur Manganese
	Fertilizers: Phosphate Aqueous ammonia Iron, copper, zinc, manganese	

slimes and filaments. In the lower Colorado River of Arizona, the predominant causes of emitter clogging and flow reduction are physical factors, followed by the combined effects of biological and chemical factors, as shown in table 3.1.3.

TABLE 3.1.3

Causes of clogging or flow reduction and relative percent occurrence in trickle irrigation emitters at Yuma, Arizona (after Gilbert et al., 1981)[a].

Causes of clogging	Percent of occurrence	
	Individual	Total
Physical factors		
Sand grain	17	
Plastic particles	26	
Sediment	2	
Body parts of insects and animals	3	
Deformed septa[b]	7	55
Biological factors		
Microbial slime	11	
Plant roots and algal mats	3	14
Chemical factors		
Carbonate precipitates	2	
Iron-manganese precipitates	0	2
Combined factors[c]		
Physical/biological	8	
Physical/chemical	2	
Chemical/biological	6	
Physical/biological/chemical	2	18
Nondetectable (probably physical)		11

[a] Results are representative of eight emitter systems and four water treatments (C, D, E, and F, table 3.1.4), which were operated for more than 4 years. There were 1200 emitters installed in these water treatments, and 119 with reduced flow or clogged conditions (50% design flow) were dissected and microscopically examined for causes of flow reductions.

[b] Silicone rubber discs were deformed by water treated with chlorine and acid treatments (D and E, table 3.1.4), which restricted flow.

[c] The observations indicated that the most likely initial cause of flow reduction was a physical factor, followed by the development of biological and chemical factors. The major physical factors involved were sand grains and plastic particles.

Detailed biological analysis of the emitters have shown that the most common bacteria are ***Pseudomonas*, *Flavobacterium*, *Vibro*, *Brevibacterium*, *Micrococcus*,** and ***Bacillus*.** The occurrence of ***Bacillus*** is enhanced by sand and screen filtration and markedly reduced by chemical conditioning of the water. No strictly anaerobic bacteria, such as ***Clostridium*** sp., is detected when the water is treated with chlorine and acid. Pigmented bacteria, ***Flavobacterium lutescens*** and ***Cytophaga hutchinsonii***, may cause the yellow coloration of the slime deposits in biologically clogged emitters, and their growth may be supported by ***Pseudomonas stutzeri***, a nonpigmented bacteria. Iron bacteria, *Sphaerotilus* spp., are absent in the samples from Arizona water.

In Florida, bacterial clogging is associated with three types of slimes: (1) iron, (2) sulfur, and (3) nonspecific filaments and nonfilamentous slimes. Ochre (filamentous iron deposits) occurs when ferrous iron in the water is precipitated as ferric iron by the activity of filamentous bacteria such as ***Gallionella*, *Leptothrix*, *Toxothrix*, *Crenothrix*,** and ***Sphaerotilus*** plus nonfilamentous aerobic slimes such as ***Pseudomonas*** and ***Enterobacter*.** The primary clogging agents are the sticky bacterial slimes that adhere to the suspended solids and not necessarily the precipitated ferric iron itself. Aerobic sulfur slimes are formed by the transformation of hydrogen sulfide to elemental sulfur by the filamentous bacteria ***Thiothrix*** and to a lesser extent ***Beggiatoa*** sp. ***Thiothrix*** requires only traces of oxygen for its development, and its optimum pH range is between 6.7 to 7.2. Hydrogen sulfide in solution can inhibit ochre and other nonsulfur slime clogging problems. Miscellaneous bacteria such as filamentous ***Vitreoscilla*** and nonfilamentous ***Pseudomonas*** and ***Enterobacter*** can also clog emitters by their sheer mass.

Iron reactions with organic complexing agents are also important in ochre formation and clogging of drain lines in Florida. Certain Fe-organic complexes can stick to glass slides, grooves of emitters, and walls of trickle irrigation tubing, even when bacteria are killed with chlorine. The adhering properties of complexed Fe are observed when well waters containing phenolics are complexed with Fe and dripped over limerock. The rocks stained reddish brown, whereas they are not stained when the water is treated continuously with chlorine (0.5 mg/L NaOCl) or hydrogen peroxides (5 mg/L H_2O_2). Treatments for 30 min each hour can prevent Fe deposits on the limerock only when the oxidizing biocides are being injected. Apparently the oxidants can separate ferric iron from the complex so that Fe^{3+} cannot stick to the limerock.

3.1.3.b System operation

Statistical studies have shown that emitter clogging can be the major cause of emitter discharge variation within a trickle system. Other conditions such as emitter construction, water temperature, and emitter aging can also cause flow variations. System operation in respect to time of day or year, water temperature, fluctuation in water quality, and the addition of chemicals, other than for water treatment, can all affect emitter clogging and discharge variability. Preventative maintenance continues to be the best solution for reducing or eliminating emitter clogging.

3.1.4. *Prevention*

Preventative maintenance practices include water filtration, field inspection, pipeline flushing, and chemical water treatment. Water filtration and field inspection are absolutely essential. Flushing laterals and pipe lines can help to minimize sediment build-up, and chemical water treatment can improve the long-term performance of a trickle irrigation system. Refer to chapter 3.2 for a complete discussion of water filtration and chemical treatments for trickle irrigation systems.

3.1.4.a Chemical water treatment research

The following discussion is limited to experimental results from research in Arizona and Florida, where most of the experimental data are available. The Arizona study, using Colorado River water, compares the effectiveness of various water treatment for preventing emitter clogging (table 3.1.4) in terms of the number of emitters operating at less than 50% of initial design at the end of the experiment. Screen filtration alone is inadequate, as the overall emitter failure rate was 68%. In contrast, a combination of sand and screen filtration (treatment C) can reduce the incidence of clogging to 23%, and chemically conditioning the water (treatments D, E, and F) reduced it even further. Emitters functioned best with continuous sulfuric acid (treatment F), under which only an 8% clogging rate is present. These results indicate that the most efficient treatment for prevention of emitter clogging is the continuous acidification for pH control to prevent carbonate formation and precipitation. However, such treatment must be coupled with sand and screen filtration. A combination of continuous (1 mg/L) or intermittent (10 mg/L) chlorine and sulfuric acid treatments is shown to be nearly as effective as the acid alone for reducing emitter clogging.

Evaluation of water treatments for slime control requires some knowledge of the interactions between treatments and slimes. Special bacterial growth chambers have been used successfully to grow emitter-clogging slime bacteria (Ford, 1977b; 1978). The growth chambers consist of 5 cm diameter corrugated polyethylene tubes 6 m in length. Bacteria can grow on the walls of the tubing, glass slides, strips of Neoprene, and stainless steel screens inserted in the lines. Various materials such as ferrous iron, bacterial energy sources, complexing agents, biocides, and acids to control pH are injected into the system. The glass slides can be monitored for the formation of any deposits. Bacterial slimes can form within 10 hours and bacterial iron deposits within 24 hours. Results using such methods indicate that chemically oxidized iron is more porous and, therefore, less of a clogging agent than biologically induced ochre. The easiest organisms to grow are ***Thiothrix*** and ***Beggiatoa*** sulfur bacteria at pH 7.2. Slimes are visible in the chambers within 2 days after treating continuously with 3 mg/L

TABLE 3.1.4

Effect of water treatment on clogging using Colorado River water at water at Yuma, Arizona (after Gilbert et al., 1981)[a].

Water treatment	Number of emitters clogged[b]	Percent clogged
A	205	68
B	205	68
C	68	23
D	53	18
E	41	14
F	25	8
Total:	597	33

Treatment letter	Water Treatment: Filtration	Water Treatment: Chemical
A	Screen (50 mesh)	None
B	Screen (50 mesh)	Chlorine[c] (10 mg/L) and acid[d] (lower pH to 7) - intermittent
C	Sand (silica No. 20) + Screen (20 mesh)	None
D	Same as treatment C	Same as treatment B
E	Same as treatment C	Chlorine (1 mg/L) and acid (lower pH to 7) - continuous
F	Same as treatment C	Acid (lower pH to 7) - continuous

[a] Pooled observations from 300 emitters for each of eight different designs with a total of 1800 emitters.

[b] Clogged and partially clogged emitters had discharge rates of less than 50% of initial design.

[c] Free residual chlorine.

[d] Sulfuric acid.

hydrogen sulfide. Biocide treatments can be rated for minimum concentrations of the active ingredient that completely inhibit ***Thiothrix*** for 7 days. Ratings for 7 biocides against ***Thiothrix*** are shown in table 3.1.5.

The chambers are also invaluable for demonstrating undesirable side effects. For example, ammonia injected at 32 mg/L completely inhibits growths of ***Thiothrix***; however, anhydrous ammonia precipitates calcium carbonate in amounts greater than could be dissolved by 4 mg/L of hydrogen sulfide. The treatment, which would be an excellent substitute for chlorine as a control for sulfur slime cannot be used commercially unless the calcium content of the water is less than 40 mg/L.

TABLE 3.1.5

Slimicide screening in a monitored bacterial growth chamber after 7 days continuous treatment for control of ***Thiothrix*** slime development in Florida (after Ford, 1977b)[a].

Slimicide treatment	Rate of application[b] (mg/L)	Slime detected	Undesirable side effects
Sodium hypochlorite	16.0	3 days	sulfur formed
Sodium hypochlorite	35.0[c]	none	none
Acrolein	0.5	none	none
Quaternary ammonium A	1.0	none	brown stain
Quaternary ammonium B	2.0	none	black sludge
Quaternary ammonium C	2.0	none	black sludge
Xylene	5.0	2 days	***Thiothrix*** slime
Isopropanol	245.0	3 days	***Thiothrix*** slime

[a] Hydrogen sulfide injected at 4 mg/L in all chambers.

[b] Expressed as amount of active ingredient.

[c] One mg/L free residual chlorine (34 mg/L sodium hypochlorite was destroyed by 4 mg/L hydrogen sulfide).

Experiments in Florida indicate that a ferrous iron content as low as 0.2 mg/L can contribute to iron deposition, and that chlorination successfully controlled ochre when iron is less than 3.5 mg/L and the pH is below 6.5. One method for iron control is to inject chlorine near the bottom of the well, and remove the precipitated iron in a sand filter. Long-term operation using water with high levels of iron, manganese, or hydrogen sulfide along with bacteria (table 3.1.1) may not be suitable for trickle irrigation.

3.1.4.b Preventive maintenance practices

Water filtration. The selection of filter type, size, and capacity depends upon water quality and emitter design. Recommendations by emitter manufacturers on the degree of filtration required should be followed; however, where no recommendations are available, the general practice is to filter to one-tenth the diameter of the emitter's smallest opening. When physical factors become severe (table 3.1.2), two or more types of filters in series may be needed. As a general rule, filtration units should be designed with at least 20% extra capacity. Pump size should also be increased accordingly to provide some reserve operating pressure and capacity for backwashing of filters and flushing of trickle lines.

Screen filters, made of stainless steel, plastic, or synthetic cloth and enclosed in a special housing, are the simplest. Aquatic algae in the water tend to cause screen blockage and can reduce the filtering capacity. Manufacturers recommend screen sizes ranging from the finer 100 or 200 mesh (150 or 75 μm) to a coarser 30 mesh (600 μm). Screen filters, as well as other filtering systems, must be routinely cleaned and inspected to insure satisfactory operation of any trickle irrigation system.

Media filters consist of fine gravel and sand of selected sizes placed in a pressurized tank. Since media filters are not easily plugged by algae, they can remove relatively large amounts of suspended solids before backwashing is needed. However, they can provide conditions favorable for increased bacterial growth. Media filters that are presently used will retain particle size in the range of about 25-100 μm. Media filters can be followed by a secondary filter or a rinse-away valve to prevent possible contaminants from going beyond the sand filter during the backwashing process.

Sand separators, hydrocyclones, or centrifugal filters remove suspended particles that have a specific gravity greater than water and that are larger than 75 μm, but these filters are ineffective in removing most organic solids. A sand separator can effectively remove a large amount of sand particles and can be used efficiently as a pretreatment located before another type of filter.

Settling basins, ponds, or reservoirs can remove large volumes of sand and silt. Unless they have a protective covering, the water is subject to windblown contamination and algae growth that must be controlled. Thus, these structures are normally used only for pretreatment for surface water sources.

<u>Chemical water treatment</u>. Sulfuric and hydrochloric acid are commonly used acids to reduce chemical precipitation. Phosphoric acid can also be used as a water treatment and fertilizer source. Chlorination is the primary means for controlling microbial activity; sodium hypochlorite (a liquid), calcium hypochlorite (a solid or powder), and chlorine gas are the basic sources. When chlorination is used, test kits designed to measure free residual chlorine, which is the excess of active chlorine over the amount required to kill bacteria, should be used. The orthotolidine indicator commonly used for swimming pools should not be used, because the chemical only gives total not free residual chlorine concentrations. Research indicates that very high applications of chlorine are necessary to injure citrus roots (Ford, 1977b). Plant damage would not be expected from applications of lower concentrations of chlorine in the field.

Another problem in using chemicals for controlling slime involves the need to obtain an Environmental Protection Agency registration for the biocide. Only those compounds with proper clearance for a specific problem should be used in the United States. Numerous potential biocides are being screened for use against various types of slime bacteria that can clog emitters. The interactions and side effects are complicated.

Some of the alternative chemicals to control bacteria and algae are acrolein, copper salts, hydrogen peroxide, iodine, and quaternary ammonium salts. Acrolein requires special handling and does not destroy certain iron complexes in the laboratory.

Copper salts have been widely used to control algae in settling basins, ponds, or reservoirs. Hydrogen peroxide can be a good bactericide for iron problems, but there is no satisfactory method to monitor concentrations. It is generally poor for sulfur and other slimes that do not contain iron. Iodine is an excellent bactericide, although it complexes with iron and is toxic to plants. Quaternary ammonium salts also kill bacteria and even snails, but they are expensive.

Chemicals for water treatments (acids, algicides, bactericides) can be safely injected through trickle irrigation systems. Several types of pumps are available for injecting the water-soluble chemical for water treatment. Final chemical concentrations for water treatments are generally low, between 0.5 and 10 mg/L. The concentration should be determined routinely after the solution has passed the primary filter and before it enters the main line. Occasionally, a test should be done at the end of the last lateral line to ensure that the treatment solution at the proper concentration has been distributed throughout the trickle irrigation system.

Acids and chlorine compounds should be stored separately, preferably in epoxy-coated plastic, or Fiberglas storage tanks. Acid can react with hypochlorite to produce chlorine gas and heat - a hazardous situation. Sodium and calcium hypochlorite will react with emulsifiers, fertilizers, herbicides, and insecticides, and destroy their effectiveness. Bulk chemicals, and diluted solutions as well, should be stored in a secure place. In preparing dilutions, concentrated stock of acid or other chemicals should be added to the water and not vice versa. A ready water source should be provided near the chemical tank and injector for washing off any chemicals that may contact the skin. Protective goggles, face shields, and clothing should also be worn when making the chemical dilutions. State and local regulations and codes must be followed in regards to type of backflow-prevention device used to ensure against contamination of an irrigation well or potable water supply. Where gas chlorinators are used, safety devices should be installed and routinely inspected to prevent the buildup of excessive pressure or contamination in the chlorine supply bottles.

3.1.5 *Reclamation*

Reclamation of partially clogged emitters has been successful in several cases. Slime deposits have been removed with superchlorination at high levels (1000 mg/L), but extreme care is required to prevent injury to plants. In Florida (Myers et al., 1976), emitters with small amounts of slime and minimal reductions in discharge rates have been cleaned by using 250 mg/L of sodium hypochlorite for at least 12 hr without causing injury to citrus trees. A 2% hydrochloric acid treatment used for 15 min removed ochre and slimes from operational emitters. Unfortunately, the soil pH was lowered from 6 to 5. In Arizona, (Nakayama et al., 1977), flushable emitters clogged with biological slime were reclaimed by treating the system for about 24 hr with 100 mg/L of chlorine and adding sulfuric acid to lower the pH to 2. The discharge rates were increased from as low as 50% back to 90 and 95% of the original design. After the reclamation treatment, a continuous 1 mg/L chlorine (NaOCl) at a pH of 7 helped to maintain the system operational for the three remaining years of the investigation.

TABLE 3.1.6

Water classification system for unfiltered water used in trickle irrigation system (after Bucks et al., 1979).

	Physical	Chemical[a]		Biological rating
Arbitrary rating[b]	Suspended solids (max. mg/L)	Dissolved solids (max. mg/L)	Total iron and/or manganese (max. mg/L)	Bacterial populations[c] (max. no./mL)
0	10	100	0.1	100
1	20	200	0.2	1000
2	30	300	0.3	2000
3	40	400	0.4	3000
4	50	500	0.5	4000
5	60	600	0.6	5000
6	80	800	0.7	10000
7	100	1000	0.8	20000
8	120	1200	0.9	30000
9	140	1400	1.0	40000
10	160	1600	1.1	50000

[a] Tentative chemical classification is based on the highest rating for either dissolved solids, soluble iron, or manganese.

[b] If water pH is 7.5 or greater, rating is increased by 2.

[c] If water is known to contain an abundant reproductive snail population, rating is increased by 4. Bacteria populations do reflect increased algae and microbial nutrients.

3.1.6 *Recommended guidelines*

Reliability in the performance of a trickle irrigation system depends upon preventive maintenance. Because water quality is of primary importance in this preventive maintenance program, a tentative water classification system was developed to establish criteria for evaluating the clogging potential of a trickle irrigation water supply (table 3.1.6). The water classification system is similar to the water quality criteria system presented previously (table 3.1.1), except that numerical ratings are developed.

These numerical ratings selected for the physical, chemical, and biological composition are arbitrary, but they give a basis for comparing different types of water. With further research, the classification will undoubtedly be improved. Each of the three factors is given a rating of zero to 10. A combined value of "0-0-0" for the water is considered excellent, whereas one of "10-10-10" is poor. Alternately, when the sum of the three factors totals 10 or less, little problem is anticipated, whereas 10 to 20 indicates some problem and 20 to 30 a severe problem. For the latter two situations, water filtration plus other preventive measures are needed.

By using such a scheme different water sources used for trickle irrigation can be classified (table 3.1.7). Except for most city waters, irrigation waters require more than minimum water treatment including filtration. Water source No. 2 with its high suspended load needed special water filtration. This was not initially chemically treated; and after two years of operation, chemical treatment was required to reclaim and prevent further clogging. Water sources No. 3 and No. 4 were used with various filtration and chemical treatments in which the clogging process was closely monitored.

TABLE 3.1.7

Example of classification of four water sources used in trickle irrigation systems (after Bucks et al., 1979).

Type of source	Physical	Chemical		Biological	Water class.
	Suspended solids[a] (max. mg/L)	Dissolved solids (max. mg/L)	Total iron or manganese (max. mg/L)	Bacteria populations[a] (max. no./mL)	(phys.-chem.-biol.)
1. City water	1	500	0.05	10	0-4-0
2. Runoff water	300	50	0.05	10,000	10-0-6
3. River water	70	900	0.10	4,000	5-8-4
4. Well water	1	1,650	0.05	40,000	0-10-9

[a] Values of suspended solids and microbial populations varied considerably over sampling period, and the worst situation was used in this case.

One of the most serious problems encountered in trickle irrigation is the clogging of emitters. Recommendations and guidelines are presented for conducting a preventive maintenance program. These include water filtration, field inspection, pipeline flushing, and chemical water treatment. A suitable combination of type, size, and capacity filter unit is required. Appropriate procedures should be followed for the field inspection and flushing of trickle irrigation systems. Chemical water treatment should be properly selected for maintaining emitter performance. Because water quality is of primary importance in the design and operation of the system, adequate water analysis should be made and evaluated on the basis of past experience such as the water classification scheme presented to evaluate the clogging potential of the trickle irrigation water source.

PROBLEMS

1. Question: Given a specific water quality, what is the water classification and recommendation for prevention of emitter clogging?

Given:

Water Quality Data

Water test	Water source 1	2	3
Suspended solids (mg/L)	3	5	250
Dissolved solids (mg/L)	300	50	900
Iron or manganese (mg/L)	0.02	0.60	0.01
Bacteria population (no./mL)	50	35	10,000
pH	6.80	7.20	8.30

Answer:

Water Source	Classification (Physical-Chem-Biol.)	Recommendation
1	0-2-0	Screen filtration and intermittent chlorination (10 mg/L) for water with high dissolved solids.
2	0-5-0	Continuous chlorination (0.5 mg/L) combined with sand filtration, automatic backflushing and holding tanks for water source high in iron content [0.4 mg/L ferrous (Fe^{2+}) iron].

3	10-9-6	Sand filtration continuous chlorination (0.5 mg/L) and continuous sulfuric acid (reduce pH to 7.0) for water source with high dissolved solids, bacteria populations and water pH> 7.5.

2. Question: What are the optimal environmental conditions for microbial growth?

   Answer: The primary environmental variables influencing microbial growth in aquatic systems include temperature, light, aeration, organic matter, pH and inorganic nutrients supply. Temperature governs all biological processes, and it is thus a prime factor of concern. Most microorganisms are mesophilic with optima ranging from 25 to 35°C and a capacity to grow from about 15 to 45°C. Growth rate is directly related to organic nutrient content so that water with high suspended loads and/or dissolved organic matter have the largest microbial numbers. Seasonal changes in microbial growth and numbers are closely related to fluctuations in temperature. Seasonal diurnal light cycles directly influence the growth of algae in the water, which in turn affect microbial growth.

3. Question: What are humic acids?

   Answer: Humic acid denotes that fraction of the soil humus which is soluble in alkali and precipitated by acid (pH 1.0 to 2.0). It is dark brown to black in color. The material which does not precipitate is often called fulvic acid. It consists primarily of humic acid-type molecules, possibly of relatively smaller molecular weights, and of polysaccharides with a very large molecular weight range. The humic acid-type molecules appear to be complex polymers of phenolic units with linked amino acids, peptides, amino sugars, and other organic constituents.

4. Question: What are iron complexing compounds?

   Answer: Iron complexing compounds are chelating agents. A chelate is an organic compound which combines with, and protects, certain metallic cations including iron, manganese, zinc, and copper. The cation-chelate combinations make complex ring structures and the metals so bound essentially lose their usual ionic characteristics. Although chelated metals are protected against soil reactions, these forms of the micronutrients are apparently assimilated fairly readily by growing plants. Thus, so long as the nutrients remain in these combinations they remain in solution and are considered as being in the form available for plant uptake and use.

5. Question: What are iron bacteria?

Answer: Iron bacteria are filamentous bacteria, such as ***Sphaerotilus***, ***Gallionella***, ***Leptothrix***, ***Toxothrix***, and ***Crenothrix***, that produce within their cells filamentous hydrophilic iron deposits (ochre) that may cause severe clogging problems in trickle irrigation systems.

6. Question: What tests should be made to characterize water quality?

Answer: Measurements for water quality can be restricted to those factors most likely to be a problem in a particular area. Most surface and well waters should be tested for iron (Fe), hydrogen sulfide (H_2S), pH, suspended solids, calcium carbonate ($CaCO_3$), dissolved solids, turbidity, and algal and bacterial numbers.

7. Question: What are ***Thiothrix*** bacteria?

Answer: ***Thiothrix*** and ***Beggiatoa*** bacteria are colorless filamentous sulfur bacteria that oxidatively utilize hydrogen sulfide (H_2S) for energy and deposit elemental sulfur. A white gelatinous sulfur slime results which contain abundant filaments of ***Thiothrix*** that clog trickle emitters and filters.

8. Question: What are biological slimes?

Answer: Bacterial slimes are responsible for most of the biological slimes that cause clogging problems in trickle irrigation systems. Bacterial slimes can be divided into 3 groups: (1) Sulfur slime; (2) iron slime; and (3) nonspecific aerobic filamentous and nonfilamentous slime.

9. Question: How are microbial counts made?

Answer: Natural environments are extremely diverse and the majority contain a wide range of microorganisms which reflect the nature of the habitat and the ability of individual members to compete successfully and coexist within that given ecosystem. In general terms, the greater the heterogeneity of the environment, the more diverse and complex will be the microflora. In natural ecosystems five major groups of microorganisms are represented: bacteria, actinomycetes, fungi, algae, and protozoa.

Of the cultural methods for determining the total microbial population, the agar-plate dilution method has been the most widely used. Other methods include direct counts by microscopic observation, culture methods using membrane filtration, and most-probable-number method for assaying certain organisms with specialized metabolism, such as nitrifiers, denitrifiers, sulfate reducers, etc. No one culture medium is adequate nutritionally for all the species present since the growth requirements for many strains are unknown, and the observed count represents only a fraction of the total.

10. Question: What are reduced conditions and how do they affect the chemistry of iron (Fe) and hydrogen sulfide (H_2S)?

Answer: Reduced conditions are common whenever the biological oxygen demand is high, as when organic substrates are actively undergoing decomposition, and the rate of oxygen entry into the system is insufficient to satisfy the demand. The creation of reducing conditions shifts the ferrous (Fe^{2+})-ferric (Fe^{3+}) iron equilibrium to favor the reduced ferrous form which is soluble, whereas the ferric is insoluble under conditions normally encountered in trickle systems. The soluble ferrous ions are quite mobile and may be leached or otherwise transported by moving water only to be spontaneously precipitated and deposited as the oxidized ferric form, sometimes in massive amounts, by biological oxidation or, most likely, when an oxygen-rich site is encountered. These types of reduction and oxidation of iron and other transition metals commonly occur in soil, lakes, rivers, estuaries, and the open sea.

Similarily, hydrogen sulfide (H_2S) accumulation is a process typically associated with reduced conditions or anaerobiosis (no oxygen present). The sources of this potent destructive product are two: (1) the sulfate reduced by ***Desulfovibrio*** or physiologically related anaerobes, and (2) the sulfur-containing amino acids attacked during proteolysis. Sulfide biogenesis is common in lake and marine sediments, flooded soils, ditches, pond mud, sewage digestion tanks, feces, and the intestine. One can recognize hydrogen sulfide by the rotten egg odor and by the white cotton-like masses of sulfur slimes, that can form in the trickle irrigation system. The slimes occur because there are certain filamentous bacteria (***Thiothrix*** and ***Beggiatoa***) that can oxidize hydrogen sulfide to insoluble elemental sulfur. During this process, the bacteria make a sulfur slime which is deposited both within and on the outside of the organism. The bacteria are very long and stringy so they can form an extensive mat which accumulates and clogs the fine tubes in emitters and microsprinklers.

REFERENCES AND SUPPLEMENTAL READING

Bralts, V.F., Wu, I.P. and Gitlin, H.M., 1981. Drip irrigation uniformity considering emitter plugging. Trans. Amer. Soc. Agric. Eng. 24(5): 1234-1240.

Bralts, V.F., Wu, I.P. and Gitlin, H.M., 1981. Manufacturing variation and drip irrigation uniformity. Trans. Amer. Soc. Agric. Eng. 24(1): 113-119.

Bresler, E., 1977. Trickle-drip irrigation: Principles and applications to soil-water management. In Brady, N.C. (Ed.), Advances in Agronomy 29: 343-393. Academic Press, New York.
Bucks, D.A. and Myers, L.E., 1973. Trickle irrigation: Application uniformity from simple emitters. Trans. Amer. Soc. Agric. Eng. 16 (6): 1108-1111.
Bucks, D.A. and Nakayama, F.S., 1980. Injection of fertilizer and other chemicals for drip irrigation. Proc. Agri-Turf Irrig. Conf., Houston, Texas, Irrigation Association, Silver Spring, Maryland, pp. 166-180.
Bucks, D.A., Nakayama, F.S. and French, O.F., 1981. Keys to successful trickle: Management and maintenance. Proc. 15th Natl. Agric. Plastics Congr., Tucson, Arizona, pp. 3-8.
Bucks, D.A., Nakayama, F.S. and Gilbert, R.G., 1977. Clogging research on drip irrigation. Proc. 4th Ann. Int. Drip Irrig. Assoc. Meet., Fresno, California, pp. 25-31.
Bucks, D.A., Nakayama, F.S. and Gilbert, R.G., 1979. Trickle irrigation water quality and preventive maintenance. Agric. Water Manage. 2: 149-162.
Bucks, D.A., Nakayama, F.S. and Warrick, A.W., 1983. Principles, practices, and potentialities of trickle (drip) irrigation. In Hillel, D. (Ed.), Advances in Irrigation 1: 219-298, Academic Press, New York.
Davis, S., 1983. Drip irrigation. In Pair, C.H., Hinz, W.W., Sneed, R.E., and Frost, K.R. (Eds.), Irrigation (formerly Sprinkler Irrigation), Fifth Edition, Irrig. Assoc., Silver Spring, Maryland, pp. 528-546.
Ford, H.W., 1975. The use of chlorine in drip irrigation systems. Proc. Florida State Hortic. Soc. 88: 1-4.
Ford, H.W., 1976. Controlling slimes of sulfur bacteria in drip irrigation systems. HortSci. 11: 133-135.
Ford, H.W., 1977a. The importance of water quality in drip/trickle irrigation systems. Proc. Int. Soc. Citriculture 1: 84-87.
Ford, H.W., 1977b. Controlling certain types of slime clogging in drip/trickle systems. Proc. 7th Int. Agric. Plastics Congr., San Diego, California, pp. 118-123.
Ford, H.W., 1978. Bacterial clogging in low pressure irrigation systems. Proc. Tech. Irrig. Conf., Palm Springs, California, Irrigation Association, Silver Spring, Maryland, pp. 239-245.
Ford, H.W. 1979a. The complex nature of ochre. Kulturtechnik und Flurbereinigung 20: 226-232.
Ford, H.W. 1979b. Characteristics of slime and ochre in drainage and irrigation systems. Trans. Amer. Soc. Agric. Eng. 22(5): 1093-1096.
Ford, H.W., 1979c. The use of surface water for low pressure irrigation systems. AREC Res. Report GS75-2-HWF-100, Univ. of Florida, Lake Alfred, Florida, 6 pp.
Ford, H.W., 1980. The problem of dripper clogging and methods for control. Proc. of Symp. on Drip Irrig. in Hortic., Skierniewice, Poland, pp. 199-206.

Ford, H.W. and Tucker, D.P.H., 1974a. Water quality measurements for drip irrigation systems. Proc. Florida State Hortic. Soc. 87: 58-60.
Ford, H.W. and Tucker, D.P.H., 1974b. Clogging of drip systems from metabolic products of iron and sulfur bacteria. Proc. 2nd Int. Drip Irrig. Congr., San Diego, California, pp. 212-214.
Ford, H.W. and Tucker, D.P.H., 1975. Blockage of drip irrigation filters and emitters by iron-sulfur-bacterial products. Hort. Sci. 10: 26-64.
Gilbert, R.G., Nakayama, F.S. and Bucks, D.A., 1979. Trickle irrigation: Prevention of clogging. Trans. Amer. Soc. Agric. Eng. 22: 514-519.
Gilbert, R.G., Nakayama, F.S., Bucks, D.A., French, O.F. and Adamson, K.C., 1981. Trickle irrigation: Emitter clogging and flow problems. Agric. Water Manage. 3: 159-178.
Gilbert, R.G., Nakayama, F.S., Bucks, D.A., French, O.F., Adamson, K.C. and Johnson, R.M., 1982. Trickle irrigation: Predominant bacteria in treated Colorado River water and biologically clogged emitters. Irrig. Sci. 3: 123-132.
Goldberg, D. and Shumeli, M., 1970. Drip irrigation - a method used under arid and desert conditions of high water and soil salinity. Trans. Amer. Soc. Agric. Eng. 13(1): 38-41.
Goldberg, D., Gornat, B. and Rimon, D., 1976. Drip Irrigation: Principles, Design, and Agricultural Practices. Drip Irrig. Scientific Publ., Kfar Shumaryahu, Israel, 296 pp.
Howell, T.A., Stevenson, D.A., Aljibury, F.K., Gitlin, H.M., Wu, I.P., Warrick, A.W. and Raats, P.A.C., 1980. Design and operation of trickle (drip) systems. In Jensen, M.E. (Ed.), Design and Operation of Farm Irrigation Systems, Monograph 3, Amer. Soc. Agric. Eng., St. Joseph, Michigan, pp. 663-717.
Marsh, A.W. and Gustafson, C.D., 1971. Irrigating a drop at a time. Crops and Soils 23: 9-11.
McElhoe, B.A. and Hilton, H.W., 1974. Chemical treatment of drip irrigation water. Proc. 2nd Int. Drip Irrig. Congr., San Diego, California, pp. 205-220.
Myers, J.M., Harrison, D.S. and Phillips, J.W., 1976. Soil moisture distribution in a sprinkler irrigated orange grove. Proc. Florida State Hortic. Soc. 89: 23-28.
Nakayama, F.S. and Bucks, D.A., 1981. Emitter clogging effects on trickle irrigation uniformity. Trans. Amer. Soc. Agric. Eng. 24(1): 77-80.
Nakayama, F.S., Bucks, D.A. and Clemmens, A.J., 1979. Assessing trickle emitter application uniformity. Trans. Amer. Soc. Agric. Eng. 22(4): 816-821
Nakayama, F.S., Bucks, D.A. and French, O.F., 1977. Reclaiming partially clogged trickle emitters. Trans. Amer. Soc. Agric. Eng. 20(2): 270-280.
Nakayama, F.S., Gilbert, R.G. and Bucks, D.A., 1978. Water treatments in trickle irrigation systems. J. Irrig. Drain. Div., Amer. Soc. Civil Eng. 104(IR1): 23-24.

Parchomchuk, P., 1976. Temperature effect on emitter discharge rates. Trans. Amer. Soc. Agric. Eng. 19(4): 690-692.

Pelleg, D., Lahav, N. and Goldberg, D., 1974. Formation of blockages in drip irrigation systems. Proc. 2nd Int. Drip Irrig. Congr., San Diego, California, pp. 203-208.

Rolston, D.E., Rauschkolb, R.S., Phene, C.J., Miller, R.J., Uriu, K., Carlson, R.M. and Henderson, D.W., 1979. Applying Nutrients and Other Chemicals to Trickle-irrigated Crops. Univ. of Calif. Bull. 1893, Berkeley, California, 14 pp.

Rutenberg, G., 1967. Trickle Irrigation. Proc. Israel Centre of Waterworks Appliances, Publ. PR-80767-E, Tel-Aviv, Israel, pp. 29-31.

Sharp, R.B. 1956. The growth of mucus-forming bacteria in drip-fed irrigation lines. J. Agric. Eng. Res. 1: 83-88.

Shoji, K., 1977. Drip irrigation. Scientific Amer. 237(5): 62-68.

Solomon, K., 1979. Manufacturing variation of trickle emitters. Trans. Amer. Soc. Agric. Eng. 22(5): 1034-1038, 1043.

White, G.C., 1972. Handbook of Chlorination. Van Nostrand Reinhold, New York, 744 pp.

Wilson, D.L., 1972. Filtration, filters and water treatment. Proc. 3rd Drip Irrig. Seminar, San Diego, California, pp. 17-23.

Wilson, D.L., 1977. Filtration in drip irrigation. Proc. 7th Int. Agric. Plastics Congr., San Diego, California, pp. 160-165.

OPERATIONAL PRINCIPLES

3.2 WATER TREATMENT

F. S. NAKAYAMA

The ideal water for use in trickle irrigation is usually municipal water which has been treated to remove suspended particles, color, odor, and pathogenic bacteria. However, municipal water maybe too expensive or not readily available for crop production. The alternate source of water suitable for farm operation then is the on-site treatment of water to a level acceptable for proper performance of the trickle system. The pioneers of trickle irrigation surely did not envision the need for an elaborate water treatment, but since emitter operation is the key to the success of the whole system this subject has become the focal point of design, operation, and maintenance.

3.2.1 *Filtration*

Very rarely is there irrigation water that does not require some sort of filtration treatment. Particles present in the water range in size from the submicrometer virus to the larger sand-size fractions as shown in table 3.2.1.

TABLE 3.2.1

Classification of screens and particle sizes.

Screen mesh no.	Equivalent diameter (micrometer)	Particle designation	Equivalent diameter (micrometer)
16	1180	Coarse sand	>1000
20	850	Medium sand	250-500
30	600	Very fine sand	50-250
40	425	Silt	2-50
100	150	Clay	<2
140	106	Bacteria	0.4-2
170	90	Virus	<0.4
200	75		
270	53		
400	38		

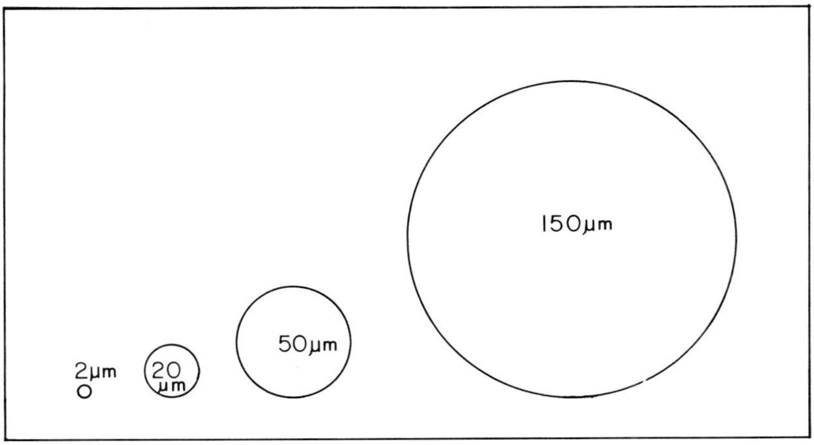

Fig. 3.2.1 Relationship of different particle sizes.

Individual silt-size particles are visible to the eye, but not the smaller clay-size particles. The relative sizes of the particles are illustrated in figure 3.2.1. The size of soft, organic particles is difficult to pinpoint accurately since these suspensions behave differently from solid particles especially in flow situations. Emitter opening varies greatly among types, but is about 250 µm and is frequently larger than this, so one wonders why we have clogging problems at all when the silt and clay particles encountered in the irrigation water are much smaller than the emitter pore dimension. The amount of suspension present in irrigation water ranges from a few parts to greater than 1000 mg/L (1000 ppm). Turbidity as an indication of suspended load can be observed in the 50 mg/L range and higher depending upon the particle size.

Example 3.2.1

Problem: Estimate the total amount of solids in 1.233×10^6 liters of irrigation water containing 10 ppm (mg/L) suspended load.

Solution: The concentration term ppm represents one unit per million units so that 10 ppm are equivalent to 10 g per 10^6 g. Assuming that the specific gravity of irrigation water is 1.0 g/cm^3, the total amount of solids is calculated as 1.233×10^3 g. The 1.233×10^6 liters were conveniently selected and represent one acre-ft of water, so that the solid weight is 27.2 lbs per acre-ft.

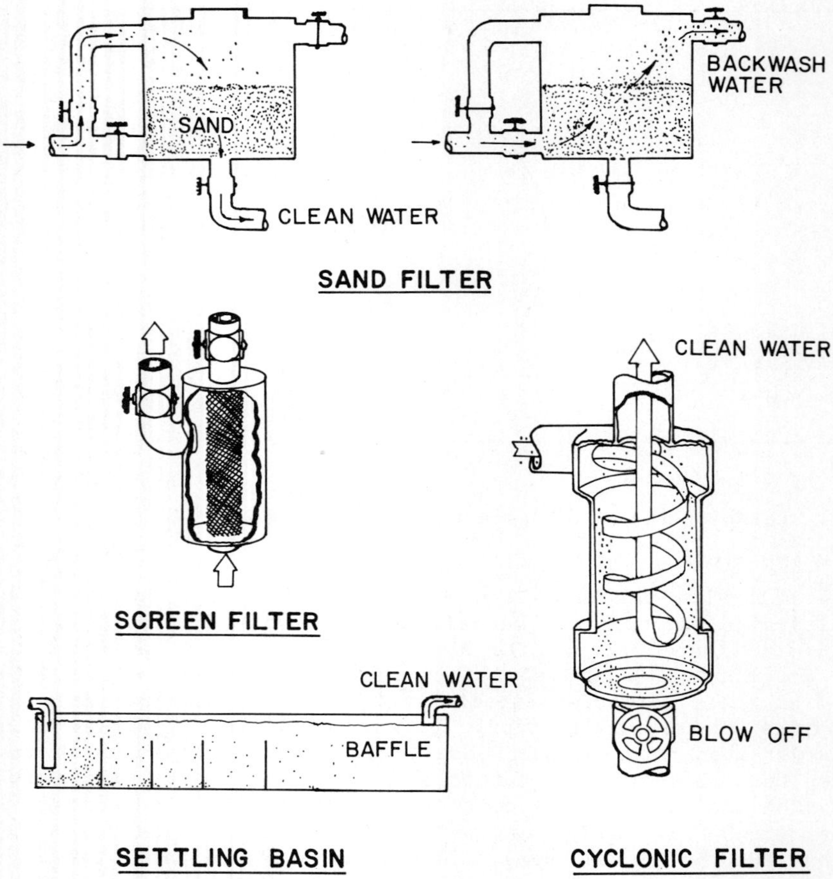

Fig. 3.2.2 Schematic description of various sediment removing devices.

The sample problem can be used as a basis for estimating further the amount of material and the potential clogging problem possible in the trickle system. In practice the suspended solid concentration is much higher than the 10 mg/L value and is equivalent to a tremendous amount of solid which the pipelines and emitters especially must contend. If the suspended materials are left in the pipelines, they can be further aggregated and cemented by microbial by-products and chemical reactions causing changes in the flow characteristics of the supply lines. Complete removal of the suspended materials from the irrigation water for use in trickle systems is impractical. In industrial applications where high pressure and flow conditions prevail, the general rule is that the particles at least 1/2 the diameter of the pore opening and larger should be removed, but field experience has shown that this does not necessarily hold for trickle emitters. Because filtering out the finer particles becomes cost prohibitive, water treatment is aimed primarily at removing the larger particle sizes and allowing the final suspended load to be in the range that the emitters and delivery system can tolerate for long-range operations. Filtration equipment used for removing the suspended solids in trickle work is based on technology developed for industrial, municipal, and domestic water treatment systems. Technological transfer in this instance has been successful with some additional modifications needed to allow for automatic cleaning and weather protection for the equipment to withstand extremes in the outdoor working environment. The filters used are self-descriptive and may be classified into the screen, media, and cyclonic or centrifuge types.

3.2.1.a Screen filters

Screen filters (figure 3.2.2) are the most frequently used equipment for removing particles. The one illustrated is used in pressurized lines. The method of water entry, circulation, and exit varies among the manufacturers. The figure shows water coming from the inside of the screen and exiting from the periphery. Others are designed with water flowing in the opposite direction. When properly sized and maintained, screen filters do an adequate job of removing suspended particles from the water, but they are limited in their load capacity. The total area has been increased to increase capacity, but there is a practical limit to the expansion. To overcome part of this limitation, holders have been designed and constructed using the cross-flow principle whereby the particle buildup on the screen is washed away by the flowing liquid and provides a self-cleaning ability without the need to dismantle the equipment for cleaning. The mesh sizes commonly in use range from 140 to 200 and are used primarily to remove the very fine sand particles. In actual operation, successively finer particles are removed because of the buildup of materials on the screen opening leading to an effective diameter smaller than the initial value. High amounts of algal debris are very difficult to handle with screen filters. The soft algal material tends to intertwine among the screen mesh and removal is difficult when the packing becomes dense.

Corrosion resistant stainless steel and plastic materials are the more common substances used in the construction of the screens. Good support for the screening component, which may be soft-textured cloth-like material or stiff wire mesh, is needed to prevent deformation caused by pressure differentials. Screens must be inspected routinely for physical integrity. Seemingly minor tear or hole enlargement drastically affects the particle removing ability of the screen. Water and particle flow is disproportionately higher through the enlarged opening than the normal screen opening. Pressure gauges at the inlet and exit sides of the filter give indication of the screen condition. Development of an abrupt pressure drop across the holder means that the screen is beginning to clog. In contrast, if no pressure change occurs for long periods, the screens or seals are broken or the mesh size is too large, assuming that the system is not the self-cleaning type where pressure changes are less noticeable than the other types of screen filters.

A slight modification of the rigid screen is the "bag" type filter which has soft, flexible sides that conform to the surrounding basket when in operation. This type is usually made of materials that can be cleaned a few times before it must be disposed. Also in the similar category are the cartridge filters that have washable and disposable inserts, and the Y-strainers both of which are used primarily in home gardens, nurseries, and landscape installation. The smaller sized units can be conveniently installed into the pipeline.

Nonpressurized, gravity screen filters are sometimes installed in an irrigation canal before the pump intake or delivery system. These are designed to remove large organic debris such as leaves and weeds. Sometimes smaller screen openings have been used to remove gravel, sand, and silt where the suspended load is extremely high. Screening devices have been constructed to operate like a looping conveyor belt to make them self-cleaning. Water jets are also directed at an appropriate angle to the screen to wash away particles that become entrapped in the mesh.

3.2.1.b Sand filters

Pressure-type, high-flow sand or mix-bed filters are the more popular ones used in clarifying irrigation water for trickle systems. Gravity operating sand filters have low flow rates and thus require large surface area to produce the equivalent volume of filtered water as pressurized filters. Almost the full depth of the sand is used in the pressurized filters compared to gravity filters where surface action is the primary filtration mechanism. In either case, the filtration process actually becomes more efficient with passage of time because successively smaller particles can be filtered out as the flow passages become smaller. Unfortunately, this cannot proceed indefinitely since the resistance increases and the flow rates are drastically reduced. Sand filtration appears to be a simple process, yet a mass of complex relation has been formulated to explain the filtration and flow characteristics in the sand bed.

Filter capacity is designated in terms of volume flow per unit time per unit bed area. Flow rates encountered in trickle systems range from 80 to 2,000 $L/min/m^2$ (2 to 50 $gal/min/ft^2$) with finer filtration occurring at the lower rates. The filter body is packed to a depth of 15 to 30 cm with silica sand or equivalent crushed material that is inert to water and chemicals. In some parts of the world, beach sand originating from coral material is available, but it should not be used since it is composed of calcium carbonate. Manufacturer's recommendation concerning the amount, type, and size of fill material to use must be followed, since the filter dimension has been optimized for a specific volume of filtering agent, pressure, and flow rate. Conceivably, sand-size particles carried in by the raw water can change the depth of the bed and this factor must be noted to avoid detrimental changes to filter performance.

The sand sizes usually used in the filters are the No. 11 (0.79 mm), 16 (0.66 mm) and 20 (0.46 mm) which give filtration of particles in the 75, 50, and 40 μm range, respectively.

Example 3.2.2

Given: Column packed with 0.60 mm sand for filtering out suspended solids from river water.

Problem: Determine the hydraulic radius of the packing assuming that the grains are spherical.

Solution: An expression relating hydraulic radius, R_h, is the ratio of the volume, V_p, to the surface area, A_p, of the particle.

$$R_h = [(4/3)\pi(D_p/2)^3]/[4\pi(D_p/2)^2] = D_p/6 \quad (3.2.1)$$

where D_p is the particle diameter. Thus, in this simplified example, the hydraulic radius for the 0.60 mm particle is 100 μm.

Actually, the sand particles are not perfect spheres and not single dimensioned but have ragged edges, with a range of sizes, so that filtration for a given particle size depends upon a variety of physical factors which make up the filter medium.

No universal guide has been developed that would help to predict accurately when filter cleaning or backwashing is necessary. Predicting the time of filter run, T, is strictly empirical, and may be related to the function

$$T = f[(A^2p^4\ H \cdot S \cdot L)/(V^{1.5}c)] \quad (3.2.2)$$

where c = concentration of suspended particles in the water, A = area of filter bed, p = porosity, H = terminal head loss, S = surge or changes occurring in the flow rate through the sand filter, L = depth of penetration, and V = filtration rate. Operators soon find by experience the time interval needed between cleaning. Decision for backwashing usually is related to the pressure differential developed between the inlet and outlet section of the filter system, in the order of 69 kPa (10 psi) and less. Filters have been able to operate properly from 1 to 12 hours or even longer without backwashing. The selection of the filter run is based primarily on convenience and cost. (See section 3.4.6 for discussion on filter removal efficiency).

Backwashing is a critical part of filter operation and performance. Reverse flow rates must be of sufficient velocity to cause the separation and suspension of sand material into individual particles. Filter design must allow for the expansion of the bed, otherwise the sand would be lost with the backwash water. Increasing the backwash flow rate beyond the suspension or fluidized rate serves no useful purpose unless there are some deeply trapped particles. Even then, water is wasted unless provision is made to use the water in other areas of the field. In the fluidized state, where particle-to-particle contact in motion is maximum, sand particles interact to rub off the loosely adhered materials. Observing the backwash water by passing it through a transparent plastic pipe that can withstand the operating line pressure is a simple method for checking the efficiency of the cleaning operation. The initial backwash water is extremely turbid, followed by a gradual clearing. Backwash operation should take from 5 to 15 minutes. Periodic inspection of the filter bed right after backwashing should be made to see whether the sand particles are loosely packed as in the original condition. Additional sand may be required if some is washed away, or a complete change may be needed when backwashing cannot remove the organic or other materials adhering to the sand grains.

The first increment of water entering the lines immediately after backwashing may be of similar quality as the untreated water when raw water is used for backwashing. To avoid this, dual filter systems are sometimes installed, where the filtered water from one is passed through the other for backwashing. Another alternative is to provide a by-pass or dump valve arrangement to get rid of the first increment of the new water. A transparent pipe for observing the quality of the filtered water entering the mains is useful. Also, screen filters should be provided after the sand filters to prevent accidental spillage of sand into the lines.

Caking or cementation of the sand can result in complete clogging of the bed or the formation of large pores that affect filter performance. In certain conditions, "rat holing" results where large pores become predominant within the filter bed so that filtration is nil and backwashing serves no beneficial purpose. Such conditions can be brought about through microbial aggregation of the sand granules where backwashing fails to disperse and remove the aggregating materials properly. The filter beds are extremely good environment for microbial activity since organic nutrients are readily available. Cementing of the sand with suspended silt and clay particles can also occur by chemical reactions due to carbonate precipitation. A particularly good example of this condition has been observed when well water high in dissolved carbon dioxide was passed through the sand filters. When the ambient pressure was decreased from that of the originating water, carbon dioxide was released into the atmosphere and calcium carbonate precipitated because of the resultant decrease in carbon dioxide concentration.

3.2.1.c Settling basins

Sedimentation or settling basins are used for clarifying waters, especially where the suspended loads of the water are extremely high and would otherwise quickly overload the sand or screen filters. Settling basins can also aid in removing the soluble sulfides and heavy metals like iron and manganese which create special problems in emitter clogging. The reduced forms of manganese and iron when allowed to oxidized to the insoluble compounds in the basins can precipitate and can be filtered out before they can get into the trickle lines. The basins need not be elaborate structures as long as they can be adequately cleaned and maintained. Land availability usually puts a limitation on the installation of the basins.

The settling behavior of suspended particles can be described by Stokes' equation as follows

$$V = g(d_s - d_w)\, D^2/18\mu \qquad (3.2.3)$$

where V = the velocity of the particle (cm/sec), g = the gravitational acceleration constant (cm/sec^2), d_s and d_w = the densities of the particle and water (g/cm^3), respectively, D = the diameter of the particle (cm) and μ = the liquid viscosity (g/cm-sec).

Example 3.2.3

Given: Particle density = 2.65 g/cm^3; liquid density = 1.00 g/cm^3; liquid viscosity = 0.008 g/cm-sec; gravitational constant = 980 cm/sec^2.

Find: Estimate the time required for a 50 μm silt-size particle to settle 30 cm (6 in) at 25°C; a 2 μm particle.

Solution: Stokes law is used to calculate the velocity of fall of the solid as

$$V = \frac{980 \times (2.65-1) \times (2.5 \times 10^{-4})^2}{18 \times 0.008} \quad \text{and } t = 30/V$$

The time of settling for 30 cm distance is then 107 seconds and for the 2 μm particle is 66,790 seconds.

From the example presented, we can see why the clay-size particle takes forever to settle out. Small currents in the water can keep these small particles in suspension almost indefinitely.

An idealized settling basin is illustrated in figure 3.2.3. A particle with settling velocity v_o at height H_o if it follows the path V will be removed once it touches the basin bottom. Other suspended particles entering the basin with heights less than h_o will follow similar parallel paths as V and will also be removed. Suspended particles with settling velocities $v_i < v_o$ will follow another path v_i and will be removed if their entrance elevation is less than h_i, and will be carried out by the main stream if the entrance height is greater than h_i.

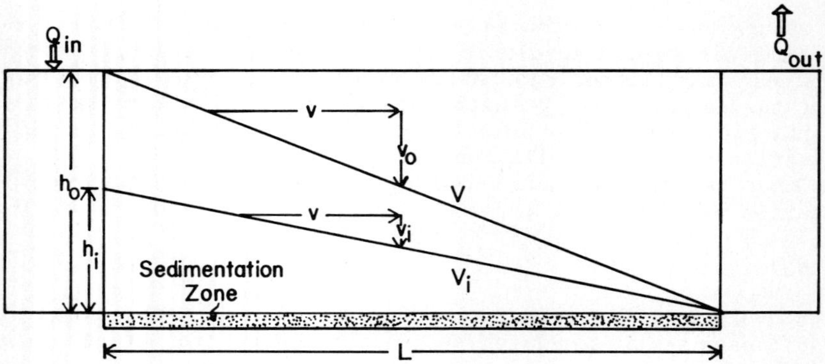

Fig. 3.2.3 Schematic of sedimentation behavior in a settling basin.

The net time for settling is $t = L/V$ and the critical settling velocity is $V_o = h_o/t_o$, but since $V = Q/h_oW$, where W is the width and the product WL the area of the basin

$$V_o = \frac{h_oW}{L} = \frac{Q}{WL} = \frac{Q}{A} \tag{3.2.4}$$

Thus, the ratio Q/A, the overflow rate or surface loading, defines the minimum particle sedimentation velocity for complete removal of materials passing through the settling zone.

Water velocities in the sediment basin cannot exceed a critical value, otherwise scouring of the settled particles will occur and the deposited materials will be carried out with the flow. This velocity V_c can be related to the various parameters by

$$V_c = [(8k/f)g(d_p - d_w)D]^{1/2} \tag{3.2.5}$$

where k = 0.04 for single particles to 0.06 for sticky materials, f = friction factor = 0.03, g = gravitational constant, d_p and d_w = the densities of the particle and water, respectively, and D = the diameter of the particle. To avoid scouring velocities, the ratio of the basin depth and/or surface area, A_s, to the cross sectional area, A_c, must be kept below a threshold value as

$$\frac{L}{h} = \frac{A_s}{A_c} = \frac{V_c}{V_s} \times K \tag{3.2.6}$$

where K = 1 for an ideal basin.

Example 3.2.4

Given: Particle diameter = 50 μm and similar variables as in example 3.2.3.

Find: The scouring velocity and the limiting length to depth ratio of the basin suitable for the removal of this particle size.

Solution:

$$V_c = \frac{8 \times 0.04}{0.03} \times 980 \times (2.65-1)(0.005)^{1/2}$$

$$= 9.3 \text{ cm/sec.}$$

For the previous example, V_s = 0.28 cm/sec so the length to depth ratio is 9.3/0.28 or 33.

By using similar approaches for other particle sizes, a practical limit can be estimated for basin dimensions and overflow velocity. Experience based on municipal sedimentation basins preceding sand filter systems indicates an average of 25 cm/sec with minimum and maximum ranging by a factor of 3. However, the velocity within the basin is not uniform throughout the cross sectional area of the basin, i.e., zero at the bottom and walls and higher closer to the water surface.

The example is applicable for an ideal situation so that uncontrollable environment factors such as wind and wave action, convection currents caused by temperature gradients, biological activities, and variable inlet velocities and liquid densities alter the flow pattern in the basin and affect the initial estimate of basin design.

Sedimentation ponds suffer from other shortcomings, the major one being algal growth and organic matter decomposition. Control is limited to the application of algacide at low concentrations to avoid damage to plants which receive the treated water. Wind-blown dust and debris also detrimentally affect the esthetic condition of the basin besides creating management problems.

3.2.1.d Cyclonic or centrifugal filters

Cyclonic filters or separators are in-line systems that are used to remove suspended materials with specific gravities greater than water. Actually particle size need not be categorized since any dense particle can be considered potentially removable. Their operational principle is entirely different from the screen or sand filters, but since the end result is similar, they are sometimes considered as filters. Normally, organic materials which are less dense than water are not removed unless they are bonded to the heavier particles. By directing water tangentially at the inlet section of the chamber, kinetic energy is converted to centrifugal force. This is equivalent to increasing the gravitational factor of Stokes' equation that was illustrated previously, causing particle separation to occur during the period that a given volume of water moves from the inlet to outlet ports. Various types of internal construction are used to aid in optimizing the flow geometry. The centrifugal force directs the solids to the outer edge or perimeter and the particles then drop to the less turbulent part of the chamber to be purged from the separator. The rest of the liquid flow with the particles removed is directed radially inward due to the lower pressure and up through the outlet port. Pressure drop is exhibited within the separator even though the water is not passed through tiny pores as in the case of sand or screen filters.

The design capacity of any type of filter installation should closely match the actual flow of the trickle system. Filters are most efficient at their rated capacity - too small capacity would require frequent backwashing or cleaning in the case of sand and screen filters and sediment basins or undesirable flow patterns in the cyclonic separators to optimize particle separation. On the other hand, overcapacity also leads to waste in material and equipment performance.

3.2.2 *Chemical treatment*

As noted in an earlier chapter, physical, chemical and biological factors are responsible for the plugging of emitters. The physical problem can be alleviated by the filtration of the water. The remaining two factors require other types of water treatment involving the addition of pH-modifying and bactericidal chemicals. Selecting the amount and kind of chemicals for water treatment requires an understanding of the composition of the water and the reactions that the chemicals undergo when added to the water.

3.2.2.a Chemical precipitation

Formation of insoluble salts as a result of chemical reactions at the orifice and internal parts of emitters has been shown to be responsible for emitter clogging. One of the primary chemical constituents identified is calcium carbonate formed from the soluble calcium and carbonates originally present in the irrigation water. Removal of the dissolved salts with ion exchange columns or reverse osmosis equipment is impractical. The other alternative is to prevent the precipitates from forming by controlling solution pH. Before this is done, however, the water analysis should be evaluated to determine whether carbonate precipitation would occur. The classical Langelier Saturation Index (LSI) concept provides a systematic approach for determining the tendency for $CaCO_3$ formation. It is based on relating a calculated pH, pH_c, to the measured pH, pH_m, of the water. The calculated pH_c is obtained from the Ca, HCO_3, and total salt concentrations of the water.

The following simplified derivation of LSI involves the definitions of the solubility product of $CaCO_3$, K_s, and the dissociation constant K_d of HCO_3:

$$Ca^{2+} + CO_3^{2-} = CaCO_{3\ (solid)} \qquad (3.2.7)$$

$$K_s = (Ca^{2+})(CO_3^{2-}) \qquad (3.2.8)$$

$$HCO_3^- = H^+ + CO_3^{2-} \qquad (3.2.9)$$

$$K_d = \frac{(H^+)(CO_3^{2-})}{(HCO_3^-)} \qquad (3.2.10)$$

Division of K_s by K_d gives

$$\frac{K_s}{K_d} = \frac{(Ca^{2+})(HCO_3^-)}{(H^+)} \qquad (3.2.11)$$

The logarithm of the various parts of the preceding equation yields

$$\log K_s - \log K_d = \log(Ca^{2+}) + \log(HCO_3^-) - \log(H^+) \qquad (3.2.12)$$

The negative log of the hydrogen ion activity is defined as pH, and under equilibrium condition is a distinct and identifiable value. For simplification it is assumed that the solution is closed or unexposed to atmospheric carbon dioxide. Langelier proposed that by calculating the pH of the solution using equation 3.2.12, and comparing this pH_c with the measured pH_m of the test solution, the difference between these two values would give an indication of the precipitation potential of the Ca and bicarbonates. If $pH_m - pH_c = 0$, the constituents are in equilibrium, if $pH_m - pH_c < 0$ no precipitation will occur, and if $pH_m - pH_c > 0$ then the precipitation of $CaCO_3$ is very likely to occur. Redefining equation 3.2.12 in terms of negative logarithms with "p" representing such a translation, the equation becomes

$$pH_c = (pK_d - pK_s) + p[Ca] + p[HCO_3] + p(ACF) \qquad (3.2.13)$$

In the original derivation, total alkalinity(= $HCO_3^- + 2CO_3^{2-} + OH^- - H^+$) was used, but under the conditions encountered, HCO_3 is the dominant species and so this concentration is usually used. The term ACF is the activity coefficient factor and includes the activity coefficients for Ca and HCO_3. This is necessary to correct for the non-ideality of the solution since the dissociation and solubility product constants defined by equations 3.2.8 and 3.2.10 are derived in terms of an ideal solution, where the activity coefficients are unity.

Both K_d and K_c values are temperature dependent and the relation to show this dependency is given by

$$pK_d - pK_s = 2.586 - 2.621 \times 10^{-2}t + 1.01 \times 10^{-4}t^2 \qquad (3.2.14)$$

where t is the solution temperature in degrees Celsius. The ACF factor is also temperature dependent, but the difference in the temperature range of 0 to 50°C is only in the order of 0.02 and, thus, can be ignored for practical situations. However, ACF is solution concentration dependent and can be related to the equation

$$p(ACF) = 7.790 \times 10^{-2} + 2.160 \times 10^{-2}\ TDS - 5.477 \times 10^{-4}\ TDS^2 + 5.323 \times 10^{-6}\ TDS^3 \qquad (3.2.15)$$

where TDS is the total dissolved ion concentration in me/L.

Example 3.2.5

Given: Irrigation water analyzed at Ca = 1.9 me/L, HCO_3 = 4.4 me/L, TDS = 9.4 me/L, measured pH = 7.70 and the solution temperature = 25°C.

Find: Langelier saturation and evaluate the water in terms of a $CaCO_3$ precipitation problem.

Solution: From the preceding equations 3.2.15 and 3.2.14 p(ACF) = 0.24, ($pK_d - pK_s$) = 2.00 at 25°C. The conversion equations to change the me/L to mole/L concentration to get p[Ca] and p[HCO_3] are p[Ca] = 3.30 - log[Ca] and p[HCO_3] = 3.00 - log [HCO_3] and leads to p[Ca] = 3.02 and p[HCO_3] = 2.36. Thus, pH_c is calculated as 7.62, and the LSI = 7.70-7.62 = 0.08. The positive difference would indicate the potential for $CaCO_3$ precipitation.

3.2.2.b Acid treatment

Two important aspects of water treatment and precipitation potential can be derived from the various equations presented relating to

$CaCO_3$ precipitation. First, adjusting the solution pH or pH_m by acid addition will force the saturation index to become negative so that precipitation can be prevented. With irrigation waters close to an LSI equilibrium, adjustment of 0.5 pH unit is usually sufficient to obtain a negative LSI. Second, since the equilibrium constants pK_d and pK_s are temperature dependent (equation 3.2.14) precipitation potential is also temperature dependent. An increase in temperature would lead to a decrease in $CaCO_3$ solubility. At 25°C, (pK_d - pK_s) was calculated to be 2.00 and at 50°C, a temperature that can be encountered in trickle lines and emitters exposed to direct sunlight, the value is 1.55. Thus, the calculated LSI would become more positive and suggests a greater potential for carbonate precipitation. In other situations, a zero or negative LSI can become positive strickly due to temperature increase. Once precipitates are formed, whether caused by temperature or other mechanisms, redissolution by reversing the condition does not take place at the same rate, but rather at a much slower rate.

A spontaneous change in pH can occur especially in well water containing large amounts of dissolved CO_2 brought about by a high hydrostatic pressure in the aquifer. A decrease in this pressure once the water encounters atmospheric pressure releases CO_2 causing a rise in pH and $CaCO_3$ precipitation if it was dissolved in the groundwater.

The saturation index value, while easily applied, cannot be used to determine the quantity of acid needed to adjust the pH of the irrigation water to the level necessary to prevent carbonate precipitation. This can only be accomplished with acid titration of the water involving the addition of known increments of standard acid to the water and measuring the pH changes occurring in the mixture. An example of such a titration curve is shown in figure 3.2.4. A rapid drop in pH is observed with the first increment of acid addition followed by a gradual change. For trickle irrigation work, interest is focused on pH decrease down to the 6.0 level. Waters vary in their response to acid because of their buffering capacity, but for most waters examined with an initial pH 8 range, the pH decreased approximately one unit with 0.5 me/L acid. In general, 1 me/L acid addition would lead to a final pH between 6 to 6.5. With continuous dilute acid treatment, carbonate precipitation can be prevented; and furthermore, if minor precipitation had already started prior to acid treatment, it can be dissolved assuming that the acidified water can make contact with the precipitated material. Iron and manganese sulfide precipitates would not dissolve at the acid levels used for carbonate control. The use of excessive acid is uneconomical and may cause other problems such as the corrosion of metallic fittings if they are present in the pipeline.

The question always rises as to what effect chemical treatment would have on the salinity of the water. A 1 me/L sulfuric acid addition is equivalent to 49 mg/L or 49 ppm dissolved solid concentration, and depending on whether the irrigation water is 350 or 1500 mg/L, could have some to very little effect on total salinity. Since one of the principal ions is sulfate, the use of sulfuric acid may be considered as the addition of gypsum ($CaSO_4$), but at a very low concentration.

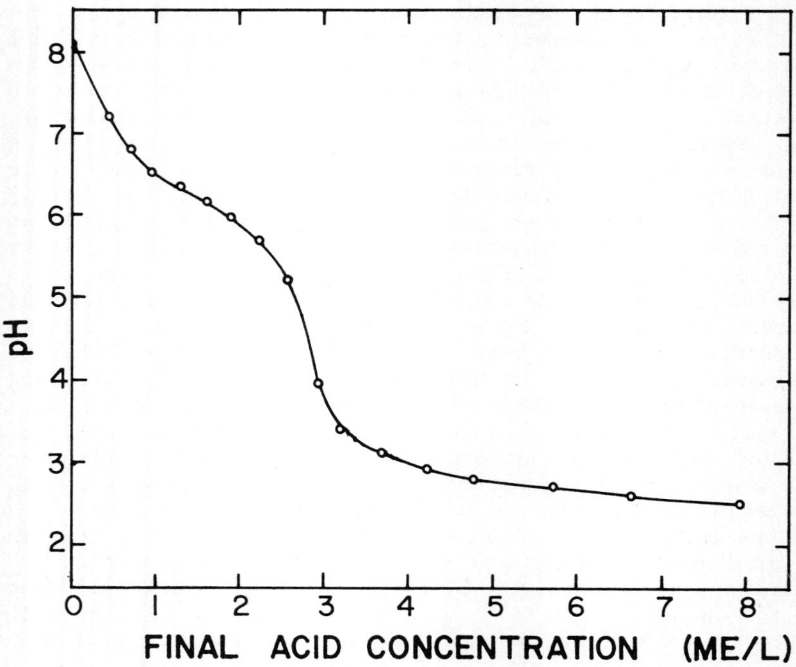

Fig. 3.2.4 Typical acid titration curve for irrigation water.

Besides the carbonate, sulfide precipitation problems are present and are usually related to a specific locale. Carbonate problems occur primarily in the West and Southwest, whereas sulfide problems predominate in the East and Southeastern regions of the United States. The sulfide reactions are

$$Mn^{2+} + S^{-} = MnS \text{ (ppt)} \quad (3.2.16)$$

$$Fe^{2+} + S^{-} = FeS \text{ (ppt)} \quad (3.2.17)$$

The precipitated MnS and FeS have black coloration. With exposure to oxygen the hydroxy oxides of these metals are formed and are characterized as "ochre". The iron and manganese sulfide precipitates have created problems in drainage systems where they form deposits that can actually obstruct large diameter pipes. The sulfides are found under reduced conditions caused by bacterial reduction of sulfate (SO_4) when no oxygen supply is available, and similarly could be present at pond bottoms where oxygen diffusion to that region is limited. Hydrogen sulfide, H_2S, with the typical "rotten egg" odor indicates the presence of the sulfide. Portable test kits are available which can be used to quantify sulfide concentration. Once formed the MnS and FeS are difficult to remove chemically.

3.2.2.c Chlorination

As noted earlier, biologically-caused clogging creates problems in trickle systems, particularly with surface water where contamination probability by bacterial materials are limitless. Well water pumped directly into the trickle lines generally presents less biological problems than surface waters because contamination sources are less. A unique exception is when iron-reducing bacteria contamination is present in the well water and the metal casing and screens can be corroded by biological activity resulting in soluble materials which eventually precipitate in the emitters and lines when exposed to oxygen.

The chemistry and application principles of chlorination for trickle irrigation water are similar to those used in home swimming pools, industrial and municipal drinking, cooling tower and wastewater treatment facilities. An understanding of chlorine chemistry is helpful for avoiding potential problems that may be encountered when using this chemical. In table 3.2.3 are listed the various forms of chlorine and the reactions commonly applicable to trickle irrigation waters.

TABLE 3.2.2

Basic forms and reactions of chlorine and its salts.

$Cl_2 + H_2O = H^+ + Cl^- + HOCl$	(3.2.18)
$HOCl = H^+ + OCl^-$	(3.2.19)
$NaOCl + H_2O = Na^+ + OH^- + HOCl$	(3.2.20)
$Ca(OCl)_2 + 2H_2O = Ca^{2+} + 2OH^- + 2HOCl$	(3.2.21)
$HOCl + NH_3 = NH_2Cl + H_2O$	(3.2.22)
$HOCl + NH_2Cl = NHCl_2 + H_2O$	(3.2.23)
$HOCl + NHCl_2 = NCl_3 + H_2O$	(3.2.24)
$HOCl + 2Fe^{2+} + H^+ = 2Fe^{3+} + Cl^- + H_2O$	(3.2.25)
$Cl_2 + 2Fe(HCO_3)_2 + Ca(HCO_3)_2 = 2Fe(OH)_3 + CaCl_2 + 6CO_2$	(3.2.26)
$HOCl + H_2S = S^\circ\downarrow + H_2O + H^+ + Cl^-$	(3.2.27)
$Cl_2 + H_2S = S^\circ\downarrow + 2H^+ + 2Cl^-$	(3.2.28)

When chlorine gas is injected into water it reacts with water to form hypochlorous acid, HOCl, hydrogen, H^+, and chloride, Cl^-, ions (equation 3.2.18). The HOCl can further dissociate to form the hypochlorous anion, OCl^- (equation 3.2.19). Note in this reaction that the hydrogen ions, H^+, formed will lower the pH. The extent of pH lowering will depend upon the amount of chlorine gas added and the buffering capacity of the water.

The sodium hypochlorite, NaOCl, of equation 3.2.20 and calcium hypochlorite, $Ca(OCl)_2$, of equation 3.2.21 are other sources of HOCl. The sodium hypochlorite is available in liquid form, more commonly called laundry bleach, and the calcium hypochlorite in solid form, which must be dissolved in water. Note in both the sodium and calcium hypochlorite compounds salts that hydrolysis in water produces hydroxyl ions, OH^-, which will raise the pH of the water.

Hypochlorous acid can react with ammonia, NH_3, or ammonium, NH_4^+ ions, and amine group, NH_2, the latter which is a integral part of organic matter. Equations 3.2.22 through 3.2.24 depict such reactions with the formation of monochloramine, NH_2Cl, dichloramine, $NHCl_2$ and trichloramine, NCl_3. The chloramines are perceptible to taste and smell and are in the order of sensitivity, $NCl_3 > NHCl_2 > NH_2Cl > HOCl$ in the concentration range of 0.02, 0.8, 5.0, 20.0 mg/L, respectively. The off-taste and odor in swimming pool and drinking waters are due to the chloramine and not necessarily the chlorine gas, Cl_2, itself. Hypochlorous acid can oxidize soluble ferrous ion, Fe^{2+}, to the ferric ion, Fe^{3+}, and in this case insoluble ferric hydroxide, $Fe(OH)_3$, can be readily formed (equations 3.2.25 and 3.2.26). Similar reactions occur with the manganous ion, Mn^{2+}. Reaction can also occur between chlorine and hydrogen sulfide forming elemental sulfur, S° (equations 3.2.27 and 3.2.28). Since the chlorine is used up by reacting with the sulfide ion, allowance must be made for the extra chlorine that is needed to permit enough residual for controlling the microorganisms.

Investigations have shown that hypochlorous acid plays the dominant role in controlling bacteria. The amount of HOCl present in solution is pH dependent (figure 3.2.5), with more of the active form occurring at the lower pH as described by equation 3.2.19 of table 3.2.2. Note that at extremely low pH or high acidity the Cl_2 form will dominate (equation 3.2.18). This is the reason for recommending that acids and hypochlorite sources be stored separately. Any accidental mixing of these chemicals, will result in the release of chlorine gas and also large amounts of heat that can cause a fire.

pH control has been strongly emphasized in swimming pool chlorination, and from a knowledge of the reaction of chlorine, we can see why alkaline solutions must be added to pools that are gas treated, whereas acid solutions must be added to the hypochlorite treated water. Sulfur dioxide (SO_2), from the burning of sulfur, instead of sulfuric acid has been used directly as an acid source for trickle systems. This is not practical when chlorination is going on because the sulfurous acid (H_2SO_3) formed by the reaction of SO_2 with water can also react with the hypochlorous compound and result in the inactivation of the chlorine source. Thus, the complete process can be represented by

$$SO_2 + H_2O = H_2SO_3, \text{ and} \quad (3.2.29a)$$

$$H_2SO_3 + HOCl = H_2SO_4 + HCl. \quad (3.2.29b)$$

Actually, sulfur dioxide has been used as a dechlorinating agent in treatment plants when undesirable high level of chlorine is present in the water.

Chlorination effectiveness is tested by measuring the concentration of HOCl. Portable units suitable for field operations are available where the test-water is treated with an indicator crystal or solution and the resultant color compared on a color chart, disk or standard reference solution. The classical swimming pool type test kits, based on the orthotolidine indicator, give the total

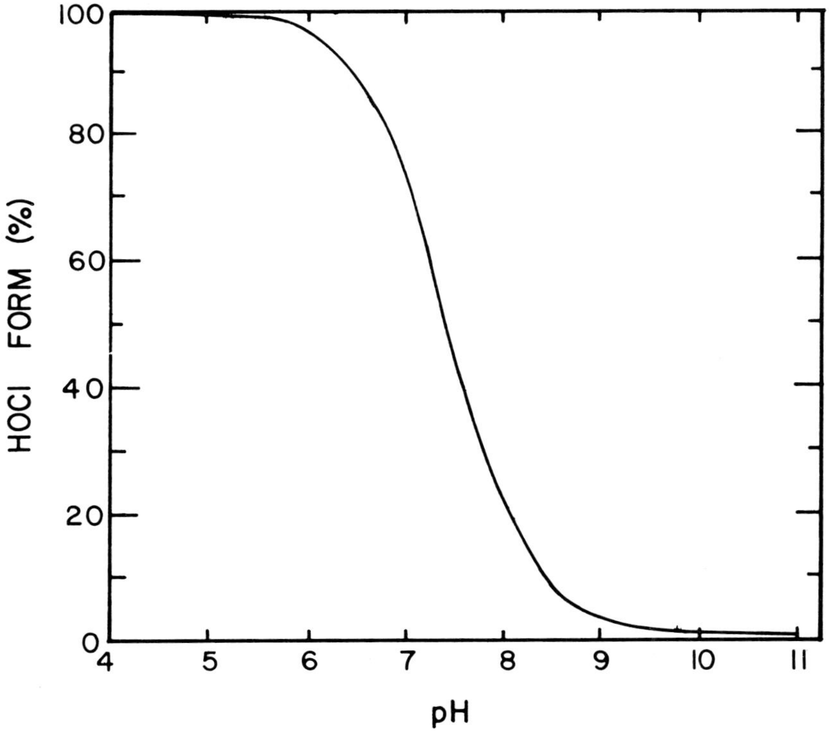

Fig. 3.2.5 Relative amounts of hypochlorous acid (HOCl) present in solution as a function of pH.

combined residual chlorine that includes the various chloramines plus the hypochlorous acid forms. More recently, test kits have become available with the N,N-diethyl-p-phenylene diamine (DPD) indicator reagent which is specific for the "free residual or available chlorine" which is the primary bactericidal agent. DPD in conjunction with pH-buffers and potassium iodide can be used to determine the individual chloramine compounds. More precise measurements of free available chlorine can be obtained by amperometric titration in the laboratory. A reducing agent is used as the titrant for the HOCl. An abrupt change in current can be observed when the end-point is reached.

Most pathogenic bacteria and viruses are inactivated at free residual chlorine concentration of 1 mg/L, with sufficient contact time in the order of 10 to 30 minutes. Municipal treatment plants must not only control bacteria, but taste and odor so that levels higher than 1 mg/L are sometimes used. Free residual chlorine is less than 1 mg/L by the time the water is distributed to the household. Studies made in municipal systems have shown that the transmission capacity of poorly performing pipe lines could be improved and maintained by chlorination. The build-up of bacterial slime was controlled by the chlorine treatment and such observations have been put to use in trickle irrigation systems.

Achieving a specified free residual chlorine content in trickle irrigation water must be done by some trial and error. To obtain a 1 mg/L final chlorine equivalent concentration would require a slightly higher injection rate or concentration than simply based on calculation. The reason for this is that most waters have an inherent chlorine demand. Chlorine can react with suspended organic matter, soil particle, or other dissolved constituents besides the bacteria. For example, if hydrogen sulfide is present, it will take 2 mg/L chlorine to react with 1 mg/L of the sulfide (equations 3.2.27 and 3.2.28); if ferrous iron is present, 1 mg/L Fe will require 0.6 mg/L chlorine (equation 3.2.25). In fact this method has been used as a possible means of solving the iron clogging problem. Chlorination would oxidize the soluble ferrous iron to the ferric form which then would lead to the formation of the insoluble ferric hydroxide. This latter material can then be filtered out from the water using the filtration system. The usual practice is to chlorinate the water just before the filter, but this close proximity to the filter may not allow enough time for complete chemical reaction to take place and for the smaller-sized $Fe(OH)_3$ precipitate to agglomerate to larger sizes which can then be removed by the filter.

Different chlorine sources are given in table 3.2.3. Relative costs can be estimated by comparing the quantity of the specific chemical needed to attain a given chlorine level and the price of the chemical. Initial cost of the injection system and availability of the chemical must also be considered in the selection of a chlorination system. Operators with systems that require large water volumes have selected chlorination with gas because it is the most economical, whereas, those with smaller needs have chosen liquid hypochlorite solutions because of its convenience.

TABLE 3.2.3

Chlorine equivalents of commercial sources and quantity needed to treat 1,233 m^3 (1 acre-ft) of water to obtain 1 mg/L Cl_2.

Chemical	Quantity equivalent to 454 g (1 lb) of Cl_2	Quantity to treat 1,233 m^3 (1 acre-ft) to 1 mg/L Cl_2
Chlorine gas	454 g (1.0 lb)	1226 g (2.7 lb)
Calcium hypochlorite 65-70% available chlorine	681 g (1.5 lb)	1816 g (4.0 lb)
Sodium hypochlorite		
15% available chlorine	2.54 L (0.67 gal)	6.81 L (1.8 gal)
10% available chlorine	3.78 L (1.0 gal)	10.22 L (2.7 gal)
5% available chlorine	7.57 L (2.0 gal)	20.44 L (5.4 gal)

3.2.2.d Chemical injections

Chemical injection for water treatment is an important operating component of the whole irrigation system. Controlled amounts of known chemical concentrations must be introduced into the main water stream to maintain the desired concentration for achieving the proper results. Fortunately, existing technology provides the operator with a choice from a variety of injection equipment that can do an adequate job of applying chemicals.

Chemical injection procedures require special considerations not necessarily related to the actual injection process. Bulk chemicals must be stored separately if they are incompatible. Storage should be in secured facilities, and storage tanks and fittings must be compatible with the chemical solution. Safety showers or ample water supply, protective clothing, respirators, and related devices must be nearby and in working condition for use in case of accidental chemical spillage and contact by personnel. Primary stock dilutions when needed should always be made by the addition of the chemical to the bulk water to avoid possible heating and splattering of concentrated solutions. Backflow prevention devices must be installed at the inlet side to prevent chemicals from getting back into the water supply lines. This is particularly important with domestic water supply connections where the water is used for human and animal consumption. Local ordinances frequently require such backflow controllers. The injection pump should be linked with the primary water flow to insure that chemicals will not be injected into the system when water is not flowing. Also, the chemical supply containers should be protected from water flowing back into the tanks to avoid overflow of the chemical solution from the storage tanks. Chemicals should be injected separately, unless there is a good reason to do so with knowledge that any reactions occurring between the injected materials will not harm the system, particularly the emitters.

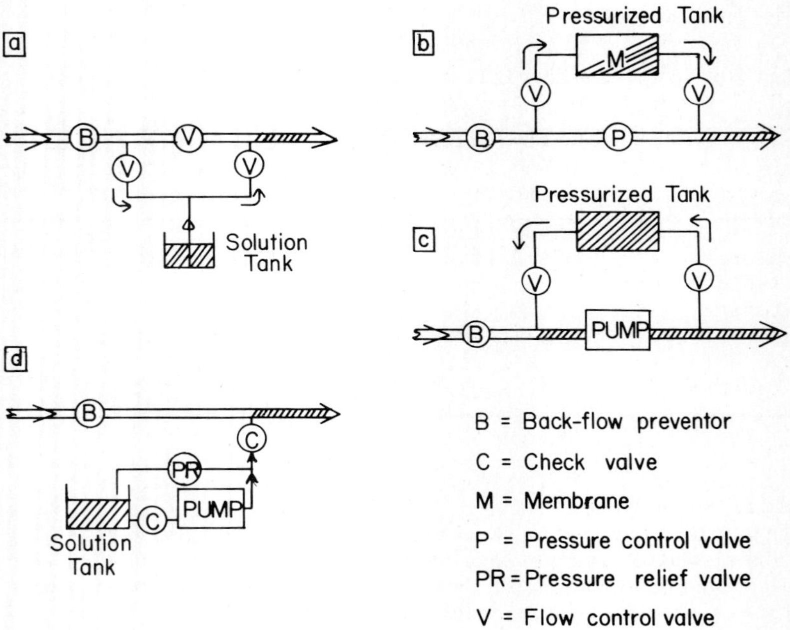

Fig. 3.2.6 Schematic description and operation of various types of chemical injectors.

Three different types of chemical injection system are used based primarily on the kind of pressure used (figure 3.2.6). In the venturi injectors (a), solution injection depends upon the creation of a reduced pressure as the water flows through a constriction in the line. The inlet of the chemical solution is placed at the low-pressure point and the solution is "sucked" into the line. Because injection is based on pressure differential, the rate of injection depends upon flow and to some extent on the level of the solution in the supply tank.

The differential pressure injection system (b) uses the line pressure and part of the water flow to introduce the chemical into the main flow stream. Regulating valves or microtubings are used to divert part of the flow into and out of the pressurized supply tank. The solution is constantly being diluted by the new water entering the container and displacing this solution into the main stream. This principle has been modified by placing a flexible membrane on the entrance side such that the water entering displaces the membrane and consequently the solution into the line. In some instances, a flow meter has been configured into the upstream side so it can activate a valve that would introduce the chemical after a given volume flow is registered by the meter resulting in a proportional injection system.

A water supply pump can be used as an injector by shorting the chemical solution source across the pump (c), but in this case the chemical should not react with the pump body.

Appropriate fittings should be installed on the supply tanks so that replacement solutions can be easily put into or drained from the tank. Pressure relief valves should be provided when necessary.

Gas chlorinators work on a combination of the two principles. Injection is controlled by the differential pressure or "vacuum" at the injection point. This is a safety feature associated with chlorinators to halt the injection immediately when water flow stops. A booster pump is used to create sufficient pressure drop at the injection site and also for the dissolving and mixing of chlorine gas with the water before the chlorinated solution is introduced into the main line. Gas flow meters are provided with the chlorinators so that the injection rate can be monitored and controlled. The chlorinators also have check valves to prevent water from entering the meter and tanks should there be a drastic change in pressure between the water line and supply tank. The chlorine tanks are actually filled with liquid chlorine, which vaporizes to provide the necessary pressure and gaseous chlorine that is injected into the water. This conversion process when done at a high rate may cause rapid cooling of the tank and could cause the freezing of the valve. Temperature of the tank affects the rate of conversion so that the tanks should be stored in a sheltered area.

The positive displacement injectors (d) rely either on a diaphragm or piston driven pump that produces a higher pressure on the chemical solution side compared to the water supply line side. Combinations of check valves are used during the compression and intake cycles to permit injection and resupply of chemicals. Displacement adjustments are provided on the pumps so the rate of injection can be controlled. The rate of injection can be monitored with a rotameter type flowmeter.

Where only a specific amount of chemical, such as fertilizer, is to be injected, the less expensive venturi or pressure differential type systems are adequate, but when a constant level of chemical concentration must be maintained in the line, then the proportional or positive displacement type pumps are needed. The more sophisticated equipment for constant feed has flowmeters, back-pressure regulators to maintain a constant pressure in the solution side between the pump and injection site, and flow control valves. The various components must be matched to the flow rate of the trickle system to which the units are attached.

Example 3.2.6

Given: Concentration of stock solution is 30,000 me/L and the chemical injector pump is set at 5 unit volume per hour, and the flow rate in the main line is at 3,000 unit volume per hour.

Find: The dilution of the concentrated stock acid needed to provide a concentration of 0.5 me/L acid in the lines.

Solution: The dilution ratio between the injector and the supply flow rate is 5:3000 or 1:600. The acid dilution to achieve the final 0.5 me/L concentration is 0.5:30,000 or 1:60,000. Therefore, the dilution of the stock solution is 600:60,000 or 1:100. One liter of the concentrated acid diluted into 99 liters of water provides the necessary secondary stock solution.

After some interval following the start of injection operation, the operator should have acquired a feel for the volume of chemicals needed for his specific needs. Monitoring the quantity of chemical in the storage tanks, by volume or weight, provides a method for checking to see that the injector is operating properly.

Water treatment is an unavoidable additional expense that the trickle irrigation operator must contend with to get the system performing properly. Prevention of clogging is much cheaper than the replacement or reclamation of emitters. Properly working emitters lead to uniform water and fertilizer application which automatically improves operation efficiency.

REFERENCES AND SUPPLEMENTAL READING

American Society of Agricultural Engineers, 1982. Safety devices for applying liquid chemicals through irrigation systems. ASAE EP409, Agricultural Engineers Yearbook, St. Joseph, Michigan, pp 534-535.

American Water Works Association, Inc., 1971. Water Quality and Treatment. McGraw-Hill, New York, 654 pp.

Betz Laboratories, Inc., 1980. Handbook of Industrial Water Conditioning. Betz Laboratories, Inc., Trevose, Pennsylvania, 437 pp.

Bower, C.A., Wilcox, L.V., Akin, G.W. and Keyes, M.G., 1965. An index of the tendency of $CaCO_3$ to precipitate from irrigation waters. Soil Sci. Soc. Amer. Proc. 29: 91-92.

Brazelton, R.W. and McHenry, W.B., 1975. How Much Chemical Do You Put In The Tank? Univ. of Calif., Cooperative Extension Leaflet 2718, Berkeley, California, 11 pp.

Bucks, D.A. and Nakayama, F.S., 1980. Injection of fertilizers and other chemicals for drip irrigation. Proc. Agri-Turf Irrig. Conf., Houston, Texas, Irrigation Association, Silver Spring, Maryland, pp. 166-180.

De Tar, W.R., 1983. Travel time for chemicals in an irrigation system. Trans. Amer. Soc. Agric. Eng. 26(2): 495-497.

Fair, G.M. and Geyer, J.C., 1958. Elements of Water Supply and Wastewater Disposal. John Wiley and Sons, New York, 615 pp.

Ford, H.W., 1976. Controlling slimes of sulfur bacteria in drip irrigation systems. HortSci. 11: 133-135.

Fereres, E. (Ed.), 1981. Drip Irrigation Management. Univ. of Calif., Cooperative Extension Leaflet 21259, Berkeley, California, 39 pp.
Goldberg, D., Gornat, B. and Rimon, D., 1976. Drip Irrigation: Principles, Design, and Agricultural Practices. Drip Irrig. Scientific Publ., Kfar Shmaryahu, Israel, 296 pp.
Langelier, W.F., 1936. The analytical control of anticorrosion water treatment. J. Amer. Water Works Assoc. 28: 1500-1521.
Miyamoto, S., Prather, R.J. and Stroehlein, J.L., 1975. Sulfuric acid for controlling calcite precipitation. Soil Sci. 120: 264-271.
Morris, A., 1975. Filtration with sand. Proc. 3rd Ann. Int. Drip Irrig. Assoc. Meet., Fresno, California, pp. 115-123.
Urquhart, L.C., 1959. Civil Engineering Handbook. McGraw-Hill, New York, 1146 pp.
White, G.C., 1972. Handbook of Chlorination. Van Nostrand Reinhold, New York, 744 pp.

Chapter 3

OPERATIONAL PRINCIPLES

3.3. AUTOMATION

C. J. PHENE

3.3.1 *Introduction*

Irrigation is the process of applying water essential for crop growth (Israelson, 1967). A well-controlled irrigation system is one which optimizes the spatial and temporal distribution of water, not necessarily to obtain the highest yield or to use the least amount of water possible, but to maximize the benefit-to-cost ratio (Hillel, 1980). Traditionally, irrigation applications are managed such that they consist of a relatively brief period of infiltration followed by an extended period of water redistribution and plant water extraction. Irrigation management to satisfy traditional concepts is based on decisions dealing with "when" and "how much" water to apply. Typically, irrigation water is applied to refill the root zone to capacity when most of the "available water" has been depleted.

Plant water uptake to satisfy growth and evapotranspiration processes follows a diurnal cycle with the water moving from a periodically replenished root zone (source) through the plant to the atmosphere (sink). At the end of a typical irrigation cycle, soil water storage becomes depleted, the hydraulic conductivity decreases drastically and the root system cannot resupply water fast enough to meet the atmospheric evapotranspiration demand of the plant, thereby creating a plant water deficit or stress condition.

Irrigation methods capable of operating frequently, such as center pivots, lateral move, mini-sprinkler, trickle, and subsurface, offer the means to maintain soil water at nearly constant levels and, thus, minimize or impose plant water stress at the desire of the irrigator (Phene, 1982). However, with frequent irrigations, control of the soil-water-root environment is critically dependent upon the irrigator, regardless whether the manager is a human or computer. Any disruption or disturbance to the irrigation schedule will quickly create detrimental water or oxygen stress on the crop. Therefore, control of high-frequency trickle irrigation must be automatic, redundant, and capable of responding to small and rapid changes in soil water, plant water, or evapotranspiration. The objectives of this chapter are to discuss: (1) basic control theory applicable to high-frequency irrigation systems, (2) automatic irrigation control systems requirements and design, and (3) instrumentation and hardware.

3.3.2 *Control theory*

Control theory or system analysis is the body of mathematical techniques used to model how one component controls the activity of another component in an interlinked system (Riggs, 1970). Usually, control systems are divided into two categories: (1) open loop systems and (2) close loop systems.

An open loop control system is defined as a system in which the results of the operation are independent of the input and an operator is needed to make decisions; for the irrigation systems two decisions are made: (1) when to irrigate and (2) how much to irrigate. Figure 3.3.1 shows examples of the open loop irrigation control systems.

In a close loop control system the input is directly dependent on the output through a feedback mechanism from the output to the input. Closing the loop via the feedback device allows for comparison of the output with some reference input signal (either constant or variable) and precise control can be achieved. For example, in the close loop feedback control system shown in figure 3.3.2, the lysimeter measures crop evapotranspiration (E_{tc}) directly via a sensor or other weighing device and this information is used to adjust the irrigation volume or time so that the depth of irrigation water applied to the field (d_i) is proportional to E_{tc} such that

$$d_i = \frac{E_{tc}}{E_i} \qquad (3.3.1)$$

where E_i is the irrigation application efficiency of the irrigation system. Within both open and close loop categories, three major control modes are available: (1) on-off control, (2) stepwise control, and (3) continuous control.

3.3.2.a Control Methods

On-off control

The on-off control system turns the system on or off, and the control condition is independent of the system. Figure 3.3.1.a shows a block diagram of this control system where the irrigation valve is either on or off. Most existing irrigation systems are controlled by this mode of control. In some cases, the operator is replaced by a timer switch or more sophisticated devices; but nevertheless, the control condition remains independent of the system.

Stepwise control

For Stepwise Control (figure 3.3.1.b), d_i may be varied by selecting different positions on a valve, a flowmeter, or a timer to give different irrigation volumes and to meet E_{tc} more precisely. For instance, early in the season when E_{tc} is low, position Lo could be used. As E_{tc} increases, position Med then Hi could be selected to

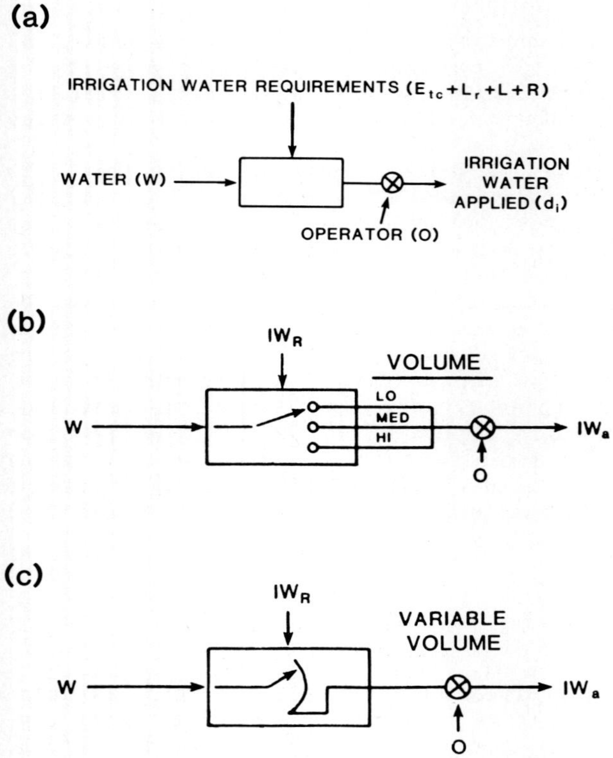

Fig. 3.3.1 (a) Open loop, on-off irrigation control system. The operator calculates the water requirement of the field and opens a valve for a given amount of time (t) to apply the amount of water required (d_i). Time as set by the operator is the only variable available.
(b) Open loop, stepwise irrigation control system. The operator calculates the water requirement of the field (E_{tc} + L_r + L + R) and can select either Hi, Med, or Lo volume of irrigation before he opens the valve to deliver the required amount of water (d_i). Three preset irrigation volumes are variables available to the operator to achieve the irrigation objective.
(c) Open loop, continuous irrigation control system. The operator calculates the water requirement (E_{tc} + L_r + L + R) of the field and can select any irrigation volume before he opens the valve to deliver the required amount of water (d_i). An infinite number of irrigation volumes are available to the operator to achieve the irrigation objectives.

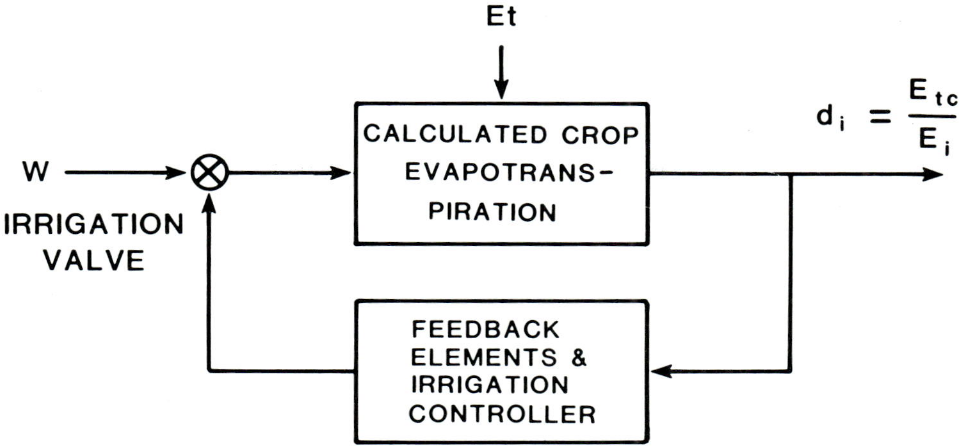

Fig. 3.3.2 Close loop, feedback irrigation control system. No operator is involved. Evapotranspiration of the crop (E_{tc}) is calculated and the irrigation valve is turned on until $d_i = \frac{E_{tc}}{E_i}$, where E_i is the irrigation application efficiency. The feedback system has the capability to modify the application by either stopping the irrigation if it is raining or increasing the irrigation set if a change in weather conditions requires it.

increase d_i and apply progressively more water with each irrigation. Note again that the volume of water applied to meet E_{tc} is not a direct function of E_{tc}, but an estimate based on other measurements. The application of stepwise control in irrigation is sometimes implemented by a time clock with fixed intervals of time control, but operated exactly like the on-off control. The irrigation time T_i can be calculated from the relation

$$T_i = 100 \frac{(E_{tc}\ I_f)}{(E_i\ I_a)} \tag{3.3.2}$$

where T_i is the irrigation time in hr, E_{tc} is the crop evapotranspiration in mm/day, I_f is the irrigation frequency in days per irrigation, E_i is the irrigation application efficiency in %, and I_a is the application rate of the irrigation system in mm/hr; E_{tc} can be multiplied by any linear factors to adjust for leaching and percolation losses.

Continuous control

For continuous control (figure 3.3.1.c), the depth of irrigation water can be selected from minimum to maximum values by adjusting time or volume of water in a continuous manner. Any value of time or volume between maximum and minimum can be achieved by varying the time or the volume setting on the flowmeter. Again, the final volume of water applied can meet the water requirement more precisely, but it is not a direct function of E_{tc}.

3.3.2.b Linear systems

A system is classified as linear if the output is directly proportional to the input, i.e., the ratio of the output to the input is constant. For irrigation in arid regions where insufficient rainfall occurs during the growing season, the input variables are the E_{tc}, the leaching requirement (L_r) for salinity control, irrigation system losses (L), and runoff (R). The output variable is d_i and the objective is to irrigate so that

$$d_i = \frac{E_{tc} + L_r + L + R}{E_i} \qquad (3.3.3)$$

If L_r required to keep the soil profile free of salts is 15 percent of E_{tc}, E_i = 0.90 and both L and R are zero, the correct amount of water to apply d_i is

$$d_i = \frac{E_{tc}(1 + 0.15)}{0.90} = 1.28\ E_{tc} \qquad (3.3.4)$$

In irrigation practices, the principles governing close loop feedback control systems are provided by linear system theory, where the input represents the command or cause and the output represents the results of the system process. In the preceding, d_i is the result of an adjustment of the irrigation system to apply just enough water to meet E_{tc} of the crop being irrigated, L_r to maintain a satisfactory salt balance in the root zone, and E_i of the irrigation system.

Example 3.3.2

Problems:
1. Give examples of control systems (other than irrigation) that describe (1) on-off control, (2) stepwise control, and (3) continuous control.
2. Differentiate between open and close loop systems and show how each could be used to control irrigation systems.
3. Describe the type of inputs which can be used to control irrigation systems in the close loop mode, (1)

where rainfall is not expected and (2) where rainfall is expected.
4. Construct a block diagram for an automatic irrigation system using an evaporation pan feedback control system.
5. Modify block diagram in problem 4 to include an automatic water filling system which refills the pan at midnight every day.

3.3.3 *Automatic control systems*

Irrigation scheduling is the means to supply water according to crop water requirements within the limitations of the irrigation system. With trickle irrigation, varying amounts of water can be supplied frequently to a crop, provided that an adequate supply of water is available. Technology is now available to tailor an irrigation schedule using real time analysis of factors such as weather, crop growth stage, desired plant water stress, soil aeration and water potential, and soil water salinity. Yet in most cases, scheduling of trickle irrigation systems has been limited to an on-off control system using time or water volume as the control variable. The computer is merely programmed to sequence solenoid valves and check flow rates, and pressures, wind, temperature, and other indirect variables.

To achieve minimal cost-to-benefit ratio and most efficient use of the water supply, water use efficiency (WUE) must be achieved simultaneously with high crop yield. The various factors of excess water applied in equation 3.3.3 must be either eliminated or reduced so that the precise irrigation application is only that needed by the crop. Trickle irrigation is usually managed so that L and R are near zero, E_i is high (90 percent or better), and L_r is achieved either by applying a large volume of preplant irrigation at the beginning of the season or by applying a percentage of water in excess of the plant requirement with each irrigation. The dominant factor of equation 3.3.3 is E_{tc}. Unfortunately, it is also the most difficult factor to estimate in irrigation scheduling because E_{tc} is variable on a seasonal and on a daily basis. In addition, E_{tc} is a function of weather, soil, plant, and sometimes irrigation systems. Factors which affect irrigation scheduling are listed in table 3.3.1. Most of these factors are interdependent and variable, both spatially and temporally.

Many crops are good integrators; however, accurate scheduling of trickle irrigation can minimize the adverse effects of some of these factors and makes maximum crop productivity attainable. Assuming that water, fertilizer, and management factors are not limiting and that we are primarily interested in automatic systems to schedule trickle irrigations, there are four basic close loop feedback methods which can be used: (1) soil water, (2) plant water, (3) evapotranspiration estimates, and (4) combinations of 1, 2, and 3. Each of these approaches has advantages and disadvantages which will not be discussed in detail here; instead, examples of some of the methods will be discussed in terms of how they can be implemented and managed.

TABLE 3.3.1

Factors affecting irrigation scheduling.

WATER FACTORS	SOIL FACTORS
water availability (amount and time)	soil structure
water quality	soil texture
	soil depth
CLIMATIC/WEATHER FACTORS	mechanical impedance
ambient temperature (day/night)	infiltration rate
solar radiation	drainage rate
wind speed	soil aeration
rainfall	water retention characteristics
humidity	hydraulic conductivity
day length	water table
length of growing season	soil salinity
	soil fertility
PLANT FACTORS	soil temperature
crop variety	soil borne organisms
rooting characteristic	
drought tolerance	MANAGEMENT FACTORS
growth stage	dates of planting/harvesting
harvestable constituent	plant population
yield and quality	irrigation system
length of growing season	critical growth stages
salt tolerance	fertilization
nutrient requirement	crop protection
	cultivation

3.3.3.a Soil water method

Soil is composed of three major components: (1) solids, (2) liquids, and (3) gases. Depending on soil type, conditions and use, the composition of these three components can vary greatly. The largest component of the liquid fraction is water and that of the gas fraction is nitrogen, although oxygen is also important. For the purpose of irrigation, one should be concerned with the energy, mass, or volume relationships between liquids and gases since the solid fraction of the soil will remain nearly constant. Active plant root systems require all three soil components, but the balance between the liquid and gas phases is most critical since it regulates root activity and plant growth processes and, it is particularly of interest in irrigation and evapotranspiration. Aeration (gas exchange) in soil is primarily a diffusion process. Diffusion coefficients are inversely proportional to the thickness of water films in soil pores. Thus, aeration increases proportionally to the concentration gradient. Adequate aeration of the root zone is necessary to maintain plant growth so that proper irrigation management helps in maintaining a balance of air and water. The potential energy of soil water in the root zone of the crop results mostly from the various force fields to which it is subjected. Water will flow or diffuse along gradients from high to low energy status. For instance, in the evapotranspiration process, water will move along the water potential gradient as the stomata of the crop leaves open and the transpiration process begins at dawn (figure 3.3.3). Hence, plant processes are directly dependent on water potential variables rather than water content.

The water holding capacity of soils differs greatly with soil type. In sandy soil, plants may be subjected to rapidly changing soil water potential ranging from saturated to wilting conditions within a day. In clays soils, water potential decreases more slowly. Therefore, depending on soil characteristics, the frequency of irrigation for optimum crop productivity may vary between several irrigations per day

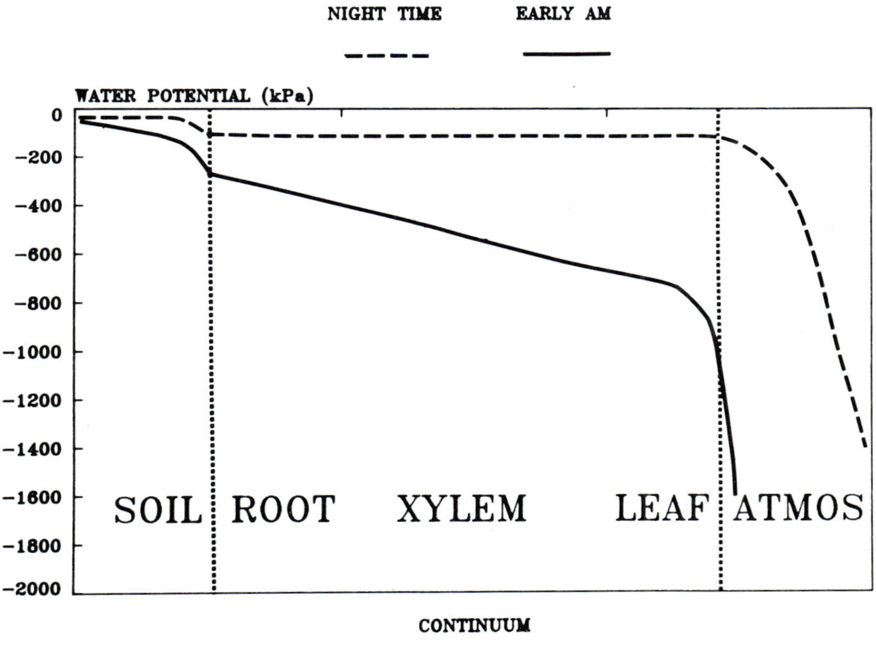

Fig. 3.3.3 Soil-plant-atmosphere water potential continuum as affected by the opening of the plant stomata.

to one irrigation every few weeks. As irrigation frequency increases, the water holding capacity of the soil become less important because water application rates match E_{tc}. Furthermore, since the frequency of the irrigation pulse can conceivably be very large, the application rate can be adjusted to fit the infiltration rate of the soil. This allows water to move through soil under unsaturated conditions, thus maintaining continually favorable conditions for gaseous diffusion and adequate aeration of the root zone.

Scheduling frequent irrigations can be accomplished with automatic feedback control based on soil water potential. Because the storage capacity of soil is de-emphasized and water is applied to supply the water potential continuum and match the evapotranspiration rate, there is less margin for error and timeliness is important.

Irrigations based on soil water potential are among the oldest irrigation scheduling techniques used. Tensiometers, (Richards, 1936), thermal methods (Shaw, 1939), gypsum blocks (Boyoucos, 1947), and thermocouple psychrometers (Richards, 1958), have all been applied successfully. Microprocessors tied to sensors can be used to simplify irrigation applications (Cary & Fisher, 1983). Sensors can provide immediate information to assist decisions made on irrigation water application. Microprocessor-based circuits can be coupled to programmable calculators to give an on-site estimate of the allowed time until the next irrigation, based on field data and an operator-supplied parameter. The program in the portable calculator can assess the soil matric potential sensor and use it to extrapolate the soil drying rate to estimate the number of days until the next irrigation.

A thermal method which measures soil matric potential independent of soil texture, temperature or salinity is based on frequent measurements of the ability of a porous ceramic sensor to dissipate a small amount of heat applied to it (Phene et al., 1973, 1984). With proper calibration the sensor can be used in any soil to monitor soil matric potential and control irrigation automatically. In addition to water availability, soil physical properties such as oxygen diffusion (aeration) and soil mechanical strength (impedance) are used to define the range of soil matric potentials (ψ_m) optimal for root growth and activity. Figure 3.3.4 shows an example of the optimal ψ_m for a Hanford fine sandy loam. Within this range, a soil matric potential value is selected at which irrigations are to be started (threshold). Data in figure 3.3.4 indicate that the optimal ψ_m should be about -25 J/kg and has a range extending approximately between -10 and -60 J/kg. The optimal ψ_m has a range which increases as soil texture becomes finer, but is extremely narrow in compacted coarse-textured soils. Therefore, physical characterization of sandy soils is usually necessary, whereas the optimal ψ_m for fine-textured soils can be approximated.

For close loop-feedback automated irrigation, the soil sensor, should be placed midway in the root zone. In this location the majority of the root zone is never allowed to dry beyond the soil matric potential threshold before the sensor detects the drying trend and triggers another irrigation.

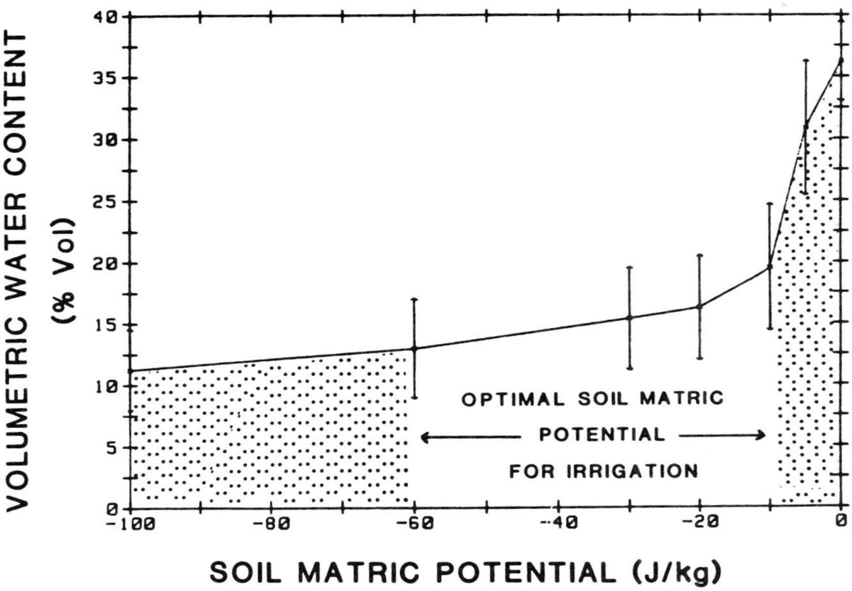

Fig 3.3.4 Water desorption curve and optimal ψ_m for irrigation of a Hanford fine sandy loam soil (Typic Xerorthents).

Figure 3.3.5 illustrates a closed loop feedback control system for an irrigation system. The computer system software developed to maintain a nearly constant soil matric potential in the root zone of the crop is outlined in figure 3.3.6.

Monitoring soil matric potential and controlling an irrigation system automatically requires equipment to (1) sample automatically several sensors sequentially, (2) compare each sensor output to the soil matric potential at which irrigation is to start (threshold), and (3) have computer outputs capable of controlling the irrigation system. Desktop computers and microprocessors have been successfully applied. Commercial equipment is also available to measure soil matric potential and to control the irrigation system automatically. The responses of three soil matric potential sensors to changes in soil matric potential in three plots irrigated simultaneously by high-frequency subsurface trickle irrigation are presented in figure 3.3.7. The computer calculates the average readings of the three soil matric potential sensors, compares the average soil matric potential measured to the threshold value at which each irrigation is to be applied, and turns the irrigation system for a preselected time period if needed. In this case, irrigations are initiated when the soil matric potential

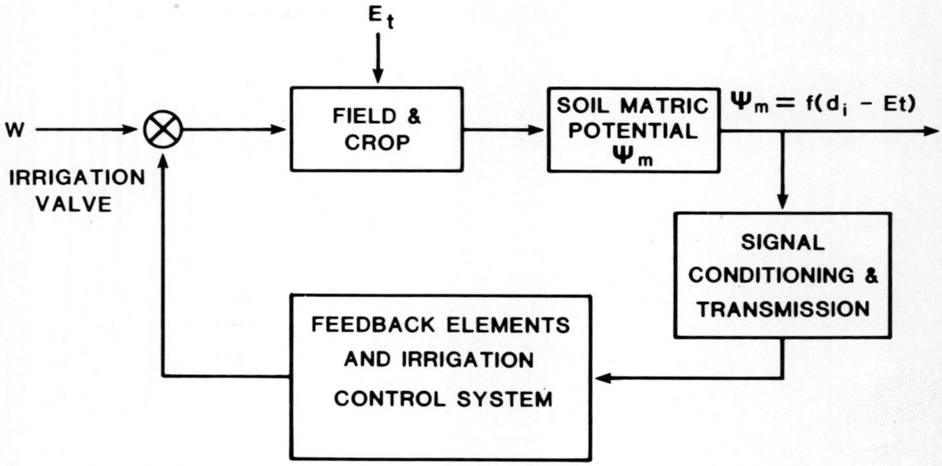

Fig. 3.3.5 Close loop, feedback irrigation control system using soil matric potential (ψ_m) as the control variable. The time variable is used to adjust the amount of water being applied.

reaches -20.0 J/kg. The number of irrigations per day, shown for each day above the x-axis, varies between 5 and 15 during the eight day period. Up to 168 individual sensor measurements are made in each plot.

Figure 3.3.8 shows the performance of another sensor used for irrigation control. On day 183, the threshold level had been changed from -35 to -45 J/kg and the next day the sensor measured the effect of no irrigation. Because the new irrigation threshold was set at a lower soil matric potential (more negative value), the system response would be somewhat erratic since the unsaturated hydraulic conductivity of the soil is much lower, and thus, the movement of water toward the sensor is slower. During the period when the threshold is -35, the mean soil matric potential is -33.5 J/kg. During the period when the threshold is -45, the mean is -40.1 J/kg. These types of data substantiate the validity of the method for scheduling irrigations precisely.

3.3.3.b Plant water methods

Water is frequently the most limiting factor in crop production; however, most of the water taken up by plants is lost to transpiration in response to the evaporative demand of the surrounding atmosphere. Less than one percent of the water absorbed is actually retained by the plant. Even this small fraction of water is sometimes used to make up the deficit between water uptake and transpiration; thus, any lack of water, causes a deficit in plant water. Total leaf water

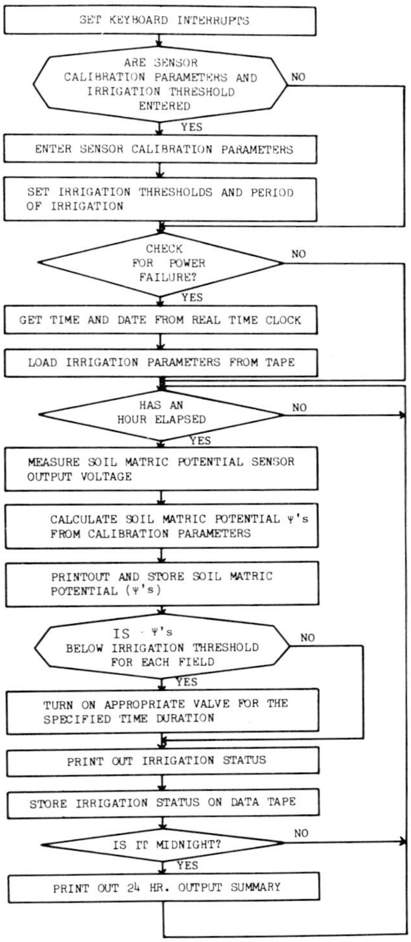

Fig. 3.3.6 Flow chart showing the logic system for an automatic irrigation controller.

potential (the sum of turgor, matric, and osmotic potentials) is used to indicate the water status of a plant. Most of the plant growth processes are affected by plant water deficit; but cell enlargement (growth), photosynthesis, pollination, and fruit setting are affected at low plant water stress levels to the point where yields can be reduced.

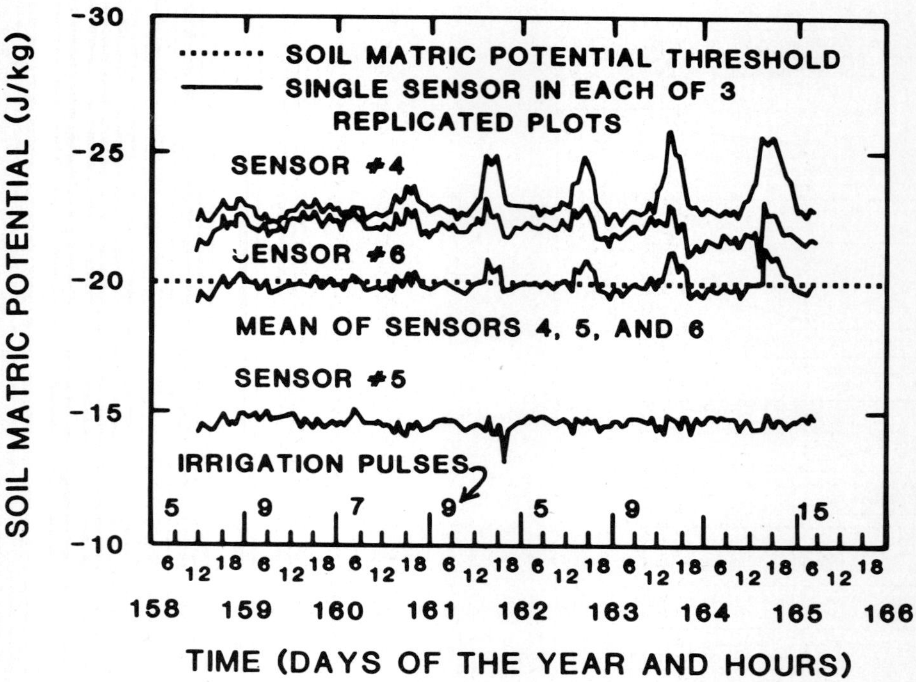

Fig. 3.3.7 Fluctuations of soil matric potential at a soil depth of 45 cm measured by sensors #4, 5, and 6 for a constant threshold of -20 J/kg (1 J/kg = 1 centibar) and the number of pulse irrigations applied automatically each day over an eight-day period in June.

Probably the plant process most sensitive to water deficit is growth by cell enlargement (Hsiao, 1973). When subjected to water deficit, the water content of the cells decrease and as the positive pressure potential (ψ_p) (also referred to as turgor pressure) approaches zero, cell enlargement stops even though all other necessary chemical and physical requirements are met.

Photosynthesis is also reduced when the plant looses its turgor pressure because, as the guard cells deflate, the stomata close reducing the diffusion pathway for CO_2 transport into the leaves. With reduced photosynthesis, the rate of dry matter production is decreased.

Pollination and fruit setting are also sensitive to water stress and fruit yield will be reduced even though the production of dry matter may not appear to be affected. Hence, a high (small negative value) total leaf water potential (ψ_L) should be maintained during pollination and fruit setting to obtain maximum fruit yield.

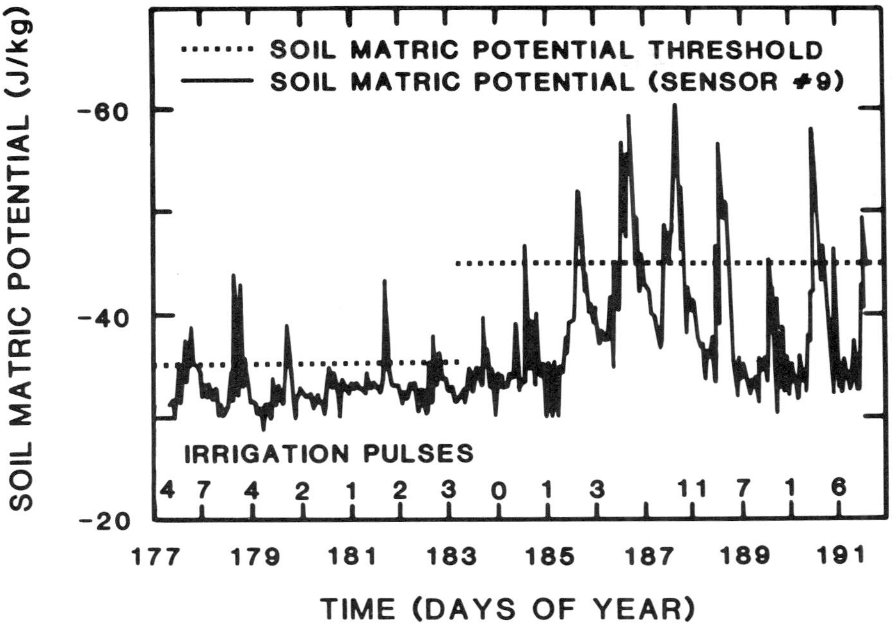

Fig. 3.3.8 Soil matric potential fluctuations at 45-cm soil depth for constant threshold values of -35 and -45 J/kg (1 J/kg = 1 centibar). Note that as the soil matric potential threshold decreases (less water in the soil), the soil matric potential fluctuates more drastically because the decrease in hydraulic conductivity slowed down the movement of water to and from the sensor.

Several methods are available to estimate plant water status. These include determination of relative water content, leaf diffusive conductance, plant water potential, and plant temperature. Plant water potential from direct or indirect measurement is probably the best indicator of plant water stress. Automatic feedback control of trickle irrigation systems can be achieved by measuring total leaf water potential using the leaf psychrometer (Hoffman and Rawlins, 1972), plant canopy temperature using the infrared thermometer (Jackson, R. D., 1982; Howell et al., 1983), and leaf water potential indirectly based on stem diameter measurements (Parsons et al., 1979).

Leaf water potential method

Direct measurements of ψ_L can be achieved by attaching thermocouple psychrometers to plant leaves. Although psychrometric measurements are made routinely in research situations, the instrumentation which needs calibration and maintenance to achieve the required accuracy, is expensive and not commercially available for irrigation scheduling.

Plant canopy temperature method

The surface temperature of a body is related to its black body radiation according to the Stefan-Boltzman equation

$$T_s = \left(\frac{R}{\varepsilon\sigma}\right)^{1/4} \qquad (3.3.5)$$

where T_s is the surface temperature in K (°C + 273), R is the emitted black body radiation in W/m^2, ε is the emissivity of the body (ratio of emitted radiation to that of a perfect black body) and σ is a constant (5.674×10^{-8} W/m^2 per K^4). Most crops are near perfect emitter in the 10 to 14 μm waveband. This principle has been applied to measure the surface temperature of a crop canopy by a noncontact infrared thermometer (IRT). The IRT accuracy in measuring plant canopy surface temperature is dependent on careful calibration. Measurements are sensitive to ambient temperature changes, and interactions from surrounding surfaces, particularly soil when the crop canopy is small.

Crop canopy temperature measurements have been introduced into the Crop Water Stress Index (CWSI) concept to estimate crop water stress (Jackson, 1982). For a given crop, the ratio of the difference between crop canopy and air temperatures to vapor pressure deficit (VPD) is bounded by two baselines (a no-stress or lower baseline and a terminal stress or upper baseline) which are determined for the specific crop by theoretical or empirical methods. The basic concept is outlined in figure 3.3.9 for a cotton crop. The CWSI is calculated by dividing the distance ($A - A_o$) by ($A_1 - A_o$) at the same VPD. The crop water stress index has a value of "zero" for no water stress and "one" for terminal stress or an essentially dead plant.

Although the CWSI has not been used to automatically schedule irrigation, to date, it can serve as the feedback to monitor and if necessary adjust the irrigation schedule. Software can be easily developed to collect data, calculate CWSI, make comparisons to irrigation threshold values, and make decisions on irrigation. For example, if a CWSI of 0.25 is set as the irrigation threshold, the system would call for irrigation at point A but not at point B. Although, the temperature difference is the same in both cases, the lower VPD in case A would create a greater evapotranspiration rate and, thus, a larger CWSI.

One type of irrigation control uses an infrared thermometer developed for continuous outdoor operation mounted on a lateral-move irrigation system (Phene et al., 1985). At specific locations within the plot, the IRT stops, rotates 180° and integrates the differences between ambient and canopy temperatures at an angle of about 30° from the horizontal. Simultaneous measurements of the wet and dry bulb temperatures above the crop are obtained and stored in the irrigation system computer to calculate the vapor pressure deficit. The VPD measurements are used to normalize the diurnal and seasonal evaporative demand.

Howell et al. (1984a) pointed out in their conclusions that "although the CWSI appears to be useful in assessing crop water stress in cotton, irrigation scheduling requires decisions for both timing and amount." Therefore, traditional irrigation scheduling models (Jensen et al., 1971) should be used to predict the irrigation application amount necessary to refill the crop root zone when an irrigation requirement is sensed by any plant indicator (either leaf water potential, CWSI or any other plant measurement). In many cases, soil water depletion can be directly measured either by soil probing or neutron methods. With irrigation systems which are frequency or rate controlled, such as center-pivot sprinkler, lateral-move sprinkler and trickle irrigation systems, the CWSI can be used to monitor the crop water stress and to indicate the need to either increase or decrease irrigation amounts or frequency.

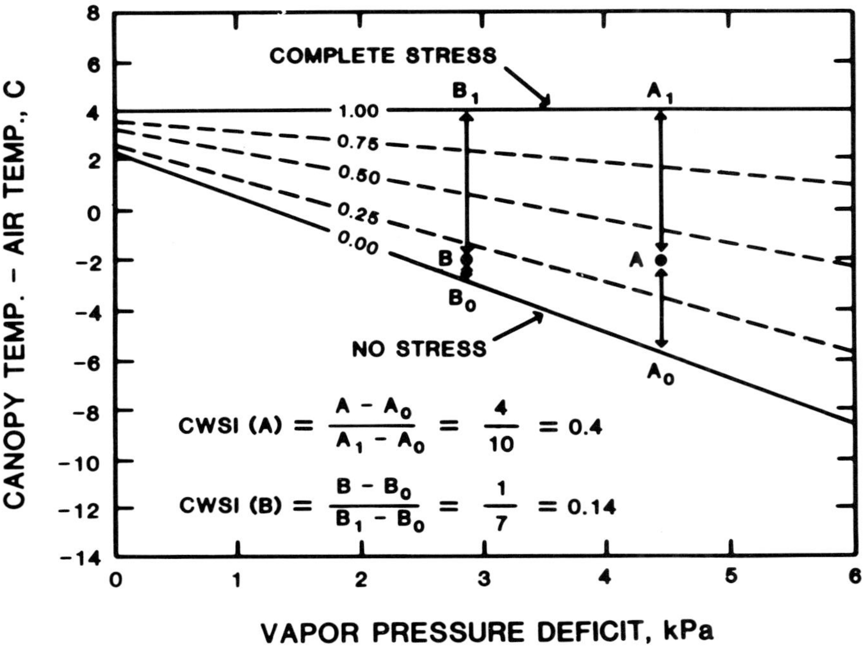

Fig. 3.3.9 Crop Water Stress Index (CWSI) relationships for a cotton crop (after Howell et al., 1984a).

Stem diameter method

Stem diameter and leaf water potential are closely related to each other (Klepper et al., 1971). Thus, stem diameter measurements can be used to monitor continuously long-term stem growth and plant water status. Two methods are available in using stem diameter to predict the diurnal variation of xylem water potential (Huck and Keller, 1977). The first and simplest procedure, the Shrinkage Modulus Method, determines an arbitrarily calibrated shrinkage modulus and relates a measured change in stem diameter to a corresponding difference in leaf water potential. The second, the Dynamic Flux Method, simulates water flow between xylem and associated phloem parenchymal tissues resulting from changes in plant water potential. Water potential differences between the xylem and surrounding tissues are assumed to induce a radial flux of water across the cambial boundary layer causing swelling or shrinking of the stem.

Stem diameter changes (ΔS) of continuously drying cotton plants can be measured with a linear variable differential transformer (Parsons et al., 1979). The reference stem diameter for computing stem diameter change is measured before sunrise throughout an experiment.

Stem diameter stress can be integrated numerically using the equation

$$ISS = \int_{t_o}^{t_1} \Delta S_{(t)} dt \qquad (3.3.6)$$

where ISS is the Integrated Stem Stress, in mm x day, t_o is the pre-sunrise time (hr), t_1 is the post-sunrise time (hr), and $S_{(t)}$ is the stem diameter change, from the nonstress stem diameter, at time t (mm).

Leaf water potentials are inferred from the hydraulic pressure necessary to cause water flow from the uncut edge of the leaf measured at sunrise and periodically each day to insure that maximum and minimum values are obtained. The relationship between the observed stem diameter changes and the minimum observed ψ_L is presented in figure 3.3.10.

This measurement technique could be used for feedback control of automatic trickle irrigation systems. Periodic calibration of stem diameter changes versus ψ_L should be obtained at least for each phenological stage of the plants. In the case of cotton, the irrigation threshold ψ_L of -1800 J/kg is based on feedback calibrated stem diameter measurements. From crop phenological stages and known water requirements of cotton, the ψ_L threshold values can be adjusted as necessary. As with the IRT method, simultaneous measurements of soil water and/or E_t should be used (at least at first) to gain confidence in the method.

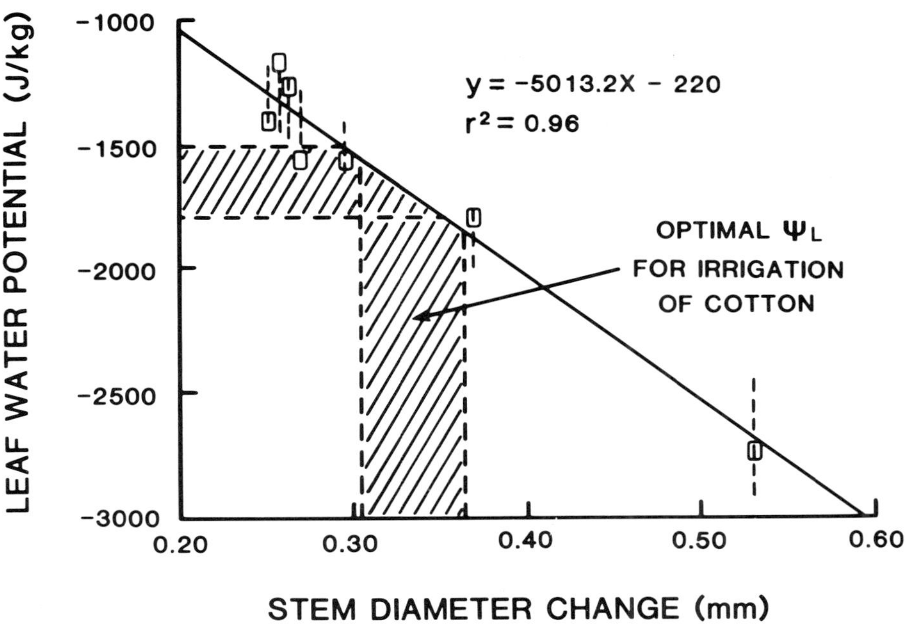

Fig. 3.3.10 Linear regression of minimum observed leaf water potential versus maximum stem diameter change from the reference stem diameter. Broken lines represent 90% confidence intervals based on the regression analysis (adapted from Parson et al., 1979).

3.3.3.c Evapotranspiration estimates

Evapotranspiration models

Irrigation scheduling models based on evapotranspiration have been widely used in the United States and worldwide (Jensen et al., 1970, 1971). Essential evapotranspiration (E_{tc}) information required for these models and the irrigation decision criteria include: (1) a climatically estimated reference evapotranspiration (E_{tr}), (2) an index for relating "expected" crop water use to E_{tr} (crop coefficient curve), (3) an index for estimating the additional soil water evaporation from a wet soil surface, (4) an index for estimating the effect of soil water depletion on the actual ET rates, (5) an estimation of extractable soil water amounts by specific crops from the specific soils, and (6) a relationship between "expected" crop yield and crop water use. Many of the input variables needed to operate the model are still not well defined and need to be estimated. Although the model can predict irrigation requirement accurately for low frequencies, it is not presently feasible for scheduling high frequency trickle irrigation available since data are limited. For instance, one of the most important inputs, the crop coefficient (item 2), is only defined for a few crops under gravity irrigation systems. Accurate weighing lysimeters and a network of weather stations are needed (Howell et al., 1984-1985). Lysimeters must provide sufficient E_{tc} resolution to permit measuring hourly crop water use rates within 0.1 mm/hr. One lysimeter should be planted to grass and used to provide E_{tr}, and the another lysimeter should be planted to the crop to be studied to measure E_{tc}. The ratio of ET's from the two lysimeters (E_{tc}/E_{tr}) is the crop coefficient with reference to grass. Irrigation scheduling models use the weather station output and a crop coefficient to compute the E_{tr} expected from a nonwater stressed reference crop such as grass to compute daily crop water use. Daily E_{tr} from integrated hourly values are calculated and used for irrigation scheduling (Doorenbos and Pruitt, 1977; Pruitt and Doorenbos, 1977).

Direct measurement of E_{tc}

An example of how E_{tc} can be used for automatically scheduling irrigation is discussed. A modified crop lysimeter serves as a feedback irrigation controller for a crop growing in and around it. A water tank is attached to the lysimeter so that the weight of the daily irrigation water is included in the weight of the lysimeter. Every time one mm of E_{tc} occurs, the lysimeter is automatically irrigated by a deep subsurface trickle irrigation system (45 cm from the soil surface) to maintain steady state soil water potential without disturbing the lysimeter weight. The lysimeter tank is automatically refilled daily at midnight to a constant tank level. Therefore, the accumulated daily change of lysimeter weight represents the crop growth and total weight. Figure 3.3.11 shows the block diagram of the feedback irrigation control system based on real time E_{tc} measurement from the lysimeter. The soil water potential is maintained nearly constant by the high-frequency irrigation. Figure 3.3.12 shows the flow chart of the algorithm used to measure the lysimeter weight, schedule the irrigation system in the lysimeter and schedule the surrounding field at two different frequencies, refill the lysimeter water tank daily at midnight, and calculate the evapotranspiration rate from the lysimeter. Grass E_{tr} is measured by the reference lysimeter and calculated by the hourly integrated Penman Equation (Pruitt and Doorenbos, 1977). Figure 3.3.13 illustrates the type of crop coefficient curve obtained for trickle irrigated tomato. Such data provides the information needed for irrigation control, as indicated at the beginning of this section, except for the relationship between the expected crop yield and water use.

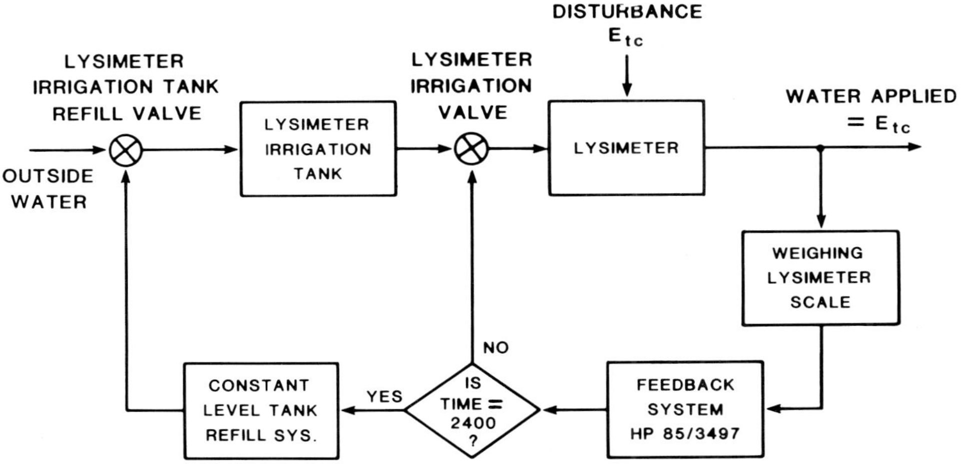

Fig. 3.3.11 Block diagram for close loop, feedback irrigation control system based on real time E_{tc} from the lysimeter which is irrigated every time one mm of E_{tc} occurs.

Example 3.3.3

Problems:

1. Using the crop coefficient curve given in figure 3.3.13 and reference evapotranspiration (E_{tr}) given in table 3.3.2, calculate the total crop water requirement of a tomato crop, and the water requirement for each of the 3 crop growth stages defined by the curve. (Assume an irrigation application efficiency (E_i) of 90%, a leaching requirement of 20% and a leakage loss of 2%).
2. Using figures 3.3.5, 3.3.6, and 3.3.11 as examples, develop a block diagram and a logic flow chart for a feedback irrigation control system using stem diameter, estimated leaf water potential and irrigation time as the control variables for irrigating a cotton crop. Use the relationship between minimum observed leaf water potential and maximum stem diameter changes given for cotton in figure 3.3.10. Assume -1800 J/kg as the threshold leaf water potential for starting irrigation and 30 minutes minimum irrigation time.
3. List factors which affect irrigation scheduling and show how these could be incorporated into a large scale irrigation model to allocate and schedule irrigation.

TABLE 3.3.2

Reference evapotranspiration calculated by the modified Penman equation, integrated hourly for irrigated grass at the University of California, West Side Field Station, Five Points, California, for the irrigation season in 1983 and 1984.

Days of the Year	Et_R 1983	Et_R 1984	Days of the Year	Et_R 1983	Et_R 1984	Days of the Year	Et_R 1984	Et_R 1985	Days of the Year	Et_R 1984	Et_R 1985
80	2.94	5.20	115	5.44	7.86	150	7.38	8.29	185	8.31	8.22
81	1.49	6.36	116	3.42	9.14	151	6.90	9.05	186	7.67	8.13
82	1.44	5.38	117	3.30	6.94	152	4.84	9.38	187	8.43	8.09
83	1.67	4.87	118	4.67	5.19	153	6.74	5.34	188	8.09	9.58
84	2.81	4.70	119	4.00	6.67	154	6.94	8.26	189	8.26	8.67
85	2.09	4.42	120	4.85	6.29	155	7.24	7.83	190	9.11	8.45
86	2.16	4.62	121	4.70	5.69	156	7.80	4.90	191	9.11	8.70
87	3.46	6.09	122	5.21	5.23	157	6.91	7.01	192	8.31	7.96
88	3.86	6.09	123	5.44	7.96	158	8.25	6.61	193	8.93	8.18
89	4.07	7.56	124	3.02	7.38	159	7.29	7.56	194	9.81	8.24
90	4.98	6.09	125	4.65	8.19	160	7.81	8.42	195	9.79	7.69
91	4.52	4.01	126	5.74	7.68	161	8.52	9.49	196	8.71	7.06
92	4.40	3.83	127	6.07	8.41	162	8.05	9.49	197	7.90	6.81
93	4.90	4.04	128	6.73	7.29	163	8.14	8.20	198	7.32	7.65
94	4.24	4.88	129	7.56	7.72	164	7.59	7.79	199	7.62	8.02
95	3.66	5.47	130	6.74	8.57	165	8.64	7.08	200	7.51	7.56
96	4.42	2.59	131	7.29	9.96	166	9.34	7.23	201	8.11	8.18
97	4.58	5.64	132	5.31	8.45	167	8.02	7.61	202	7.71	9.02
98	4.74	5.49	133	5.76	7.94	168	9.54	8.13	203	8.23	7.98
99	3.88	3.97	134	6.79	8.72	169	8.84	8.64	204	7.52	5.18
100	1.76	5.93	135	7.47	8.46	170	9.04	8.52	205	7.51	6.82
101	2.96	5.09	136	8.30	6.41	171	8.12	8.18	206	6.22	7.19
102	3.10	5.99	137	7.15	7.33	172	7.93	7.94	207	7.14	7.68
103	3.68	5.98	138	6.73	8.30	173	8.08	8.33	208	7.66	7.76
104	4.44	6.32	139	7.25	7.87	174	8.25	9.93	209	7.56	8.02
105	3.59	6.09	140	8.37	8.22	175	7.89	9.04	210	7.35	7.80
106	5.09	7.27	141	7.87	8.91	176	8.43	8.68	211	7.66	7.66
107	3.84	6.44	142	8.10	9.50	177	9.26	8.70	212	6.94	7.93
108	3.77	5.96	143	8.27	9.45	178	6.74	9.36	213	8.73	8.22
109	3.26	2.68	144	7.70	10.20	179	7.04	7.86	214	7.96	8.25
110	3.30	5.05	145	7.56	9.04	180	8.39	10.15	215	8.08	8.15
111	3.09	7.15	146	7.88	9.75	181	7.95	9.56			
112	3.16	6.91	147	8.16	9.68	182	7.95	8.50			
113	4.89	6.27	148	7.09	8.73	183	6.69	8.41			
114	4.46	7.43	149	7.23	8.93	184	8.28	8.64			

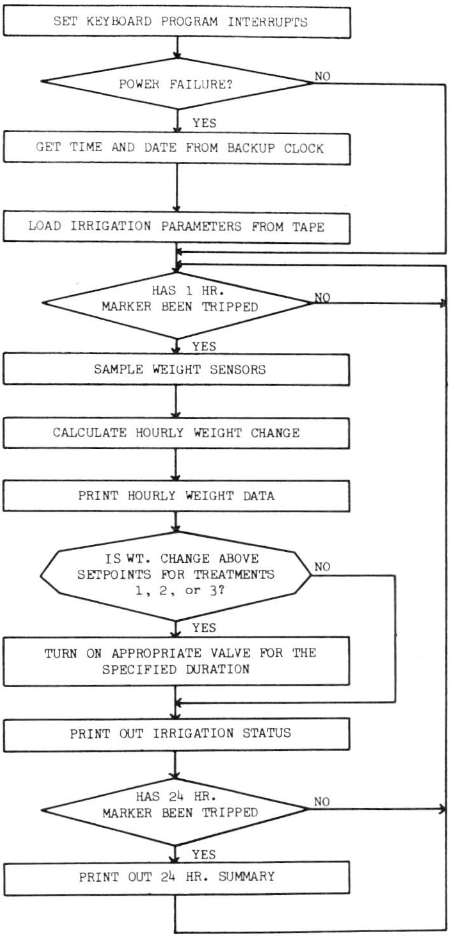

Fig. 3.3.12 Logic flow chart for lysimetric weight sensor measurements and irrigation control sequence for irrigating a crop at three irrigation control levels.

3.3.4 *Instrumentation and hardware*

The automation of a pressurized trickle irrigation system can potentially provide optimum crop yield and use of agricultural water. An automated irrigation control system should use feedback sensors to monitor on a real-time basis important functions such as: water quantity, flow rate, water pressure, and environmental conditions such as wind speed, air temperature, soil moisture, solar radiation, rainfall, crop canopy temperature, etc. Continuous monitoring and control of system performance with flowmeters, pressure transducers, solenoid valves and pressure regulators at strategic locations will enable irrigation operation at maximum efficiency (Phene et al. 1982). Data or control functions can be transmitted by electrical wires, hydraulic lines, radio frequency signals, microwave, laser or infrared devices.

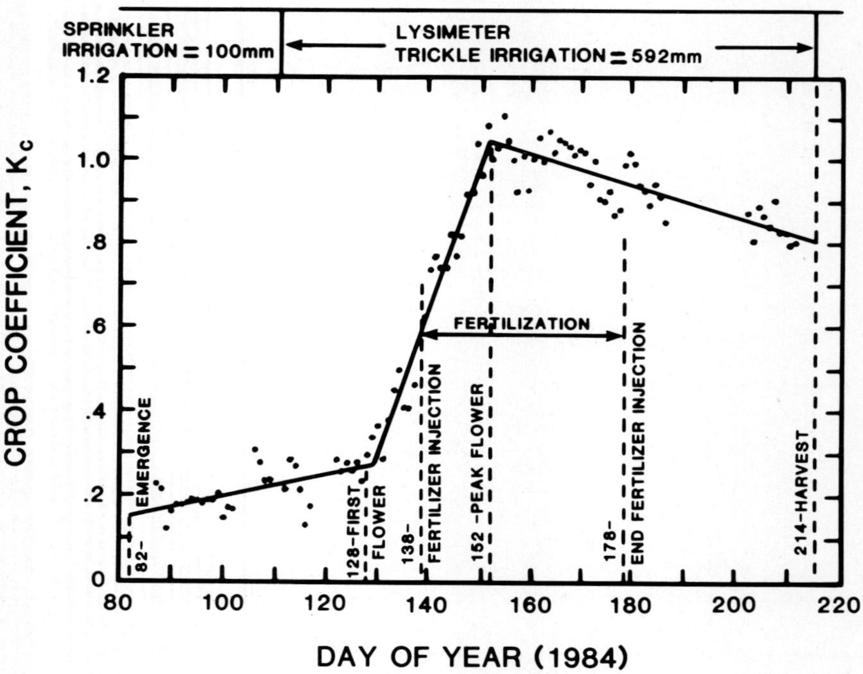

Fig. 3.3.13 Crop coefficient curve for a subsurface trickle irrigated tomato crop grown in Panoche clay loam soil. Crop phenological stages are shown for reference purposes.

The interest in automation of trickle irrigation systems has resulted in increased research and development in the field of instrumentation and hardware needed to accomplish the task. A large variety of instrumentation and hardware with varying characteristics are available commercially. These can be subdivided into six major categories:

(1) controller
(2) valve
(3) flowmeter
(4) filter
(5) chemical injector
(6) environmental sensor

Details of installation, function and maintenance are covered under other chapters in this book. The main functions described in this chapter are in terms of their respective mode of control.

Controller

The controllers receive feedback information about the volume of water per field, line pressure, flow rate, weather data, soil moisture, plant water stress, etc., from sensors in the field. This information is then compared with desired limits, and the irrigation cycle is modified accordingly. A controller issues (automatically) or is set to issue (manually) commands for operating of water valves, boosters, fertilizer or water treatment injectors, cleaning of filters, etc., according to the modified irrigation cycle.

Valve

Automated valves are activated either electrically, hydraulically, or pneumatically and are used to switch water on or off, flush filters, mains and laterals, sequence water from one field or segment to another, and regulate flow or pressure in mains, submains, or laterals. Valve differences are dependent upon their function. The controller issues commands for valve operation and receives feedback information to verify correct operation.

Flowmeter

Flow measuring feedback devices allow the computer to determine the rate and volume of water applied for estimating whether the irrigation scheduling algorithm and recommendations are followed. Two most commonly used flowmeters for monitoring flow in irrigation pipes are the propeller and the turbine types. Usually the output from these meters are digitized and calibrated in counts per unit volume of water applied (totalizing meter) or in counts per unit volume per unit time (flowrate meter).

Environmental sensor

Various types of soil moisture instruments (tensiometer, gypsum block, heat dissipation sensor, soil psychrometer), weather instrumentation (weather station, automated evaporation pan, etc.), plant water stress or crop canopy temperature (leaf psychrometer, stomatal diffusion porometer, infrared thermometer, and stem diameter sensor) are available and can be used in feedback mode for irrigation management. Soil moisture sensing devices are commonly used to override a system controller. If the soil at a particular station is "too wet," the sensor device disables part of the valve circuitry and the station is bypassed.

Filter

Plugging of emitters caused by physical, chemical, or biological contaminants is universal and at one time was considered to be the largest maintenance problem with drip irrigation systems. Filtration is accomplished with various types of sand filters or cartridge filters and screens. Suspended materials eventually build up on the filters and decrease efficiency. Either manual or automatic backwashing operations are available for sand media and screen filters for improving filter function.

Chemical injector

Methods used in the injection of fertilizer, algicides, and other chemicals into the lines are (1) pressure differential, (2) venturi (vacuum), and (3) positive displacement pumps. In the pressure differential system, a pressure difference is created by a valve or pressure regulator, installed between the tank inlet and outlet, causing a flow of water through the tank. Precise control valves maintain a preset injection rate. In the case of venturi type injectors, a rapid change of water velocity created by a reduction in the diameter of the pipe creates a pressure drop (vacuum) across an orifice to draw chemicals from the tank. The third method uses a rotary, gear, or piston pump to inject the fertilizer solution from the supply tank into the pipeline. In all cases, digital flowmeters can be used in a feedback mode to adjust the injection of chemicals proportionally to the flow rate of water to maintain a constant concentration of chemical in the irrigation water. Injectors should be made inoperable whenever the main water flow is stopped.

Example 3.3.4

Problems:

1. Draw a block diagram for the control system schematized in figure 3.3.14.
2. Draw a flow chart diagram for the logic required to operate the control system schematized in figure 3.3.14.
3. List the six major hardware categories necessary to fully automate a trickle irrigation system. Explain their function and differentiate between feedback and nonfeedback elements that would be needed to operate this system.
4. Discuss the use of various feedback sensing techniques and explain how they can best be used for control of high-frequency trickle irrigation.

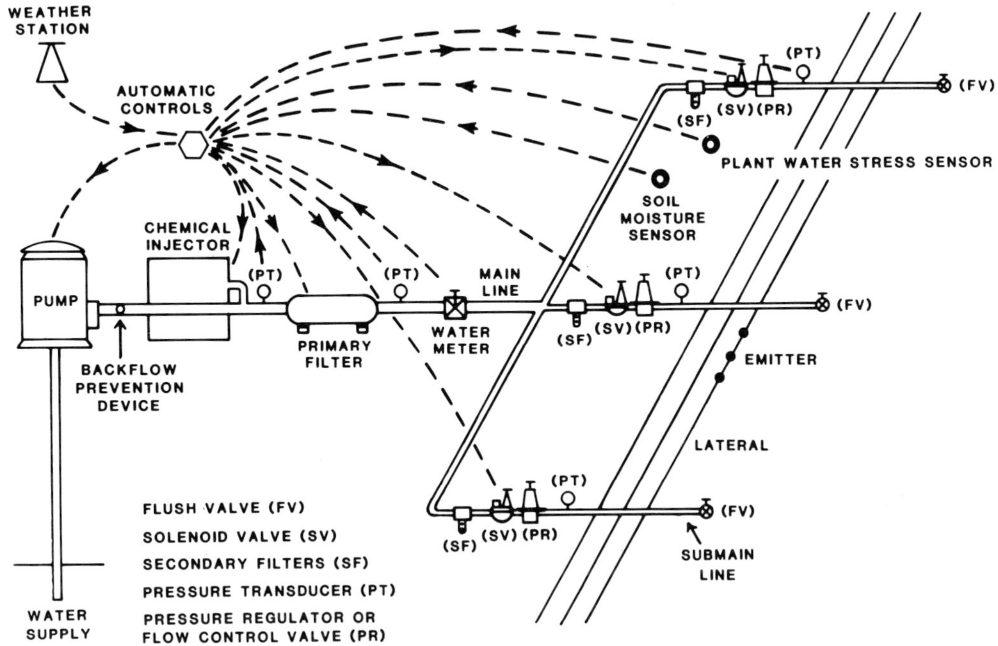

Fig. 3.3.14 Schematic of fully automated trickle irrigation system showing various input variables from transducers and output controls from computerized feedback controller (adapted from Bucks et al., 1983).

REFERENCES AND SUPPLEMENTAL READING

Bouyoucos, G.J. and Mick, A.H., 1940. An electrical resistance method for the continuous measurement of soil moisture under field conditions. Michigan Agric. Exp. Sta. Tech. Bull. 172, East Lansing, Michigan.

Bouyoucos, G.J. and Mick, A.H., 1947. Improvements in the plaster of Paris absorption block electrical resistance method for measuring soil moisture under field conditions. Soil Sci. 63: 455-465.

Bucks, D.A., Nakayama, F.S., and Warrick, A.W., 1982. Principles, practices, and potentialities of trickle (drip) irrigation. In Hillel, D. (Ed.), Advances in Irrigation 1: 219-198, Academic Press, New York.

Cary, J.W. and H.D. Fisher., 1983. Irrigation decisions simplified with electronics and soil water sensors. Soil Sci. Soc. Amer. J. 47: 1219-1223.

Doorenbos, J. and Pruitt, W.O., 1977. Crop water requirements. Food and Agric. Organization. Irrigation and Drainage Paper 24, United Nations, Rome, Italy. 144 pp.

Hillel, D., 1980. Applications of Soil Physics. Academic Press, New York, 385 pp.

Hoffman, G.J. and Rawlins, S.L., 1972. Silver-foil psychrometer for measuring leaf water potential in situ. Sci. 177: 802-804.

Howell, T.A., Hatfield, J.L, Rhoades, J.D., and Meron M., 1983. Response of cotton water stress indicators to soil salinity. Irrig. Sci. 5: 25-36.

Howell, T.A., Hatfield, J.L., Yamada, H., and Davis, K.R., 1984a. Evaluation of cotton canopy temperature to detect crop water stress. Trans. Amer. Soc. Agric. Eng. 27(1): 84-88.

Howell, T.A., McCormick, R.L., and Phene, C.J., 1985. Design and installation of large weighing lysimeters. Trans. Amer. Soc. Agric. Eng. 28(1): 106-112, 117.

Howell, T.A., Meek, D.W., Phene, C.J., Davis, K.R., McCormick, R.L., 1984b. Automated weather data collection for research on irrigation scheduling. Trans. Amer. Soc. Agric. Eng. 27(2): 386-391, 396.

Hsiao, T.C., 1973. Plant responses to water stress. Annu. Rev. Plant Physiol. 24: 519-570.

Huck, M.G. and Klepper, B., 1977. Water relations of cotton. II. Continuous estimates of plant water potential from stem diameter measurements. Agron. J. 69: 593-597.

Israelsen, O.W. and Hansen, V.E., 1967. Irrigation Principles and Practices, 3rd Edition, John Wiley and Sons, Inc., New York, 447 pp.

Jackson, R.D., 1982. Canopy temperature and crop water stress. In Hillel, D.I. (Ed.) Advances in Irrigation 1: 43-85, Academic Press, New York.

Jensen, M.E., Robb, D.C.N., and Franzoy, C.E., 1970. Scheduling irrigations using climate-crop-soil data. J. Irrig. Drain. Div., Amer. Soc. Civ. Eng. 96(IR1): 25-38.

Jensen, M.E., Wright, J.L., and Pratt, B.J., 1971. Estimating soil moisture depletion from climate, crop, and soil data. Trans. Amer. Soc. Agric. Eng. 19(5): 954-959.

Klepper, B., Browning, V.D., and Taylor, H.M., 1971. Stem diameter in relation to plant water status. Trans. Amer. Soc. Agric. Eng. 48(3): 683-683.

Parsons, J.E., Phene, C.J., Baker, D.N., Lambert, J.R., and McKinion, J.M., 1979. Soil water stress and photosynthesis in cotton. Physiol. Plantarum 47: 185-189.

Penman, H.L., 1948. Natural evaporation from open water, bare soil, and grass. Proc. Roy. Soc. (London) A193: 120-145.

Phene, C.J., Hoffman, G.J., and Austin, R.S., 1973. Controlling automated irrigation with soil matric potential sensor. Trans. Amer. Soc. Agric. Eng. 16(4): 733-776.

Phene, C.J. and Howell, T.A., 1984. Soil sensor control of high-frequency irrigation. Trans. Amer. Soc. Agric. Engin. 27(2): 386-391, 396.

Phene, C.J., Howell, T.A., Ayars, J.E., Meron, M., Litwin, Y.J., Saluja, H., Barbour, W.H., and Miller, R.S., 1982. Evaluation of Automation and Instrumentation for Optimal Water Management in Irrigation. California State Department of Water Resources, Technical Report Agric. B53853, Sacramento, California, 149 pp.

Phene, C.J., Howell, T.A., and Sikorski, M.D., 1985. A traveling trickle irrigation system. In Hillel, D. (Ed.) Advances in Irrigation, Vol. III (in press).

Phene, C.J., Radulovich, R.A., Rose, J.L., Blume, M.F., 1982. The effect of high-frequency trickle irrigation on water stress of tomatoes. ASAE paper 82-2521, Amer. Soc. Agric. Eng., St. Joseph, Michigan, 21 pp.

Pruitt, W.O. and Doorenbos, W.O., 1977. Empirical calibration, a requisite for evapotranspiration formula based on daily or longer mean climatic data. In Proc. "Evapotranspiration" Hungarian Natl. Comm., International Commission on Irrig. and Drain., New Delhi, India, 20 pp.

Richards, L.A. and Gardner, W., 1936. Tensiometers for measuring the capillary tension of soil water. J. Amer. Soc. Agron. 28: 352-358.

Richards, L.A. and Ogata, G., 1958. Thermocouple for vapor pressure measurement in biological and soil systems at high humidity. Sci. 128: 1089-1090.

Riggs, D.S., 1970. Control Theory and Physiological Feedback Mechanisms. The Williams and Williams Company, Baltimore, Maryland, 599 pp.

Shaw, B. and Baver, L.D., 1939. An electrothermal method for following moisture changes of the soil in situ. Soil Sci. Soc. Amer. Proc. 4: 78-83.

Van Bavel, C.H.M., 1966. Potential evaporation: The combination concept and its experimental verification. Water Resources Res. 2(3): 445-467.

Zeines, B., 1972. Automatic Control Systems. In Kosow, I.L. (Ed.), Prentice-Hall, Englewood Cliffs, New Jersey, pp. 1-11.

Chapter 3

OPERATIONAL PRINCIPLES

3.4 FIELD PERFORMANCE AND EVALUATION

V. F. BRALTS

A trickle irrigation system consists of a control station (water supply, pumps, filters, injectors, pressure regulators, water meters, time clocks, etc.) followed by a network of plastic pipes and fittings (main and submain lines), plastic lines or tubing (laterals), and emitters (see figure 3.4.1). The mainline is the primary artery for the delivery of water to the various irrigation zones or blocks. Within each zone there are usually a number of submain units. The unit size may be 1 to 5 hectares and is a function of the hydraulic design and topography. The zone size, in contrast, is usually a function of crop type and system capacity.

Despite the success of trickle irrigation systems in the United States and elsewhere, several problems related to optimal water and fertilizer management still remain. The theory behind trickle irrigation for conserving water and fertilizer is sound, but the implementation in the field may not always be practical. While the method has great potential for high irrigation efficiencies, poor system design, management, or maintenance, can lead to low efficiencies. In some instances the trickle irrigation systems were installed with little concern for basic engineering hydraulic principles and resulted in nonuniform emitter discharges throughout the irrigated field. Irrigators in order to overcome this lack of uniformity found it necessary to over-irrigate. Over-irrigation can lead to the waste of water, nutrients, and energy as well as the possibility of groundwater contamination due to excessive leaching.

The nonuniformity of emitter discharge is the result of several factors. The more important of these are the hydraulic and emitter discharge variations (Bucks et al., 1982). The hydraulic variation along the lateral line, submain, or manifold is a function of slope, pipe length and diameter, and emitter-discharge relations. Emitter variation at a given operating pressure is caused by manufacturing variability, emitter plugging (complete or partial), water temperature changes, and emitter wear. At the present time, a design equation which includes all the factors affecting emitter uniformity in a trickle system has not been developed.

3.4.1 *Need for evaluation*

The field evaluation of trickle irrigation systems is important for several reasons. It is important to (a) the design engineer in

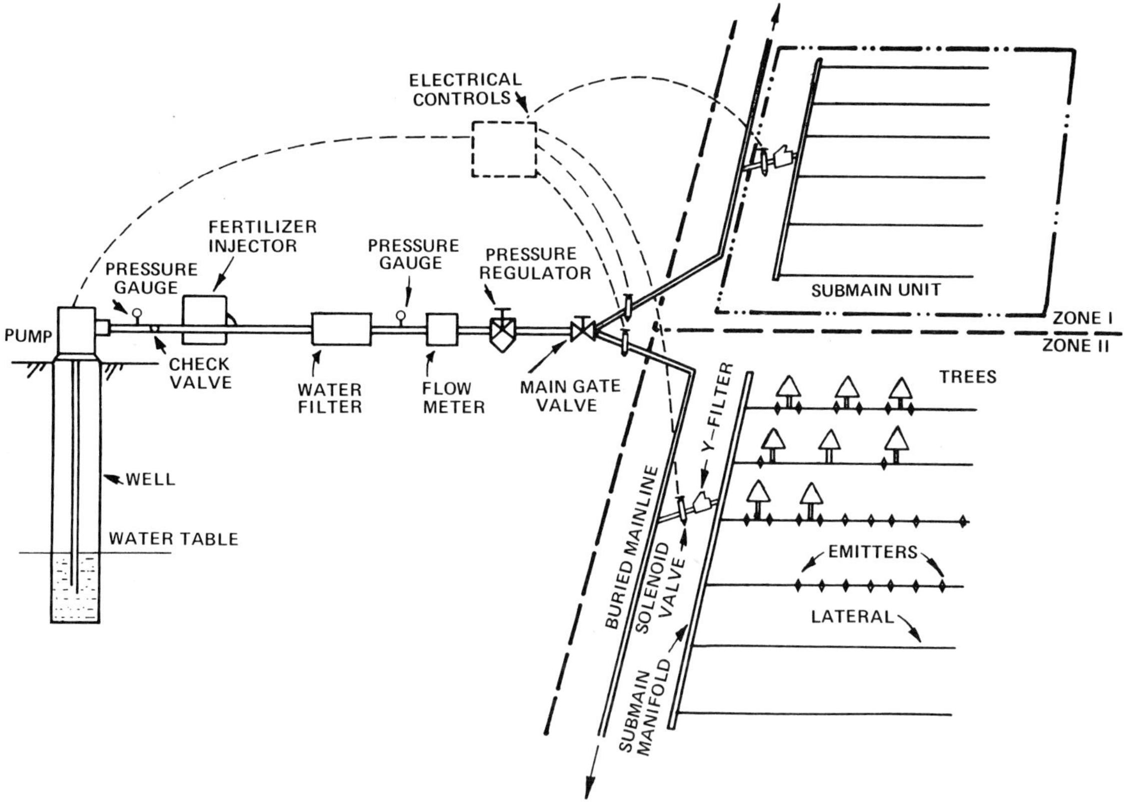

Fig. 3.4.1 Trickle irrigation system components.

establishing whether the desired emitter discharge uniformity specifications are being met, (b) the irrigator in deciding whether the system can be operated efficiently, and (c) the maintenance personnel as a diagnostic tool for determining the proper operation of the system and its components and to take remedial action as required.

Proper hydraulic design is needed to insure the successful operation of a trickle irrigation system. Several methods have been developed to aid in the design of such a system. These include the Emission Uniformity Concept (Keller, 1984), Emitter Flow Variation Concept (Wu and Gitlin, 1974), and Statistical Uniformity Concept (Bralts et al., 1981a, 1981b). All these methods are based upon classical hydraulic theory. Field evaluation techniques for each will be summarized in this chapter.

3.4.2 *Emitter discharge uniformity estimation*

The main purpose of trickle irrigation is to supply water and nutrients to the plants at frequent and low volumes adequate to meet their consumptive use and fertility requirements. With this in mind, it is essential that emitter flow variation or uniformity of water distribution for the system is known, particularly because the irrigation duration and amount are ultimately based upon the flow rate. In this section the factors affecting emitter flow variation, measurement of emitter flow and uniformity, and field evaluation procedures will be discussed.

3.4.2.a Factors affecting emitter flow variation

The variation of emitter flow rate is affected by hydraulic, manufacturing, and field conditions (Solomon, 1976). Emitters vary in design from elaborate pressure compensating types to long flow path and simple orifice types. The flow characteristics of most emitters can be described by the equation

$$q_e = kH^x \qquad (3.4.1)$$

where q_e = emitter discharge rate, k = emitter discharge coefficient, H = pressure head at the emitter, and x = emitter discharge exponent (Karmeli, 1977; Wu et al., 1979).

Equation 3.4.1 can be derived from a combination of the Bernoulli energy and continuity equations. The emitter discharge coefficient k, contains the variables such as the coefficient of discharge, emitter geometry, and the acceleration of gravity.

The value of x characterizes the type of emitter and/or the flow regime of the emitter. For example, orifice-type emitters are fully turbulent and have an emitter discharge exponent of 0.5. With long path emitters, x = 0.5 for fully turbulent flow and x = 1.0 for laminar flow (Karmeli, 1977). An emitter with a x-value of less than 0.5 would be a pressure compensating type (Wu et al., 1979), and when x = 0 would be fully pressure compensating.

The variation in flow for various types of emitters as a function of pressure head is illustrated in figure 3.4.2 (Karmeli, 1977; Bralts, 1978).

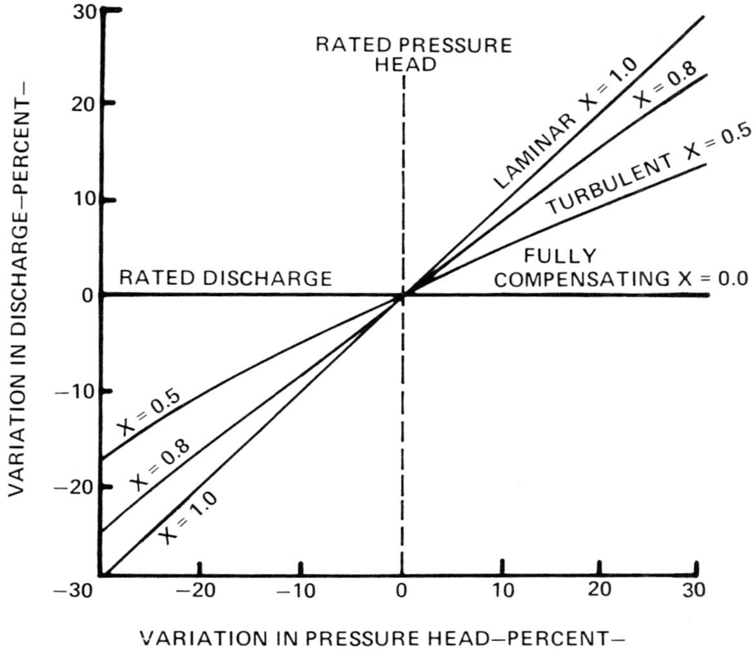

Fig. 3.4.2 Relation between discharge variation and pressure change for emitters with various discharge exponents.

The equation describing emitter flow at any point along the lateral is given by the general form (Wu, 1976)

$$q_i = q_o[1 - R_i(\frac{\Delta H}{H_o}) \pm R'_i(\frac{\Delta H'}{H_o})]^x \tag{3.4.2}$$

where q_i = emitter flow at a given length ratio, q_o = emitter flow at the input pressure H_o, $R_i \Delta H$ = headloss due to friction, and $R'_i \Delta H'$ = headloss or gain due to slope. This equation illustrates that the emitter flow is dependent upon the emitter discharge coefficient x for any given specific hydraulic condition. Furthermore, equation 3.4.2 can be used to develop the coefficient of variation due to hydraulics in the form

$$V_h x = \frac{S_h x}{\overline{h^x}} \tag{3.4.3}$$

where V_hx = coefficient of variation due to hydraulics, S_hx = standard deviation of emitter flow due to hydraulics, and $\overline{hx}$ = mean emitter flow due to hydraulics (Bralts, 1978).

Another important factor which affects emitter discharge uniformity is emitter manufacturing (Keller and Karmeli, 1974; Karmeli and Keller, 1975; Solomon, 1976, 1977). Proposals have been made to use the coefficient of manufacturer's variation (V_m) as the measure of emitter flow variation due to construction variability. For a specific head, this variation is given by the relation

$$V_m = \frac{S_q}{\overline{q}} \tag{3.4.4}$$

The values of V_m range from 0.02 to 0.2 for the various types of emitters (Solomon, 1977). The variability can be adjusted downward by increasing the number of emitters around a plant and is given by the relation

$$V_e = V_m(e)^{-1/2} \tag{3.4.5}$$

where V_e = coefficient of variation per plant, V_m = manufacturer's variation, and e = number of emitters per plant (Keller and Karmeli, 1975).

The effect of emitter plugging on flow variation can be treated statistically (Bralts et al., 1981b). For completely and partially plugged emitters, the coefficient of variation of emitter flow, V_q, is given by

$$V_q = \frac{S_q}{\overline{q}} \tag{3.4.6}$$

and further expanded to the form

$$V_p = [\frac{n(\theta + p'\ (1 - A')^2}{(\theta + p'\ (1 - A'))^2}(V_q^2 + 1) - 1]^{1/2} \tag{3.4.7}$$

where S_q = standard deviation of emitter flow, $\overline{q}$ = mean emitter flow rate, V_p = coefficient of variation of emitter flow with emitter plugging, n = total number of emitters on the lateral line, θ = number of emitters openly flowing, p' = number of partially plugged emitters, and A' = degree of plugging for the partially plugged emitters. Thus, when information is available on the coefficient of variation without emitter plugging and the percent plugging, the resultant coefficient of variation including emitter plugging can be estimated.

3.4.2.b Emitter flow variation and uniformity estimation

Emitter flow variation and uniformity equations developed by other types of irrigation practices can be readily adapted to trickle irrigation systems.

Sprinkler irrigation uniformity is defined in terms of the depth of irrigation water applied, y, as

$$U_s = 100\ (1 - \frac{\overline{\Delta y}}{\bar{y}}) \qquad (3.4.8)$$

where U_s = uniformity coefficient percentage, $\overline{\Delta y}$ = absolute value of the mean deviation of irrigation depth, and $\bar{y}$ = mean depth of irrigation (Christiansen, 1942). By substituting emitter flow rate q_e for y, the uniformity for trickle irrigation lateral lines can be estimated (Wu, 1976).

The uniformity of sprinkler irrigation can also be described using common statistical parameters such as the coefficient of variation (V_y) of the depth of irrigation water, y (Wilcox and Swailes, 1947). The statistical uniformity coefficient is thus defined as

$$U_s = 100\ (1 - V_y) = 100\ (1 - \frac{S_y}{\bar{y}}) \qquad (3.4.9)$$

where U_s = statistical uniformity coefficient as a percentage, and V_y = coefficient of variation of the depth of irrigation water, $\bar{y}$, or as previously defined the standard deviation (S_y) over the mean ($\bar{y}$). A similar statistical approach can be developed for trickle irrigation systems where the random variable y, the depth of water in sprinkler irrigation is replaced by q so that equation 3.4.9 becomes

$$U_s = 100\ (1 - V_q) = 100\ (1 - \frac{S_q}{\bar{q}}) \qquad (3.4.10)$$

(Bralts, 1981a). By using the statistical treatment, all of the various factors such as emitter manufacturing variation, lateral line friction, elevational differences, and emitter plugging are included in the final uniformity estimate. For single chamber tubing in the orifice type emitters, the coefficients of hydraulic (friction and elevation changes) and construction variation are independent of each other and can be combined into the total coefficient of variation for the lateral line by the equation

$$V_{ql} = (V_{hl}^2 + V_m^2)^{1/2} \qquad (3.4.11)$$

where V_{ql} = total coefficient of variation for the lateral line, V_{hl} = hydraulic coefficient of variation for the lateral line, and V_m = emitter manufacturer's coefficient of construction variation (Bralts, 1981a). The coefficient of total variation is then used in equation 3.4.1 and the statistical uniformity of a trickle irrigation lateral line results.

A simple way to show emitter flow variation for trickle irrigation is based on lateral line hydraulics which takes the form

$$q_{var} = 100\ (1 - \frac{q_{min}}{q_{max}}) \tag{3.4.12}$$

where q_{var} = variation of emitter flow as a percentage, q_{max} = maximum emitter flow, and q_{min} = minimum emitter flow (Wu, 1975). The relationship between q_{var} and H_{var} is

$$q_{var} = 100\ [1 - (1 - H_{var})^x] \tag{3.4.13}$$

in which

$$H_{var} = \frac{H_{max} - H_{min}}{H_{max}} \tag{3.4.14}$$

where H_{var} = variation of emitter pressure, H_{max} = maximum pressure in the line, and H_{min} = minimum pressure in the line. Thus, the term q_{var} can usually found after first determining the lateral line hydraulics. Note that the relationship between q_{var} and H_{var} is dependent upon the emitter type (Keller and Karmeli, 1974; and Karmeli and Keller, 1975). (See section 2.2.2.d on Design criteria).

Initially, the preceding measures of emitter flow variation and uniformity were applied exclusively to trickle irrigation lateral lines. Further development (Karmeli and Keller, 1975) indicated that emitter flow uniformity can also be applied to system uniformity using the following equation

$$EU' = 100\ (\frac{q_{min}}{q_{avg}}) \tag{3.4.15}$$

where EU' = field emission uniformity as a percentage, q_{min} = minimum discharge rate computed from minimum pressure in the system, and q_{avg} = average of all the field data emitter discharge rates.

The equation has been redefined and modified to include the emitter coefficient of manufacturing variation, V_m, and the number of emitters per plant, e. Thus, to estimate the emission uniformity for a proposed trickle irrigation system design, the following equation can be used:

$$EU = 100\ (1 - \frac{1.27\ V_m}{e^{0.5}})\ \frac{q_{min}}{q_{avg}} \qquad (3.4.16)$$

where EU = design emission uniformity, and all other variables have been defined previously.

The statistical uniformity coefficient equation can also be used in in the design of trickle irrigation submain units based on statistical considerations (Bralts, 1983) and will be presented in section 3.4.4 of this chapter.

3.4.2.c Field uniformity estimation

The evaluation of a trickle irrigation system encompasses significantly more than the uniformity of emitter discharges and the accurate estimation of system uniformity is probably the single most important factor in obtaining system performance. At present several methods are used for estimating field uniformity. A modified form of equation (3.4.16) is to consider the absolute emission uniformity (EU_a) as follows:

$$EU_a = 100\ (\frac{q_{min}}{q_{avg}} + \frac{q_{avg}}{q_x})\ (\frac{1}{2}) \qquad (3.4.17)$$

where EU_a = absolute uniformity as a percentage, q_{min} = minimum discharge rate, q_{avg} = average emitter flow rate, and q_x = average of the highest 1/8 of emitter flow rates. Calculations based on equations 3.4.16 and 3.4.17 are available in the publications by Karmeli and Keller (1975), Merriam and Keller (1978) and the Soil Conservation Service National Engineering Handbook (1983). An example of the type of information collected from the field is shown in figure 3.4.3, and the procedure for collecting such data from point source emitters is as follows:

Step 1. Choose four lateral lines and four emitters on each lateral per location given on the table.

Step 2. Measure the emitter discharge and calculate the emitter flow rate at two adjacent emitters (A and B) at each collection point by collecting the discharge (mL) for one minute in a graduated cylinder.

Step 3. Calculate the average emitter discharge (gph or lph) for each of the 16 locations.

Step 4. Calculate Emission Uniformity using equation 3.4.16 or 3.4.17 as desired.

At least 8 pressure and 32 discharge volume readings are recommended (Merriam and Keller, 1978). For line-source emitters, a one meter collection trough can be used for determining the discharge rates.

General criteria for EU and EU_a values are: 90% or greater, excellent; 80 to 90%, good; 70 to 80%, fair; and less than 70%, poor. The

Location on Lateral		Lateral Location on the Manifold							
		inlet end		1/3 down		2/3 down		far end	
		mL	gph	mL	gph	mL	gph	mL	gph
inlet end	A B								
	Ave								
1/3 down	A B								
	Ave								
2/3 down	A B								
	Ave								
far end	A B								
	Ave								

Fig. 3.4.3 Sample form for recording emitter discharge rates in the field for calculating field emission uniformity.

primary disadvantage of this method is its nonstatistical base. For this reason, obtaining confidence limits and breaking down the components of emitter flow variation are not possible.

Example 3.4.1

Problem: Estimate the emission uniformity EU_a for a trickle irrigation submain unit with the following field data. Volume (mL) collected in 1 minute from adjacent emitters (A/B); 80/90, 70/89, 78/88, 85/79, 84/75, 75/90, 91/88, 93/81, 78/85, 76/81, 83/64, 93/64, 75/60, 67/71, 75/83, 73/85.

Solution: Step 1. Determine the mean emitter discharge rate for each of the 16 locations. For example, (80 + 90)/2 = 85 mL/min.

Step 2. Calculate the emission uniformity using equation 3.4.17, EU' = 90.5%.

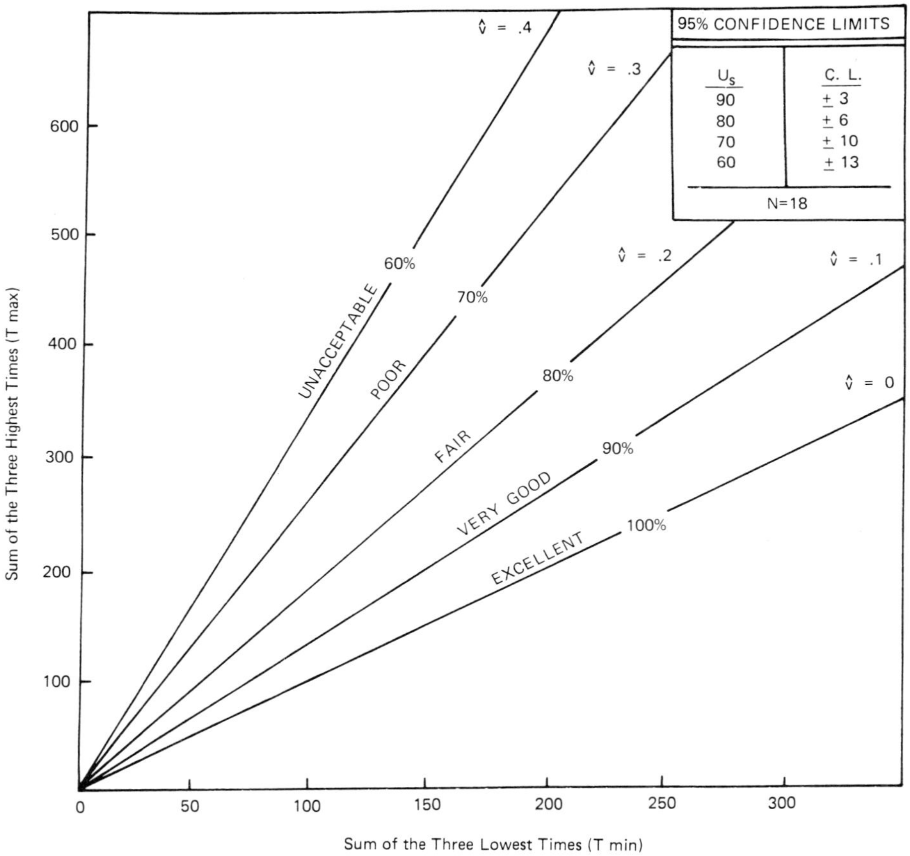

Fig. 3.4.4 Nomograph for estimating statistical uniformity based on emitter discharge rates.

Another method for field evaluation of emission uniformity relies on the design procedure based on estimating emitter flow variation (Wu and Gitlin, 1974 and 1975). In this approach, the evaluation consists in finding the minimum and maximum pressures in the submain unit and then calculating the emitter flow variation q_{var} as given in equations 3.4.13 and 3.4.14. General criteria for q_{var} values are: 10% or less, desirable; 10 to 20%, acceptable; and greater than 20%, not acceptable. A major limitation of this procedure is that it does not include emitter flow variation due to manufacturing or plugging, which can be significant factors in the system uniformity.

A third method for estimating system uniformity is based upon the statistical uniformity coefficient given by equation 3.4.10 (Bralts and

Kesner, 1982). The advantages of this method are that field measurements are simple and confidence limits of the uniformity values can be set. A graphical technique for field determination of system or statistical uniformity can be used based upon the time to fill a container rather than the emitter flow rate. This method assumes that the distribution of emitter flows is normal. It uses the highest one-sixth and lowest one-sixth of the times needed to fill a container to get the statistical uniformity. These modifications, although unjustified by an increase in statistical efficiency, were justified on the basis of practical field applications. An example of the graphical method is illustrated in figure 3.4.4 and the procedure for the use of the nomograph is as follows:

Step 1. Select a container for making the flow rate determination (100 or 200 mL).
Step 2. Choose 18 emitters at random in the submain and measure the time it takes to fill the container.
Step 3. T_{max} is the sum of the three highest times it takes to fill the container.
Step 4. T_{min} is the sum of the three lowest times it takes to fill the container.
Step 5. Find T_{max} on the vertical axis and draw a horizontal line.
Step 6. Find T_{min} on the horizontal axis and draw a vertical line.
Step 7. The intersection of the two lines gives the statistical uniformity.

Example 3.4.2

Problem: Estimate the total statistical uniformity U_s for a trickle irrigation submain unit with the following field data. Time (sec) to fill a 100 mL container: 64, 79, 67, 71, 75, 81, 68, 85, 75, 69, 85, 77, 89, 68, 81, 90, 65, and 61.

Solution: Step 1. Calculate T_{max} and T_{min} using the three highest and three lowest times so that $T_{max} = 85 + 89 + 90 = 264$, and $T_{min} = 64 + 65 + 61 = 190$.
Step 2. Use equation 3.4.10 or figure 3.4.4 to determine the statistical uniformity $U_s = 90\% \pm 3\%$.

The general criteria for an acceptable statistical uniformity coefficient are: 90% or greater, excellent; 80 to 90%, very good; 70 to 80%, fair; 60 to 70%, poor; and less then 60%, unacceptable. The main advantage of the statistical uniformity coefficient is its ability to make analysis of the variables. For example, when U_s and the percent of emitter plugging are obtained from random sampling techniques, the statistical uniformity without emitter plugging can be determined using equation 3.4.7 or using equation 3.4.5 when the number of emitters per plant is known.

3.4.3 *Hydraulic system and emitter performance*

The variation of emitter flow in a trickle irrigation system is caused by various factors. As mentioned previously, the primary factors are slope, pipeline length and diameter, emitter type, emitter manufacturing, emitter plugging, and number of emitters per plant. The complete evaluation of the system must include the separation of problems caused by hydraulics and emitter performance and a method is available to do this (Bralts and Edwards, 1983). This method is based on a constant odds statistical analysis of the general emitter flow equation 3.4.1 and is in the form

$$V_q = (V_k^2 + x^2 V_h^2)^{1/2} \qquad (3.4.19)$$

where V_q = coefficient of variation for emitter discharge, V_k = coefficient of variation for the emitter discharge coefficient k, x = emitter discharge exponent, and V_h = coefficient of variation of the pressure head h.

The coefficient of variation V_k was previously defined as the emitter coefficient of variation due to manufacturing V_m. The coefficient of hydraulic variation V_h is a measure of the hydraulic pressure variation. The major advantage of the preceding equation is that V_h can be calculated directly from the hydraulic pressure and is independent of the emitter type.

The effects of emitter type and head can best be seen by maintaining the variable k constant, and varying the pressure h and discharge exponent x. By using the general emitter flow equation and the error formulation for a single variable, the relation simplifies to the form

$$V_q = xV_h \qquad (3.4.20)$$

This expression shows that the coefficient of variation for emitter flow is directly proportional to the emitter discharge exponent x when the total system discharge is held constant. For example, if the coefficient of hydraulic variation has been determined from field pressure measurements, the coefficient of emitter flow variation due to hydraulics alone can be readily calculated for a specific emitter type using equation 3.4.20. When the coefficient of variation due to hydraulics is too high for an acceptable uniformity (see equation 3.4.10) then a new emitter with a lower emitter discharge coefficient, but a similar flow rate should be installed. The expected reduction in the coefficient of variation would be equal to the proportion of the respective emitter coefficients. In the extreme case where an emitter is chosen with a discharge coefficient of zero, the coefficient of variation due to hydraulics is also zero. In other words, the emitter is completely pressure compensating.

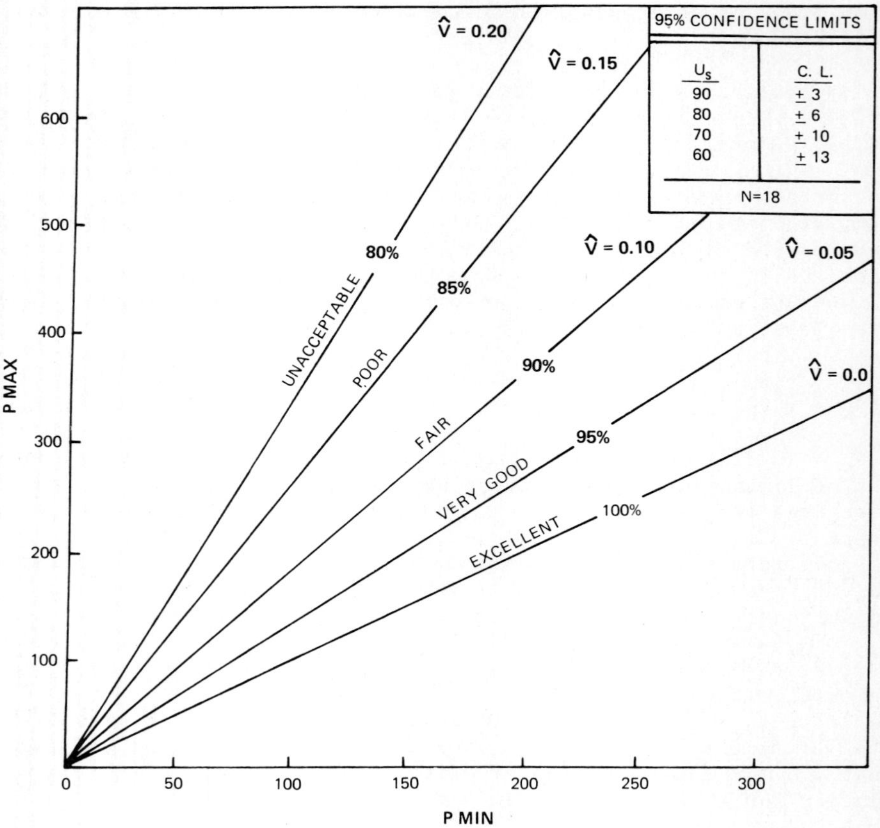

Fig. 3.4.5 Nomograph for estimating hydraulic uniformity based on pressure distribution measurements (sum of three highest and lowest).

A graphical method for determining the coefficient of variation due to hydraulics xV_h, or the statistical uniformity of emitter flow due to hydraulics can also be used (Bralts and Edwards, 1983). The nomograph (figure 3.4.5) is based upon the same statistical principles as the uniformity calculator. As illustrated the nomograph is based on an emitter discharge exponent of x = 0.5; corrections for other emitter types can be made using the equation

$$x_2V_h = 2x_2\ (x_1V_h) \qquad (3.4.21)$$

where x_2V_h = corrected coefficient of variation due to hydraulics when $x = x_2$, x_2 = actual emitter discharge exponent, and x_1V_h = coefficient of variation due to hydraulics when x = 0.5.

Example 3.4.3

Problem: Estimate the statistical uniformity due to hydraulics for a trickle irrigation submain unit with an emitter exponent of x = 0.4 and the following field data: Pressure head in meters of water of 10.20, 9.98, 8.80, 8.50, 10.18, 9.01, 7.59, 10.25, 9.43, 9.75, 10.07, 8.37, 7.92, 10.92, 9.25, 10.50, 9.43, and 8.97.

Solution: Step 1. Calculate P_{max} and P_{min} using the three highest and the three lowest pressures as P_{max} = 10.50 + 10.20 + 10.18 = 30.88 and P_{min} = 7.59 + 7.92 + 8.37 = 23.88.

Step 2. Use equation 3.4.3 or figure 3.4.5 to determine the statistical uniformity due to hydraulics for x = 0.5, U_s = 91%.

Step 3. Use equation 3.4.21 or figure 3.4.5 to determine the statistical uniformity due to hydraulics for x = 0.4, U_s = 92.8%.

The real utility of the information obtained from figure 3.4.4 is most evident when it is used in conjunction with figure 3.4.5. Referring back to equation 3.4.19, it becomes evident that when two of the coefficients of variation for the submain unit are known, the third can be determined by rearranging equation 3.4.19 to

$$V_{pf} = (V_{qs}^2 - x^2 V_{hs}^2) \tag{3.4.22}$$

where V_{pf} = coefficient of variation due to emitter performance in the submain, V_{qs} = total coefficient of variation in the submain unit, x = coefficient of discharge for the specific emitter, and V_{hs} = coefficient of variation due to hydraulics in the submain unit.

The coefficient of variation due to emitter performance, V_{pf}, was defined because it includes a variety of factors such as manufacturer's variation, emitter number, emitter plugging, and emitter number per plant. Equation 3.4.22 permits the differentiation between hydraulic and emitter performance related variables.

Example 3.4.4

Problem: Given the total statistical uniformity of 90% as determined in example 3.4.2 and the statistical uniformity due to hydraulics of 92.8 as determined in example 3.4.3, estimate the percentage variation due to emitter performance.

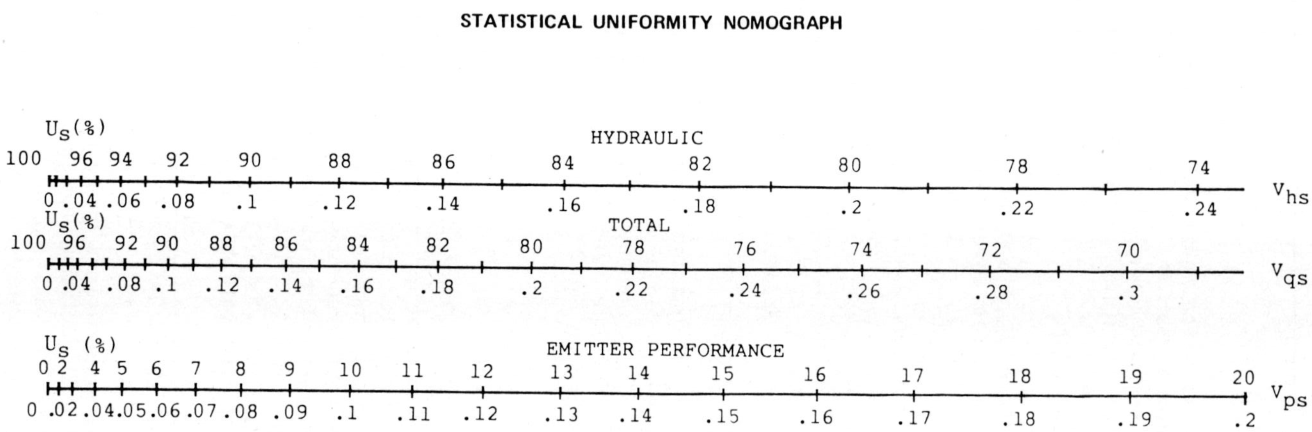

Fig. 3.4.6 Nomograph for determining emitter performance.

Solution: Use equation 3.4.22 or figure 3.4.6 to determine the emitter performance variation, V_{pf} = 0.055 or 5.5%.

The general criteria for an acceptable U_s as reported earlier are 90% or greater, excellent; 80 to 90%, very good; 70 to 80%, fair; and 60 to 70%, poor; and less then 60%, unacceptable. An acceptable U_s due to hydraulics depends on all of the various factors which affect uniformity. In general, an U_s due to hydraulics of 90% or better is recommended for a trickle system. Any system with less than this value requires a design change, or replacement of existing emitters with the pressure compensating type.

An acceptable U_s due to emitter performance is 85%. This value is based upon the desirability of maintaining a combined statistical uniformity of 80% and greater. A statistical uniformity of 80% is the minimum acceptable value when fertilizer injection is used. If the U_s due to emitter performance is less than 85%, cleaning or replacing of plugged emitters or placing additional emitters at each plant will improve the statistical uniformity.

3.4.4 *Soil water distribution and application efficiency*

Evaluation of the adequacy of soil water distribution and the resultant application efficiency, two very important components of any irrigation system, is an essential part of trickle systems. Since water in trickle irrigation operation is usually applied at rates close to the plant's consumptive water use, soil water considerations are much more critical in trickle than in sprinkler or flood irrigation. In the past, a 30 to 50% root zone wetting has been recommended for trickle irrigation. This value has been exceeded in tree crops, particularly in the southeastern United States, where additional emitters have been added to improve uniformity. Under row crops, the soil wetted volume has sometimes gone as high as 70 to 80%.

The soil type is important because it affects the extent of available water and the distribution of water under an irrigation regime (figure 3.4.7). The maximum application depth based upon soil type can be estimated from the relation

$$I_{max} = y\,(AWC)\frac{zp}{100} \qquad (3.4.23)$$

where I_{max} = maximum net depth of irrigation, y = proportion of AWC to be depleted before irrigation, AWC = available water capacity, z = soil depth (root zone), and p = wetted volume as a proportion of the root zone, expressed as a decimal fraction of 0.01 to 1 (Karmeli and Keller, 1975). For field evaluation purposes, the maximum net irrigation depth is converted to the required irrigated volume V_r by multiplying by the irrigated area.

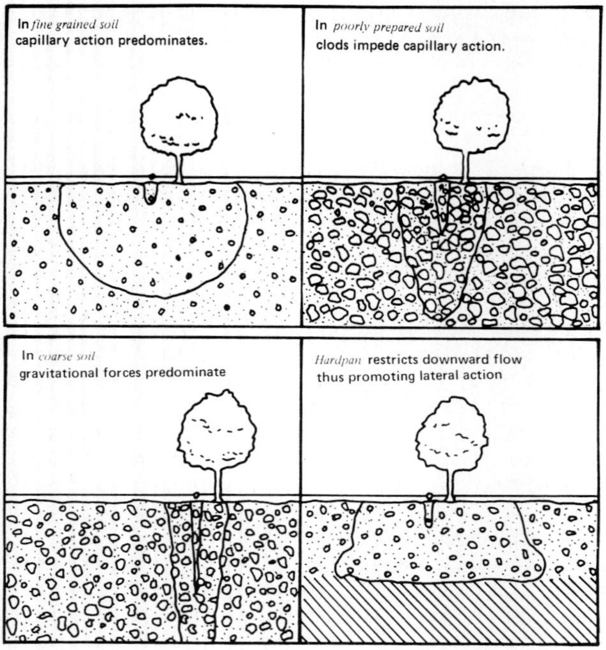

Fig. 3.4.7 Soil water distribution under different soil textures and conditions.

The application efficiency of an irrigation system is defined as the percentage of water applied that is actually stored in the crop root zone in relation to the total water applied. When the root zone is not fully irrigated, or a specific irrigation deficit as a percent of the total volume of water required is allowed, the irrigation application efficiency is defined by

$$E_a = 100[\frac{V_r(1 - P_D)}{V_a}] = 100[\frac{V_r(1 - P_D)}{3600Q_aT}] \tag{3.4.24}$$

where V_r = amount of water applied, P_D = irrigation deficit (expressed as a decimal), Q_a = actual discharge to the submain per second, and T = irrigation time in hours (Wu and Gitlin, 1983). Equation 3.4.24 can be used to determine the application efficiency when the statistical uniformity is 100% and the irrigation deficit is known.

The relationships among statistical uniformity, irrigation deficit and application efficiency are complex and can be based upon probability and normal statistical distribution. For the special case

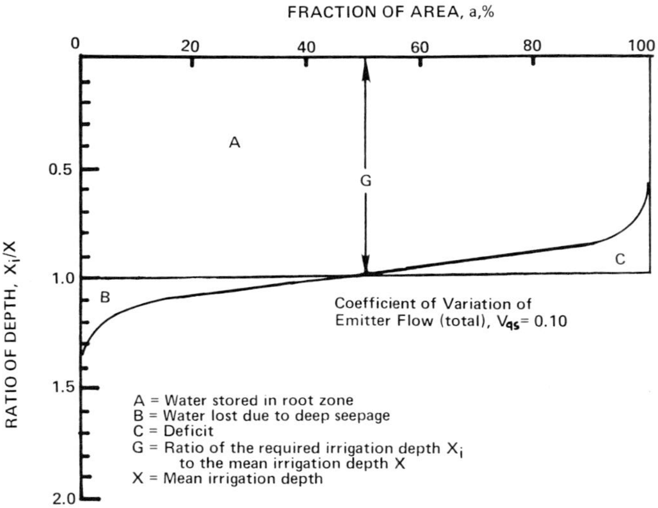

Fig. 3.4.8 Soil water distribution for a coefficient of variation of emitter discharge at $V_{qs} = 0.10$.

where the irrigation volume applied, V_a, is equal to the irrigation volume required, V_r, the irrigation deficit is equal to approximately 0.4 times the coefficient of variation for the submain unit, V_{qs}. In this instance, the application efficiency can be determined by the equation

$$E_a = 100 \left[\frac{V_r(1 - P_D)}{V_a}\right] = 100 \ (1-0.4V_{qs}) \qquad (3.4.25)$$

A dimensionless plot of the cumulative frequency curve is given in figure 3.4.8 which shows the required irrigation depth in the root zone for a coefficient of variation of 0.1 or a statistical uniformity of 90% and a deficit of approximately 4%.

The relationship of statistical uniformity, irrigation deficit, and application efficiency is graphically presented in figure 3.4.9. It is evident from the figure that over-irrigation to compensate for non-uniformity of irrigation is strictly at the expense of application uniformity. One can assume a 0.5 percent deficit to be small enough to be considered as full irrigation. Note that an irrigation deficit of zero is meaningless when related to irrigation application efficiency.

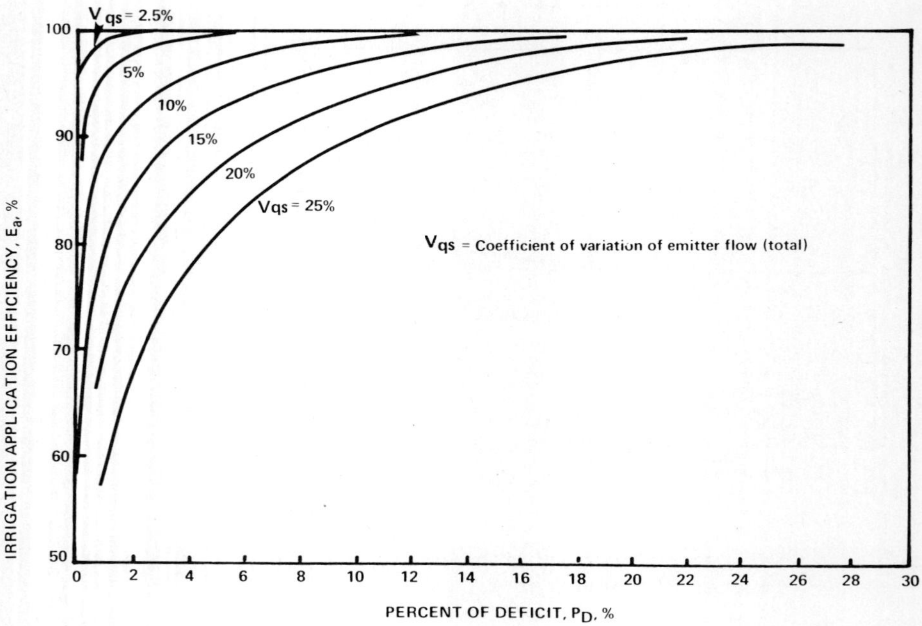

Fig. 3.4.9 Relation between irrigation application efficiency and irrigation deficit for various coefficient of variation of emitter flow (after Wu and Gitlin, 1983).

Example 3.4.5

Problem: Given the total statistical uniformity for a trickle system submain is 80% (V_{qs} = 0.2). Determine the application efficiency and the percent deficit when the irrigation application is equal to the required irrigation volume.

Solution: Step 1. Calculate the application efficiency using equation 3.4.25, E_a = 92%.

Step 2. From figure 3.4.9 the irrigation deficit for this situation can be found where E_a = 92% and V_{qs} = 0.4 as P_D = 8%.

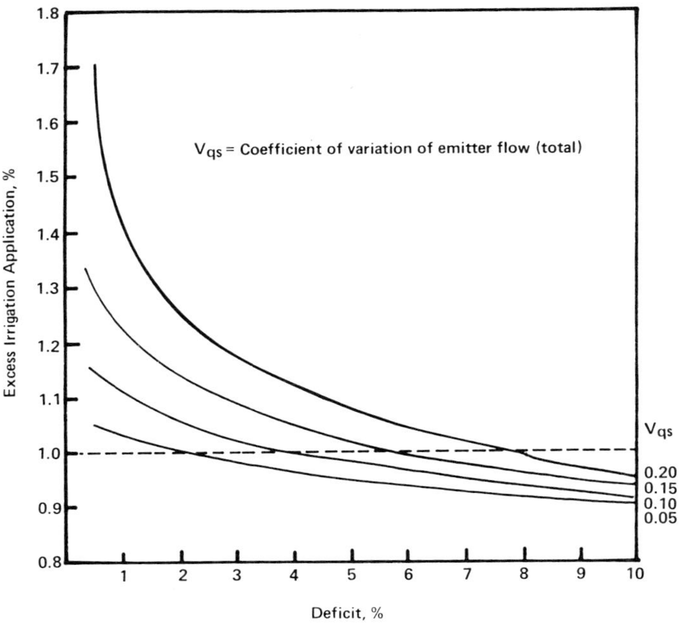

Fig. 3.4.10 Relation between excess irrigation application and irrigation deficit for various coefficient of variation of emitter flow (after Wu and Gitlin, 1983).

Example 3.4.6

Problem: Given the total statistical uniformity for a trickle irrigation submain unit as 80% (V_{qs} = 0.20). Determine the percent deficit and the application efficiency when the irrigation application is 1.2 times the required irrigation volume. Compare the results with example 3.4.5.

Solution: Step 1. Determine the percent deficit for V_{qs} = 0.20 and an irrigation application 1.2 times the required irrigation application using figure 3.4.10, P_D = 3%.

Step 2. Determine the application efficiency of the irrigation using figure 3.4.9, E_a = 82%.

Note: The application efficiency in this example is less than that in example 3.4.5 because over-irrigation to make up for a large irrigation deficit or a low system uniformity decreases the application efficiency.

The relationship of excess irrigation, irrigation deficit and coefficient of variation or statistical uniformity can be analyzed graphically (figure 3.4.10). The utility of this type of presentation can be illustrated by calculating the application efficiency. By assuming the irrigation volume to be 1.2 times the volume of the root zone and a statistical uniformity of 15%, and using figures 3.4.10 and 3.4.9 in succession, the application efficiency of approximately 83% is obtained.

From a field evaluation perspective, if the ratio of actual irrigation volume, V_a, to the maximum irrigation required, V_r, and the statistical uniformity are known, the application efficiency can be determined using figures 3.4.9 and 3.4.10. If the resultant application efficiency is less than 90%, i.e., unsatisfactory, the irrigator can improve the statistical uniformity by the methods discussed in the previous section or by increasing the acceptable deficit percentage. Of course, the irrigator could reduce the irrigation to such a point as to obtain an application efficiency of 100%. However, this would put practically the whole crop in jeopardy and is not a recommended practice. An allowable deficit in the range of 3 to 6% is recommended when the statistical uniformity is greater than 85% (Wu and Gitlin, 1983).

3.4.5 *Fertilizer injection evaluation*

Many benefits are obtained by injecting fertilizers through the trickle irrigation system (see chapter 4.3 on Fertilization). The uniformity of fertilizer application is approximately equal to the uniformity of the water application. Thus, an acceptable emitter discharge uniformity is a prerequisite to fertilizer injection. From a system standpoint, maximum application efficiency can be achieved by establishing flow equilibrium before chemical injection is started. The equilibrium period can be approximated by assuming an average flow velocity and dividing this rate into the the furthest distance from the injection point. For final evaluation, the total injection time which includes the start up, injection and shut down periods, should not exceed the available water capacity in the root zone. Otherwise, the fertilizer may be moved beyond the region of the absorbing roots.

3.4.6 *Filter performance*

Some form of filtration is required for virtually all trickle irrigation systems. Depending upon the water quality, the filtration system may include a single or combination of screen, media or centrifugal type filters. At least three important components must be considered in performance evaluation of the filters. These include (a) removal efficiency, (b) pressure differential, and (c) emitter plugging. Removal efficiency is defined as

$$E_r = 100\ (1 - \frac{S_1}{S_2}) \qquad (3.4.26)$$

where E_r = removal efficiency, S_1 = concentration in the filter outlet, mg/L, and S_2 = concentration in the filter inlet, mg/L (Oron et al., 1980). The concentration may included suspended materials or dissolved constituents. E_r can be used as a diagnostic tool when filtration problems are suspected. Water samples should be taken at least 30 minutes after the system is turned on to give it a chance to equilibrate.

Example 3.4.7

Problem: Determine the removal efficiency (E_r) of a sand and a 200 mesh screen filter system used in trickle irrigation for the removal of very fine sand. The field test results indicate the unfiltered water as containing 20 mg/L greater than 75 micrometer material and the filtered water as containing 15 mg/L greater than 75 micrometer materials.

Solution: Using equation 3.4.26 the removal efficiency can be calculated. E_r = 25%.

The second aspect of the field evaluation of the trickle irrigation filter system is the acquisition of information on differential pressures (inlet and outlet) before and after filter backwashing takes place. If the differential pressure is too great, a loss of flow can occur which affects the downstream pressure and performance of the rest of the flow system. Automatic backwash controls are based on such differential pressures and short-cycling can occur when high pressure differences develop too frequently, indicating high sediment load in water or malfunctioning control components. Conversely, if the differential pressures are constantly low or nonexistent, structural failures in the filter such as broken screen, worn seals, and nonoperating backwash valves may be present.

The third approach for assessing the effectiveness of the filter system is the determination of the extent of emitter plugging, which may not wholly be caused by failure in filter operation. Inspection of clogged emitters may give clues as to the cause of emitter failure (see chapter 3.1 on Emitter Clogging). Even a small percentage of emitter clogging can drastically affect the uniformity of water application. This is illustrated in figure 3.4.11 which can be used to estimate such effects. Thus, with as little as 5% emitter plugging, the statistical uniformity can be reduced to approximately 75%.

3.4.7 *Water Meter Evaluation*

The importance of installing water meter devices on every trickle system has been emphasized (Bucks et al., 1982). A water meter is used to check initial designs, manage or schedule irrigations, and monitor system operation to spot maintenance problems. Continuous

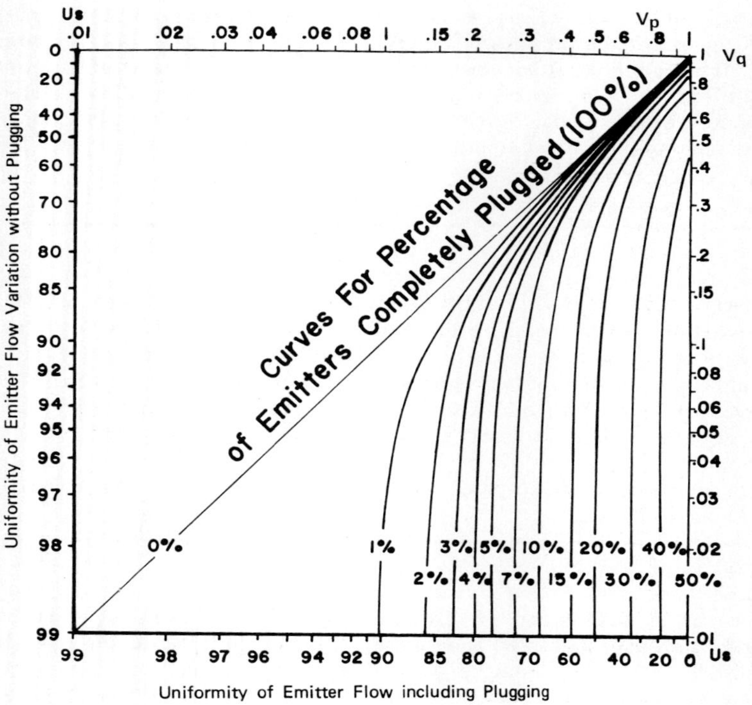

Fig. 3.4.11 Relation between the degree of emitter plugging and statistical uniformity (after Bralts, 1981b).

monitoring of the flow rate to the various trickle irrigation subunits can help to detect and locate problems before they become serious. For example, a gradual decrease in flow rate may indicate a clogged filter or the beginning of emitter plugging. In contrast, an increase in flow rate may be the result of a broken distribution line or some type of emitter failure.

Because discharge rates for the trickle emitters are normally low, the lower capacity and less expensive type of water meter can be used compared to the sprinkler system. Meter operations must also be checked and in this instance two meters are placed in series and if the flow rate of the meters is different by 5%, the low reading meter is usually replaced and repaired.

By knowing the number of emitters per subunit, an average emitter flow rate can be calculated from the water meter readings. Granted this will not give an indication of water distribution uniformity of the unit, it is useful for comparing the system emitter flow rate to the average flow obtained from individual emitters measured in the field. The meter-based emitter flow rate should be within 10% of the averaged individual emitter rate and if this is not the case, the number of emitters sampled in the field must be increased. Experience has shown that trickle irrigation systems will typically have an average discharge rate that is 90 to 95% of the initial design discharge rate. When the average discharge is below 85% of the initial, a thorough evaluation of the system performance is needed.

REFERENCES AND SUPPLEMENTAL READING

Bralts, V.F., 1978. The effects of emitter flow variation on the design of single and dual chamber drip irrigation systems. Unpub. M.S. Thesis, Univ. of Hawaii, Honolulu, Hawaii, 221 pp.

Bralts, V.F., 1983. Hydraulic Design and Field Evaluation of Drip Irrigation Submain Units. Unpub. Ph.D. Dissertation, Michigan State Univ., East Lansing, Michigan, 310 pp.

Bralts, V.F. and Edwards, D.M., 1983. Field evaluation of drip irrigation submain units. ASAE Paper 83-2596, Amer. Soc. Agric. Eng., St. Joseph, Michigan, 33 pp.

Bralts, V.F. and Kesner, C.D., 1982. Drip irrigation field uniformity estimation. Trans. Amer. Soc. Agric. Eng. 26(5): 1369-1374.

Bralts, V.F., Wu, I.P. and Gitlin, H.M., 1981a. Manufacturing variation and drip irrigation uniformity. Trans. Amer. Soc. Agric. Eng. 24(1): 113-119.

Bralts, V.F., Wu, I.P. and Gitlin, H.M., 1981b. Drip irrigation uniformity considering emitter plugging. Trans. Amer. Soc. Agric. Eng. 24(5): 1234-1240.

Bucks, D.A., Nakayama, F.S. and Warrick, A.W., 1982. Principles, practices and potentialities of trickle (drip) irrigation. In Hillel, D. (Ed.) Advances in Irrigation 1: 219-248, Academic Press, New York.

Chesnes, J.L. and Couvillon, G.A., 1980. Peach tree responses to trickle application of water and nutrients. ASAE Paper 80-2079, Amer. Soc. Agric. Eng., St. Joseph, Michigan, 19 pp.

Goldberg, D., Gornat, B., and Rimon, D., 1976. Drip Irrigation: Principles, Design and Agricultural Practices. Drip Irrig. Scientific Publ., Kfar Shmaryahu, Israel, 296 pp.

Howell, T.A. and Hiler, E.A. 1974a., Trickle irrigation lateral line design. Trans. Amer. Soc. Agric. Eng. 17(4): 902-908.

Howell, T.A. and Hiler, E.A., 1974b. Designing trickle irrigation laterals for uniformity. J. Irrig. and Drain. Div., Amer. Soc. Civil Eng. 100(IR4): 442-453.

Karmeli, D. 1977. Classification and flow regime analysis of drippers. J. Agric. Eng. Res. 22: 165-173.

Karmeli, D. and Keller, J., 1975. Trickle Irrigation Design. Rain Bird Sprinkler Manufacturing Corp. Glendora, California, 133 pp.

Keller, J. and Karmeli, D., 1974. Trickle irrigation design parameters. Trans. Amer. Soc. Agric. Eng. 17(4): 678-684.

Merriam, J.L. and Keller, J., 1978. Farm Irrigation System Evaluation: A Guide for Management. Agric. and Irrig. Eng. Dept., Utah State University, Logan, Utah, 271 pp.

Nakayama, F.S., Bucks, D.A. and Clemmens, A.J., 1979. Assessing trickle emitter application uniformity. Trans. Amer. Soc. Agric. Eng. 22(4): 816-821.

Oron, G., Shelef, G., and Zur, B., 1980. Stormwater and reclaimed effluent in trickle irrigation. J. Irrig. and Drain. Div., Amer. Soc. Civil Eng. 106(IR4): 299-310.

Soil Conservation Service, 1983. Trickle Irrigation. Chapter 5, Sec. 15, National Engineering Handbook, U.S. Dept. of Agric., Washington, DC, 129 pp.

Sokal, R.R. and Rohlf, F.J., 1969. Biometry. W.H. Freeman and Company. San Francisco, California, 776 pp.

Solomon, K.H., 1975. Distribution of emission rates from trickle irrigation systems. Unpub. M.A. Thesis, Claremont Graduate School, Claremont, California, 192 pp.

Solomon, K.H., 1976. Emission uniformity--An introduction. Proc. 4th Ann. Int. Drip Irrig. Assoc. Meet., Fresno, California, pp. 56-64.

Solomon, K.H., 1977. Performance comparison of different emitter tubes. Proc. 7th Int. Agric. Plastics Congr., San Diego, California, pp. 97-102.

Solomon, K.H., 1979. Manufacturing variation of emitters in trickle irrigation systems. Trans. Amer. Soc. Agric. Eng. 22(5): 1034-1038.

Solomon, K.H. and Keller, J., 1978. Trickle irrigation uniformity and efficiency. J. Irrig. and Drain. Div., Amer. Soc. Civil Eng. 104(IR3): 293-306.

Wilcox, J.C. and Swailes, G.E., 1974. Uniformity of water distribution by some undertree orchard sprinklers. Scientific Agric. 27: 565-583.

Wu, I.P. and Gitlin, H.M., 1974. Design of Drip Irrigation Lines. Hawaii Agric. Exp. Sta. Tech. Bull. 96. Univ. of Hawaii, Honolulu, Hawaii, 29 pp.

Wu, I.P. and Gitlin, H.M., 1975. Energy gradient line for drip irrigation laterals. J. Irrig. and Drain. Div., Amer. Soc. Civil Eng. 101(IR4): 321-326.

Wu, I.P. and Gitlin, H.M., 1983. Drip irrigation application efficiency and schedules. Trans. Amer. Soc. Agric. Eng. 26(1): 92-99.

Wu, I.P., Howell, T.A. and Hiler, E.A., 1979. Hydraulic Design of Drip Irrigation Systems. Hawaii Agric. Exp. Sta. Tech. Bull. 105. Honolulu, Hawaii, 80 pp.

CHAPTER 4

MANAGEMENT PRINCIPLES

4.1 IRRIGATION SCHEDULING

T. A. HOWELL, D. A. BUCKS, D. A. GOLDHAMER, AND J. M. LIMA

The success of any irrigation method, particularly trickle (drip) irrigation, depends to a large degree on the management of the irrigation system. Management of trickle irrigation systems is unique in many respects. Irrigating by small, frequent quantities is quite different from traditional sprinkler and surface irrigation methods where large, infrequent applications are normally applied. The management strategy changes from an extraction dominance of the soil water balance to one where water infiltration and redistribution are of primary importance. With trickle irrigation, precise information on the amount of water that the crop is using is required to determine adequately the irrigation amount. Control strategies using feedback information on soil water or plant water status can be used to determine if the irrigation applications are either too large or too small. This section will summarize the information regarding estimation of crop water requirements, irrigation scheduling methods, and implementation of effective irrigation scheduling.

4.1.1 *Evapotranspiration concepts*

Irrigation provides the crop with water sufficient for its physiological processes. Photosynthesis is the physiological process for crop growth and occurs only when sunlight, CO_2, and water are present and the leaf stomata are open. In general, when plants are well hydrated, stomata are open during the day. When plant water stress occurs, the stomata partially close thus restricting the loss of water and also the uptake of atmospheric CO_2. The process of water loss from the plant is called transpiration. In general, a linear relationship exists between transpiration and the dry matter accumulation by crops, although differences are known to exist between crop species and even between different environments. The ratio of transpiration to dry matter growth or some other yield parameter is called "transpiration ratio", and the ratio of yield to transpiration is called "water use efficiency". Thus, to achieve a specified yield, enough water to at least meet the transpiration amount for that yield level must be supplied in some manner. Other crop management factors, such as control of pests and diseases, fertility, etc., also directly influence yield and may indirectly affect transpiration.

Besides the water required for transpiration, additional water evaporates directly from plant and soil surfaces. In most trickle irrigation systems, the plant surfaces are not wetted (an exception would be low-volume spray system) so the amount of water evaporating from plant surfaces can be neglected with the exception of intercepted rainfall. However, evaporation from the soil can be significant with trickle irrigation systems. Although only a small portion of the soil surface may actually be wetted, the frequency of the wetting may result in cumulative evaporation losses as large as those commonly associated with sprinkler or surface irrigation.

As with other irrigation methods, trickle irrigations can allow water to move beneath the crop root zone where it becomes largely unavailable to the crop (note that some water can move upward toward the root zone from either wetter soil zones or from water tables). Although this drainage beneath the root zone is considered a loss in terms of the soil water balance, it may be necessary for control of the salinity within the root zone (see chapters 2.4 and 4.4). Where drainage is necessary for "leaching", it becomes an important part of the irrigation requirement. However, where drainage is not required for "leaching", soil water should be managed through irrigation scheduling to minimize drainage. Some amount of soil water storage capacity may be deliberately set aside to store rainfall and to minimize runoff. In areas with shallow water tables, crops can directly extract water from the groundwater to meet transpirational demands depending on its salinity.

Stored soil water from rainfall, groundwater use, and water harvesting from runoff complete the positive nonirrigation inputs to the crop water balance. The "irrigation requirement" is the additional net amount of water required to meet crop water demands plus leaching requirements. Depletion of soil water, drainage below the root zone, runoff from the field, evaporation from the soil surface, and transpiration use by the crop are considered the negative inputs (losses) to the crop water balance. The crop "evapotranspiration" or ET is considered to be the total of the crop water use from transpiration and the water normally evaporated from the soil. Usually, the soil water evaporation is related to the method of irrigation. Although

---

Example 4.1.1

Problem: Tanner (1981) reported that the transpiration of potatoes in Wisconsin during 1976 was about 350 mm, the dry matter yield was about 16 Mg/ha and the tuber yield was about 14 Mg/ha. What is the dry matter "transpiration ratio" and what is the tuber "water use efficiency"? How much transpiration would be expected for a potato tuber yield of 10 Mg/ha?

Solution: The transpiration ratio would be 21.9 mm/(Mg/ha)(350 mm/16 Mg/ha). The tuber water use efficiency would be 0.04 (Mg/ha)/mm (14 [Mg/ha]/350 mm). The transpiration would be 250 mm for a tuber yield of 10 Mg/ha [10 (Mg/ha)/0.04 (Mg/ha)/mm].

---

additional water is stored within the living tissue of the crop, the amount is negligible when expressed on a volume per unit land area basis (depth). Traditional definitions of "consumptive use", however, have included both water stored within the crop and water used in evapotranspiration.

4.1.1.a Climatic parameters

Many climatic variables have either direct or indirect effects on crop water use through their influence on evaporation and transpiration. Solar radiation, temperature, air relative humidity or other thermodynamic properties of water vapor, and wind are the primary meteorological variables influencing evapotranspiration. Other parameters such as barometric pressure can indirectly influence evapotranspiration. The crop can be considered the "integrator" of the growing environment both the atmosphere and soil. In many respects, the crop reacts rather passively to the environment, particularly under "full" irrigation or under adequate rainfall. However, extremes with respect to atmospheric parameters of temperature, wind, drought, etc., can result in dramatic losses in crop production. Irrigation provides one of the most practical and widely used cultural practices for the moderation of atmospheric extremes, particularly drought. Under drought or inadequate irrigation, the crop reacts more actively with the environment. The initial plant response involves reduced expansive growth due to lower cell turgor pressure. As plant water stress continues, partial stomatal closure causes a corresponding reduction in transpiration. This reduction will result in an increase in crop temperature and can induce canopy changes such as wilting, leaf rolling or curling, bud or flower abscission, leaf drop, fruit drop, etc.

Various methods for estimating crop water use from meteorological information have been proposed and are currently used. Many of these methods are reviewed by Jensen (1974), Pair et al. (1983), Doorenbos and Pruitt (1977), and Burman et al. (1980) as they relate to predicting crop irrigation requirement. Most evapotranspiration predictive methods use four factors: (1) an estimate of reference evapotranspiration (ET_r) based on a specific type of crop, commonly grass and alfalfa, using mainly climatic information, (2) a crop factor (K_{cb}) that describes both the dynamic seasonal and developmental change in the crop evapotranspiration in relation to ET_r, (3) a soil factor (K_{cs}) which describes the effect of low soil water content on transpiration and having a close interrelationship to such crop properties such as rooting, stomatal sensitivity, etc., and (4) a soil factor (K_{so}) which describes the increase in evapotranspiration from either recent rainfall or irrigations. Thus, the crop water use is represented by the following equation:

$$ET = ET_r\ [(K_{cb}\ K_{cs}) + K_{so}] \qquad (4.1.1)$$

where ET is the estimated crop water use amount in the same units as ET_r and all the K factors are expressed as unitless values.

One of the basic problems in estimating crop water use under trickle irrigation is the expression of water use in units of depth per unit time. Normally, crop water use is expressed as a flux density [quantity (in this case volume) per unit area per unit time]. The resulting dimension of evapotranspiration flux is depth per unit time. Most trickle irrigation systems are designed, however, to deliver a precise volume per unit time to either a particular plant, several plants or a location in the field. When the ground cover is complete, as with a mature orchard or a fully developed row crop, the area can be considered to be the whole area; thus, the flux equals the volume per unit area per unit time. However, with developing crops the area per plant may not be considered as fully contributing to evapotranspiration. Hence, the K_{cb} factor has to account for both the annual cycle of crop water use and the development cycle for full "effective" cover.

Example 4.1.2

Problem: The evapotranspiration rate of a mature almond orchard is 8 mm/day. The tree spacing is 8 m by 8 m. Considering the orchard to be fully mature and at full development, what is the evapotranspiration flux and what is the volume evapotranspiration rate? If there are four emitters per tree with flow rates of 8 liters/hr, how much time the irrigation application require to fully meet the evapotranspiration?

Solution: The evapotranspiration volume rate is 512 liters/day [(1000 liters/m^3)(64 m^2)(8 mm/day)/1000 mm/m]. The ET flux is 8 liters/m^2 per day [512 liters/day/64 m^2]. The total application rate is 32 liters/hr [4 emitters at 8 liters/hr]; therefore, the irrigation application time is 16 hours/day [(512 liters/day)/(32 liters/hr)].

Many methods are available for estimating the reference evapotranspiration [Et_r]. Four methods will be discussed in this section: (1) modified Blaney-Criddle method [FAO-BC] (Blaney and Criddle, 1950; Doorenbos and Pruitt, 1977); (2) modified Jensen-Haise method [M-JH] (Jensen and Haise, 1963; Jensen, 1974); (3) combination equation method [CE] (Penman, 1948; Van Bavel, 1966; Doorenbos and Pruitt, 1977); and (4) pan evaporation method [PE] (Jensen, 1974; Doorenbos and Pruitt, 1977). Since water use from crops not subjected to water stress and with "full" evaporative covers has been called by many terms in the past such as potential evaporation, potential evapotranspiration, maximum evapotranspiration, equilibrium evapotranspiration, reference evapotranspiration, etc., Burman et al. (1980) proposed that the terminology of "reference" evapotranspiration be uniformly adopted to indicate the expected evapotranspiration from a particular crop (hence the reference) that is "well" supplied with water and "fully" covers the soil. Traditionally, crops such as alfalfa and grass have been widely used world-wide for referencing the water use of various crops to a more universal standard. The reference evapotranspiration then is a calculated (modelled) parameter which in principle represents the expected evapotranspiration from the reference crop for the particular weather parameters that exist at the site in question. All of the above methods of evaluating ET_r are simply estimates of a complex biophysical process and contain considerable empiricism. Each method should be used with some degree of caution until local experience has validated the estimates. Additional guidelines are available in Jensen (1974), Doorenbos and Pruitt (1977) and Burman et al. (1980).

The FAO-BC method of estimating ET_r uses air temperature measured at the site or nearby locale and general information regarding daytime wind, humidity and sunshine from either a climatic atlas or local climatological data. The ET_r is given by

$$ET_r = a + bf \tag{4.1.2}$$

$$f = p(0.46T_a + 8) \tag{4.1.3}$$

where ET_r is in mm/day, p is the percentage of daytime hours in a day compared to the entire year (table 4.1.1) and T_a is the mean monthly air temperature in °C. The coefficients a and b represent the intercept and slope of a linear relationship between ET_r and f. A graphical estimate of ET_r for grass using daytime winds, minimum humidity and sunshine fraction (n/N) adapted from Doorenbos and Pruitt (1977) is given in figure 4.1.1. The daytime wind speed is used here since it is more representative of wind influences on the evapotranspiration process than daily (24 hour) wind speed. The daytime wind may be estimated from the daily wind using the ratio of day to night winds based on general climatic information for the area as follows:

Day/Night Wind Ratio	1.0	1.5	2.0	3.0	3.5	4.0
Daytime/Daily Wind Ratio	1.0	1.5	1.33	1.5	1.56	1.6

The corrected daytime wind speed is the product of the "Daytime/Daily Wind Ratio" for the estimated or recorded "Day/Night Wind Ratio" times the recorded daily (24 hour) wind speed. The minimum relative humidity can be estimated as the ratio of the saturated vapor pressure at the average dew point temperature to the saturated vapor pressure at maximum air temperature. The FAO-BC method of estimating ET_r is best applied to ten-day to one-month periods, and may be of limited use for trickle irrigation.

The M-JH method of estimating ET_r has been referred to as a solar radiation based method. The M-JH has been widely used in the Western United States for computer irrigation scheduling programs. The M-JH estimate of ET_r is given by

$$ET_r = C_t(T_a - T_x)R_s/L \tag{4.1.4}$$

TABLE 4.1.1

Mean daily percentage (p) of annual daytime hours for different latitudes (adapted from Burman et al., 1980).

Latitude North South[a]	Jan Jul	Feb Aug	Mar Sep	Apr Oct	May Nov	Jun Dec	Jul Jan	Aug Feb	Sep Mar	Oct Apr	Nov May	Dec Jun
60 deg	0.15	0.20	0.26	0.32	0.38	0.41	0.40	0.34	0.28	0.22	0.17	0.13
58	0.16	0.21	0.26	0.32	0.37	0.40	0.39	0.34	0.28	0.23	0.18	0.15
56	0.17	0.21	0.26	0.32	0.36	0.39	0.38	0.33	0.28	0.23	0.18	0.16
54	0.18	0.22	0.26	0.31	0.36	0.38	0.37	0.33	0.28	0.23	0.19	0.17
52	0.19	0.22	0.27	0.31	0.35	0.37	0.36	0.33	0.28	0.24	0.20	0.17
50	0.19	0.23	0.27	0.31	0.34	0.36	0.35	0.32	0.28	0.24	0.20	0.18
48	0.20	0.23	0.27	0.31	0.34	0.36	0.35	0.32	0.28	0.24	0.21	0.19
46	0.20	0.23	0.27	0.30	0.34	0.35	0.34	0.32	0.28	0.24	0.21	0.20
44	0.21	0.24	0.27	0.30	0.33	0.35	0.34	0.31	0.28	0.25	0.22	0.21
42	0.21	0.24	0.27	0.30	0.33	0.34	0.33	0.31	0.28	0.25	0.22	0.21
40	0.22	0.24	0.27	0.30	0.32	0.34	0.33	0.31	0.28	0.25	0.22	0.21
35	0.23	0.25	0.27	0.29	0.31	0.32	0.32	0.30	0.28	0.25	0.23	0.22
30	0.24	0.25	0.27	0.29	0.31	0.32	0.31	0.30	0.28	0.26	0.24	0.23
25	0.24	0.26	0.27	0.29	0.30	0.31	0.31	0.29	0.28	0.26	0.25	0.24
20	0.25	0.26	0.27	0.28	0.29	0.30	0.30	0.29	0.28	0.26	0.25	0.24
15	0.26	0.27	0.27	0.28	0.29	0.29	0.29	0.28	0.28	0.27	0.26	0.25
10	0.26	0.27	0.27	0.28	0.28	0.29	0.28	0.28	0.28	0.27	0.27	0.27
5	0.27	0.27	0.27	0.28	0.28	0.28	0.28	0.28	0.28	0.27	0.27	0.27
0	0.27	0.27	0.27	0.27	0.27	0.27	0.27	0.27	0.27	0.27	0.27	0.27

[a] Southern latitudes: apply 6 month difference as shown.

where ET_r is reference evapotranspiration in mm/day, T_a is mean daily temperature in °C, R_s is total daily solar radiation in MJ/m^2 and L is the latent heat of vaporization [2.45 MJ/kg]. Note that R_s/L must be in the same units as ET_r [1 kg/m^2 = 1 mm]. The values of C_t and T_x can be estimated by

$$C_t = 1/(C_1 + 7.3C_h) \quad (4.1.5)$$

$$C_h = 5/(e_2 - e_1) \quad (4.1.6)$$

$$C_1 = 38 - (E_1/152) \text{ [for alfalfa]} \quad (4.1.7a)$$

$$C_1 = 45 - (E_1/137) \text{ [for clipped grass]} \quad (4.1.7b)$$

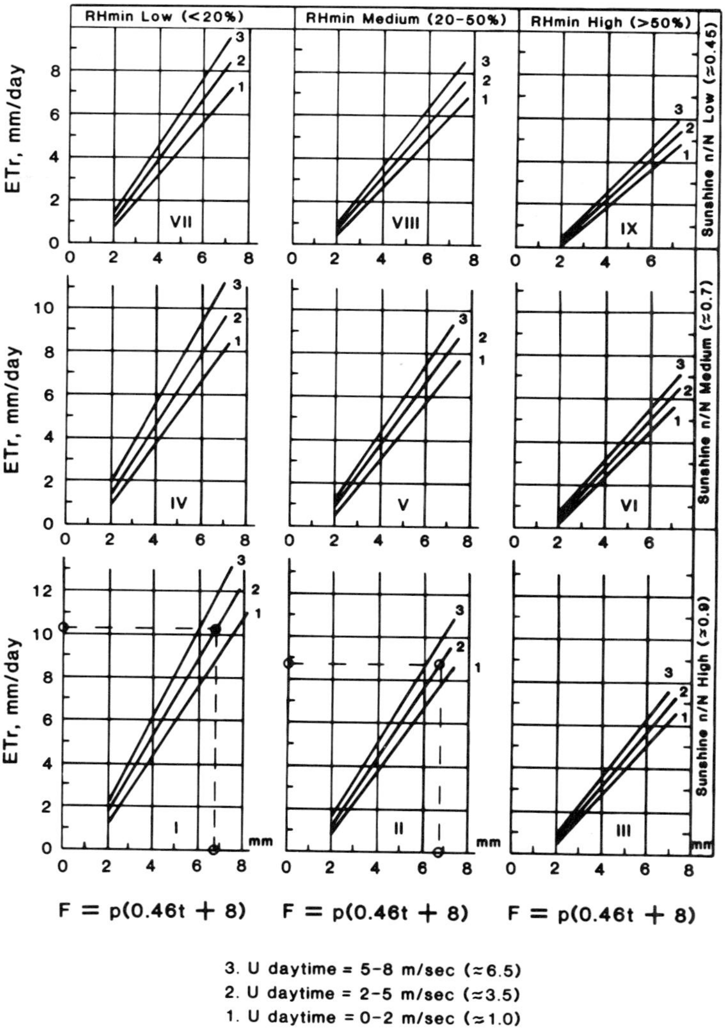

Fig. 4.1.1 Prediction of reference ET_r for grass from Blaney-Criddle f factor for different conditions of minimum relative humidity, sunshine duration and daytime wind (adapted from Doorenbos and Pruitt, 1977).

$$T_x = -2.5 - 1.4\,(e_2 - e_1) - (E_1/550) \qquad (4.1.8)$$

where e_2 is the saturated vapor pressure in kPa at the mean monthly maximum air temperature of the warmest month of the year (from long-term climatological information), e_1 is the saturated vapor pressure in kPa at the mean monthly minimum air temperature of the warmest month of the year and E_1 is the site elevation above sea level in m. The coefficients C_t, C_h, C_1, and T_x are computed once and equation 4.1.4 is used with daily values of R_s and T_a to estimate ET_r. The M-JH method is recommended for periods of 5 days to one month.

The combination equation was first suggested by Penman (1948) and represents the combination of the energy balance with the aerodynamic equation. Van Bavel (1966) discussed the critical assumptions of the combination method and illustrated a more theoretical relationship for the aerodynamic term. However, subsequent research has shown that the combination equation probably requires some "local" calibration, particularly for the aerodynamic "wind function" and in estimating the net radiation. Nevertheless, the CE method has been shown to be widely applicable and accurate over a wide range of climatic environments.

The combination equation (CE) can be presented as

$$ET_r = [\Delta/(\Delta + \gamma)][(R_n + G)/L] + [\gamma/(\Delta + \gamma)]f(U)(e_s - e_a) \qquad (4.1.9)$$

where ET_r is the reference evapotranspiration in mm/day, Δ is the slope of the saturated vapor pressure-temperature relationship $[de_s/dT_a]$ in kPa/C, γ is the psychrometric constant in kPa/C, R_n is the daily net radiation in MJ/m^2, G is the soil heat flux towards the surface in MJ/m^2, L is the latent heat of vaporization in MJ/kg, f(U) is the "wind function" in mm/day per kPa, U is the mean daily wind speed in m/s, and $(e_s - e_a)$ is the mean daily vapor pressure deficit in kPa. The relationship between $[\Delta/(\Delta + \gamma)]$ and $[\gamma/(\Delta + \gamma)]$ and elevation and temperature is given in table 4.1.2. The saturated vapor pressure-temperature curve can be represented by the following

$$e_s = 0.1 \exp[18.7209 - 3806/(T_a+273.1) - 222153/(T_a+273.1)^2] \qquad (4.1.10)$$

where e_s is the saturated vapor pressure in kPa and T_a is the temperature in °C. Then Δ can be given as

$$\Delta = e_s\,[3806/(T_a+273.1)^2 + 444306/(T_a+273.1)^3] \qquad (4.1.11)$$

The psychrometric constant can be represented by

$$\gamma = [(P_b/L)/625] \qquad (4.1.12)$$

where P_b is the barometric pressure in kPa, which can be estimated using a straight line approximation for the United States standard atmosphere as

$$P_b = 101.3 - (E_1/95) \quad (4.1.13)$$

where E_1 is the site elevation above sea level in m. The latent heat of vaporization can be estimated by

$$L = 2.501 - (T_a/423) \quad (4.1.14)$$

TABLE 4.1.2

Relation of $\Delta/(\Delta + \gamma)$ to elevation and temperature (adapted from Burman et al., 1980).

Air temp. °C	Elevation, m					
	0	500	1,000	1,500	2,000	2,500
0.0	0.401	0.414	0.428	0.433	0.458	0.475
5.0	0.477	0.491	0.505	0.520	0.536	0.552
10.0	0.551	0.564	0.578	0.593	0.608	0.624
15.0	0.620	0.632	0.645	0.659	0.673	0.688
20.0	0.681	0.693	0.705	0.717	0.730	0.743
25.0	0.735	0.745	0.756	0.767	0.778	0.790
30.0	0.781	0.790	0.799	0.809	0.818	0.828
35.0	0.820	0.828	0.835	0.844	0.852	0.860
40.0	0.852	0.858	0.867	0.872	0.879	0.886
45.0	0.878	0.884	0.889	0.895	0.901	0.907
50.0	0.900	0.904	0.909	0.914	0.919	0.924

Note that $\gamma/(\Delta + \gamma) = [1 - \Delta/(\Delta + \gamma)]$

Various methods are used to estimate the daily mean vapor pressure deficit. The value of $(e_s - e_a)$ could be approximated by taking e_s at the mean air temperature (mean of T_{max} and T_{min}) and e_a as the saturated vapor pressure at the mean daily dew point temperature (Doorenbos and Pruitt, 1977). However, in more arid areas and higher elevations where large diurnal temperature changes occur, e_s may be estimated as

$$e_s = [e_s(T_{max}) + e_s(T_{min})]/2 \quad (4.1.15)$$

where $e_s(T_{max})$ is the saturated vapor pressure at the maximum daily air temperature and $e_s(T_{min})$ is the saturated vapor pressure at the minimum daily air temperature.

The net radiation can be estimated following the procedures of Hill et al. (1983) as

$$R_n = [(1 - \alpha)R_s] - R_{bo}\,[a_1(R_s/R_{so}) + a_2] \qquad (4.1.16)$$

where R_n is the daily net radiation in MJ/m^2, α is the albedo (short-wave reflection coefficient) [0.20 to 0.23 for most "full" cover green crops], R_s is the daily solar radiation in MJ/m^2, and a_1 and a_2 are empirical coefficients. The "clear day" solar radiation curve can be developed from measurements from several years at a location. For example, Howell et al. (1983) presented the following equation for Fresno, California

$$R_{so} = 32.2\,\exp\{-[(\mathrm{Day} - 172)/167]^2\} \qquad (4.1.17)$$

where Day is the day of the year. The net outgoing long-wave daily radiation on a "clear" day can be estimated as

$$R_{bo} = \varepsilon\,(4.9\mathrm{x}10^{-9})\,(T_a + 273.1)^4 \qquad (4.1.18)$$

where ε is the emissivity which can be estimated as

$$\varepsilon = [a_3 + a_4\,(e_a)^{0.5}] \qquad (4.1.19)$$

where a_3 and a_4 are empirical coefficients or ε can be estimated by a simpler relationship (Idso and Jackson, 1969) based solely on air temperature as

$$\varepsilon = -0.02 + 0.261\,\exp[-7.77\mathrm{x}10^{-4}\,(T_a)^2] \qquad (4.1.20)$$

Table 4.1.3 lists a summary of several of the radiation equation coefficients (Jensen, 1974). In general practice, the soil heat flux (G) has been neglected. Over a one day time period, the gain or loss of heat from the soil is usually less than 5% of R_s.

TABLE 4.1.3

Coefficients for estimating radiation parameters (adapted from Jensen, 1974).

Region	(a_1,	a_2)	(a_3,	a_4)
Davis, California	(1.35,	-0.35)	(0.35,	-0.046)
Southern Idaho	(1.22,	-0.18)	(0.325,	-0.044)
England	(not available)		(0.47,	-0.065)
England	(not available)		(0.44,	-0.080)
Australia	(not available)		(0.35,	-0.042)
General	(1.2,	-0.2)	(0.39,	-0.05)

The wind function [f(U)] has been determined from empirical studies conducted mainly with precision weighing lysimeters. However, the wind function is estimated with all the compound errors associated with the estimation of net radiation and the mean daily saturated vapor pressure deficit. Consequently, a wide diversity has developed in the interpretation of the various wind functions. The difference between reference crops should only be in the net radiation (albedo and emissivity) and the wind function (aerodynamic properties of the crop). Table 4.1.4 presents several of the estimated wind functions for the combination equation for various reference crops based on standard anemometer height of 2 m. Linear functions have, generally, been used to represent the wind function; however, power functions have occasionally been used. In figure 4.1.2, a procedure is given for correcting the estimated ET_r of grass for daytime to nighttime wind [average daytime wind (U_d)/average nighttime wind (U_n)], solar radiation, and maximum daily relative humidity when using the wind function [f(U) = 2.70 + 2.33 (U)]. The method used to estimate (e_s - e_a) will also influence this correction.

TABLE 4.1.4

Empirical wind functions for the combination equation.

Source	Reference crop	Wind function f(U) in mm/day per kPa [a + b(U)] Intercept	Slope
Penman (1948)	short grass	2.63	1.38
Penman (1963)	clipped grass	2.63	1.45
Wright and Jensen (1972)	alfalfa	1.97	2.61
Kincaid and Heermann (1974)	alfalfa	2.89	2.40
Doorenbos and Pruitt (1977)	grass	2.70	2.33
Stigter (1980)		3.70	2.00
Van Bavel (1966)	various	0.00	$105.6/[\ln(z/z_o)]^2$

z is anemometer height in m and z_o is a roughness length in m.

The CE method has been widely used to estimate ET_r over time periods from one day to one month. This method has been found to be highly reliable when used with good weather information and sound judgements in the selection of the various coefficients. The CE method does require the most detailed calculations and the most direct weather information. However, neither requirement should be critical with modern microcomputers which can both acquire the data and perform the calculations. Automated weather data collection system which can

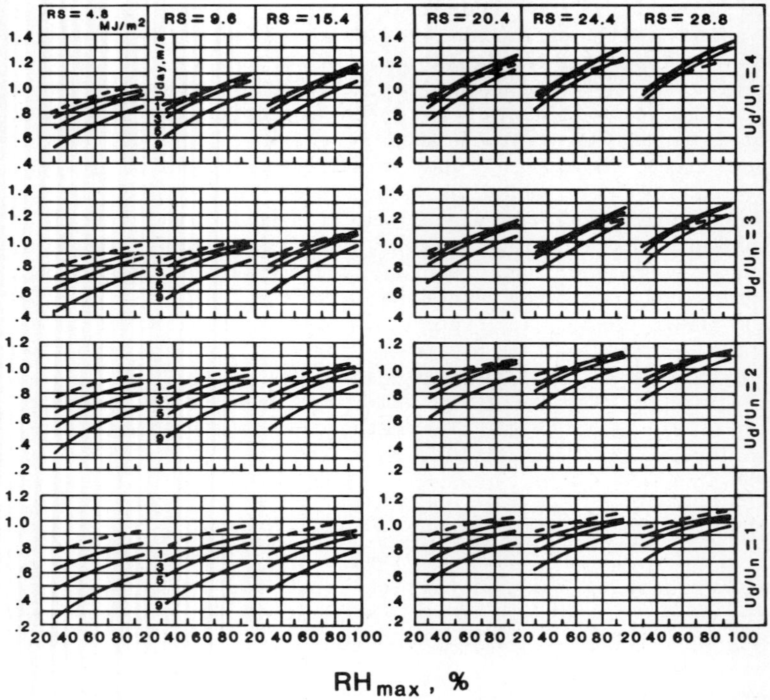

Fig. 4.1.2 Adjustment factor for ET_r of grass using the wind function of Doorenbos and Pruitt (1977) in the combination equation to correct for daytime to nighttime (U_d/U_n) wind, solar radiation [note that R_s in MJ/m^2 can be converted to mm/day by dividing MJ/m^2 by L, the latent heat of vaporization in MJ/kg (approximately 2.45 MJ/kg)] and maximum daily relative humidity (from personal communication with W. O. Pruitt, 1980).

be integrated with the calculation of reference evapotranspiration has been developed (Howell et al., 1984).

Evaporation pans of various designs have been widely used throughout the world as an index of reference evapotranspiration and crop water use. Even though the evaporation of water from a pan only vaguely resembles the biophysical process of evapotranspiration from crops, water loss from pans has been found to be correlated to the crop water use. The PE method for estimating ET_r uses a "pan coefficient" to adjust the measured evaporation from a specific type of evaporation pan as follows

$$ET_r = K_pE_p \tag{4.1.21}$$

where K_p is the pan coefficient and E_p is the pan evaporation in mm/day. The K_p factor will depend on the surface conditions around the pan (e.g., either bare soil, dry grass, irrigated grass or crop), the upwind fetch, and the crop used for reference (alfalfa, grass, etc.). Guidelines are provided for correcting the pan coefficient for local conditions of pan exposure, mean daily relative humidity, and daily wind run for grass water use (Doorenbos and Pruitt, 1977).

The most widely used evaporation pan is the United States Class A Pan (U. S. Weather Bureau, now National Weather Service, NOAA) which is 1.21 m in diameter, 0.255 m in depth, and constructed of either galvanized steel or Monel alloy metal. The pan is placed on a wooden platform about 0.15 m above the ground. The water level is maintained within a range of 50 to 75 mm from the rim of the pan and should not fluctuate more than 50 mm. Evaporation is determined by the change in water elevation measured with a vernier hook gage placed in a stilling well in the evaporation pan. The reading schedule is determined by the evaporation rate and the requirement to maintain the specific water level range. Daily readings of the water elevation are normally taken although intermittent readings, twice weekly during summer and weekly during winter, can be used. The pan location should be free from the influence of animals and birds if possible. Wire screen covers are sometimes used to prevent water consumption by wildlife; however, screens can affect the evaporation loss rate.

Example 4.1.3

Problem: Estimate the ET_r for grass at Fresno, California, for day 177 of the year (June 25) by the "uncorrected" combination equation of Doorenbos and Pruitt (1977), the M-JH, and FAO-BC equations. The following weather information is available:

R_s = 30 MJ/m^2 T_{max} = 38.0°C T_{min} = 18.0°C   U = 2.7 m/sec

T_{dew} = 10.0°C E_l = 80 m   Latitude = 36 deg North

The grass albedo (α) can be assumed to be 0.23, the monthly mean maximum temperature of the warmest month at Fresno is 36°C and the mean monthly minimum temperature of the warmest month at Fresno is 14°C.

Solution:

<u>Combination equation</u>

$$T_a = [(38 + 18)/2] = 28°C$$

$L = [2.501 - (28/423)] = 2.44\ MJ/m^2$

$e_s = 0.1 \exp[18.7209 - 3806/(28+273.1) - 222153/(28+273.1)^2]$
$= 3.77$ kPa

$e_a = 0.1 \exp[18.7209 - 3806/(10+273.1) - 222153/(10+273.1)^2]$
$= 1.22$ kPa

$(e_s - e_a) = [3.77 - 1.22] = 2.55$ kPa

$\Delta = 3.77[3706/(28+273.1)^2 + 444306/(28+273.1)^3] = 0.22$ kPa/C

$P_b = [101.3 - (80/95)] = 100.5$ kPa

$\gamma = [(100.5/2.44)/625] = 0.0659$ kPa/C

$\Delta/(\Delta + \gamma) = [0.22/(0.22 + 0.0659)] = 0.769$

$\gamma/(\Delta + \gamma) = [1 - 0.769] = 0.231$

$\varepsilon = [0.35 - 0.046(1.22)^{0.5}] = 0.299$ (assumed Davis, California, radiation parameters)

$R_{bo} = [0.299(4.9\times10^{-9})(28 + 273.1)^4] = 12.1\ MJ/m^2$

$R_{so} = [32.2 \exp\{-[(177 - 172)/167]^2\}] = 32.2\ MJ/m^2$
(assumed clear day curve from Fresno, California)

$R_n = [(1 - 0.23)30] - 12.1\ [1.35(30/32.2) - 0.35]$
$= 12.1\ MJ/m^2$ (assumed Davis, California, radiation parameters)

$f(U) = [2.70 + 2.33(2.7)] = 9.02$ mm/day per kPa

$ET_r(CE) = [0.769(12.1/2.44)] + [0.231(9.02)(2.55)]$
$= 9.1$ mm/day

Modified Jensen-Haise equation

$e_2 = 0.1 \exp[18.7209 - 3806/(36+273.1) - 222153/(36+273.1)^2]$
$= 5.93$ kPa

$e_1 = 0.1 \exp[18.7209 - 3806/(14+273.1) - 222153/(14+273.1)^2]$
$= 1.59$ kPa

$C_h = [5/(5.93 - 1.59)] = 1.15$

$C_1 = [45 - (80/137)] = 44.4$

TABLE 4.1.6

Crop coefficient (K_{cb}) for full grown deciduous fruit and nut trees (adapted from Doorenbos and Pruitt, 1977, for grass ET_r).

	With ground cover crop[a]									Without ground cover crop[b] (clean cultivated, weed free)								
	Mar	Apr	May	Jun	Jul	Aug	Sep	Oct	Nov	Mar	Apr	May	Jun	Jul	Aug	Sep	Oct	Nov
COLD WINTER WITH KILLING FROST: GROUND COVER STARTING IN APRIL																		
Apple, cherry																		
humid, light to mod. wind	-	.5	.75	1.0	1.1	1.1	1.1	.85	-	-	.45	.55	.75	.85	.85	.8	.6	-
humid, strong wind	-	.5	.75	1.1	1.2	1.2	1.15	.9	-	-	.45	.55	.8	.9	.9	.85	.65	-
dry, light to mod. wind	-	.45	.85	1.15	1.25	1.25	1.2	.95	-	-	.4	.6	.85	1.0	1.0	.95	.7	-
dry, strong wind	-	.45	.85	1.2	1.35	1.35	1.25	1.0	-	-	.4	.65	.9	1.05	1.05	1.0	.75	-
Peach, apricot, pear, plum																		
humid, light to mod. wind	-	.5	.7	.9	1.0	1.0	.95	.75	-	-	.45	.5	.65	.75	.75	.7	.55	-
humid, strong wind	-	.5	.7	1.0	1.05	1.1	1.0	.8	-	-	.45	.55	.7	.8	.8	.75	.6	-
dry, light to mod. wind	-	.45	.8	1.05	1.15	1.15	1.1	.85	-	-	.4	.55	.75	.9	.9	.7	.65	-
dry, strong wind	-	.45	.8	1.1	1.2	1.2	1.15	.9	-	-	.4	.6	.8	.95	.95	.9	.65	-
COLD WINTER WITH LIGHT FROST: NO DORMANCY IN GRASS COVER CROPS																		
Apple, cherry, walnut[c]																		
humid, light to mod. wind	.8	.9	1.0	1.1	1.1	1.1	1.05	.85	.8	.6	.7	.8	.85	.85	.8	.8	.75	.65
humid, strong wind	.8	.95	1.1	1.15	1.2	1.2	1.15	.9	.8	.6	.75	.85	.9	.9	.85	.8	.8	.7
dry, light to mod. wind	.85	1.0	1.15	1.25	1.25	1.25	1.2	.95	.85	.5	.75	.95	1.0	1.0	.95	.9	.85	.7
dry, strong wind	.85	1.05	1.2	1.35	1.35	1.35	1.25	1.0	.85	.5	.8	1.0	1.05	1.05	1.0	.95	.9	.75
Peach, apricot, pear, plum, almond, pecan																		
humid, light to mod. wind	.8	.85	.9	1.0	1.0	1.0	.95	.8	.8	.55	.7	.75	.8	.8	.7	.7	.65	.55
humid, strong wind	.8	.9	.95	1.0	1.1	1.1	1.0	.85	.8	.55	.7	.75	.8	.8	.8	.75	.7	.6
dry, light to mod. wind	.85	.95	1.05	1.15	1.15	1.15	1.1	.9	.85	.5	.7	.85	.9	.9	.9	.8	.75	.65
dry, strong wind	.85	1.0	1.1	1.2	1.2	1.2	1.15	.95	.85	.5	.75	.9	.95	.95	.95	.85	.8	.7

[a] For young orchards with tree ground cover of 20 and 50%, reduce mid-season K_{cb} values by 10 to 15% and 5 to 10%, respectively.

[b] For young orchards with tree ground cover of 20 and 50%, reduce mid-season K_{cb} values by 25 to 35% and 10 to 15%, respectively.

[c] For walnut Mar-May possibly 10 to 20% lower values due to slower leaf growth.

Table 4.1.5 presents citrus crop coefficients for predominately dry areas with light to moderate wind with 70, 50 and 20% ground cover. Table 4.1.6 lists crop coefficients of deciduous fruit and nut trees. Values are given both for clean cultivation and for cover crop conditions assuming full canopy development of at least 70% ground cover. Table 4.1.7 presents crop coefficients for grape vines for three types of growing climates.

TABLE 4.1.7

Crop coefficients (K_{cb}) for grape with clean cultivation, infrequent irrigation and dry soil surface most of the season (adapted from Doorenbos and Pruitt, 1977, for grass ET_r).

Conditions[a]	Mar	Apr	May	Jun	Jul	Aug	Sep	Oct	Nov
Mature grape grown in areas of killing frost; initial leaves early May, harvest mid-September; ground cover 40-50% at mid season.									
1	——	——	0.50	0.65	0.75	0.80	0.75	0.65	——
2	——	——	0.50	0.70	0.80	0.85	0.80	0.70	——
3	——	——	0.45	0.70	0.85	0.90	0.80	0.70	——
4	——	——	0.50	0.75	0.90	0.95	0.90	0.75	——
Mature grape in areas of only light frost; initial leaves early April, harvest late August to early September; ground cover 30-35% at mid-season.									
1	——	0.50	0.55	0.60	0.60	0.60	0.60	0.50	0.40
2	——	0.50	0.55	0.65	0.65	0.65	0.65	0.55	0.40
3	——	0.45	0.60	0.70	0.70	0.70	0.70	0.60	0.35
4	——	0.45	0.65	0.75	0.75	0.75	0.75	0.65	0.35
Mature grape grown in hot dry areas; initial leaves late February to early March, harvest late half of July; ground cover 30-35% at mid-season.									
3	0.25	0.45	0.60	0.70	0.70	0.65	0.55	0.45	0.35
4	0.25	0.45	0.65	0.75	0.75	0.70	0.55	0.45	0.35

[a] Conditions: 1 humid, light to moderate wind; 2 humid, strong wind; 3 dry, light to moderate wind; 4 dry, strong wind.

The K_{cb} values for the last two growing conditions in table 4.1.7 must be reduced when ground cover is less than 35%. Sugarcane crop coefficient values are presented in table 4.1.8. Since sugarcane cultural and cultivar variations result in different growing seasonal lengths, large differences can exist in the crop coefficients. Total growing season length will depend on the climate and on whether the crop is virgin or ratoon. The 12-month crop in table 4.1.8 refers to a ratoon crop and the 24-month crop refers to a virgin crop.

Crop coefficient (K_{cb}) values for several types of row crops are presented in table 4.1.9. Two values of K_{cb} are presented for mid-season and for harvest or maturity, respectively, for both high, low minimum relative humidity and both high, low wind conditions. The seasonal pattern or curve of the crop coefficient can be developed with these two points and one value for the initial stage from planting and germination to early growth. Basically, this K_{cb} value represents evaporation from a dry soil. In most cases where rainfall is infrequent (recurrence interval of significant amounts is greater than 10 days to two weeks), K_{cb} will be about 0.25 and decrease slightly as the value of ET_r increases. The three values of K_{cb} can then be plotted with the appropriate time to reach the midseason and maturation values. Figure 4.1.3 shows an example crop coefficient curve developed using this procedure for cotton growing near Fresno, California. A more detailed discussion of these crop coefficients is found in Doorenbos and Pruitt (1977) and Burman et al. (1980). These examples are presented to illustrate some values of K_{cb} and should be used in practice only after careful field validation.

The soil water stress coefficient (K_{cs}) illustrates the effect of limited soil water on crop transpiration. Since the effects of

TABLE 4.1.8

Crop coefficients (K_{cb}) for sugarcane (adapted from Doorenbos and Pruitt, 1977, for grass ET_r).

Crop age months			RH_{min} > 70%		RH_{min} < 20%	
			light to moderate wind	strong wind	light to moderate wind	strong wind
12	24	Growth stages				
0-1	0-2.5	planting to 1/4 canopy	0.55	0.60	0.40	0.45
1-2	2.5-3.5	1/4 to 1/2 canopy	0.80	0.85	0.75	0.80
2-2.5	3.5-4.5	1/2 to 3/4 canopy	0.90	0.95	0.95	1.00
2.5-4	4.5-6	3/4 to full canopy	1.00	1.10	1.10	1.20
4-10	6-17	peak use	1.05	1.15	1.25	1.30
10-11	17-22	early senescence	0.80	0.85	0.95	1.05
11-12	22-24	ripening	0.60	0.65	0.70	0.75

TABLE 4.1.9

Crop coefficients (K_{cb}) for field and vegetable crops for mid-season and maturity (harvest) stages of growth under prevailing climatic conditions and estimated time for various stages of development (adapted from Doorenbos and Pruitt, 1977, for grass ET_r).

Crop	Growth[a] stage	Climatic conditions: $RH_{min} > 70\%$, 0-5 m/sec	$RH_{min} > 70\%$, 5-8 m/sec	$RH_{min} < 20\%$, 0-5 m/sec	$RH_{min} < 20\%$, 5-8 m/sec
Sweet Corn (spring)	3	1.05	1.10	1.15	1.20
20/30/35/10[b]	4	0.95	1.00	1.05	1.10
Cotton (late spring)	3	1.05	1.15	1.20	1.25
30/50/60/50	4	0.65	0.65	0.65	0.70
Crucifer (late winter)	3	0.95	1.00	1.05	1.10
25/35/25/10	4	0.80	0.85	0.90	0.95
Cucumber (spring)	3	0.90	0.90	0.95	1.00
25/35/50/20	4	0.70	0.70	0.75	0.80
Lettuce (late winter)	3	0.95	0.95	1.00	1.05
25/35/30/10	4	0.90	0.90	0.90	1.00
Melon (spring)	3	0.95	0.95	1.00	1.05
25/35/40/20	4	0.65	0.65	0.75	0.75
Onion [dry] (spring)	3	0.95	0.95	1.05	1.10
15/25/70/40	4	0.75	0.75	0.80	0.85
Pepper [fresh](spring)	3	0.95	1.00	1.05	1.10
30/35/40/20	4	0.80	0.85	0.85	0.90
Squash (spring)	3	0.90	0.90	0.95	1.00
20/30/30/15	4	0.70	0.70	0.75	0.80
Tomato (spring)	3	1.05	1.10	1.20	1.25
30/40/45/30	4	0.60	0.60	0.65	0.65

a Stages: 3 mid-season; 4 harvest or maturity.

b Days for initial development, crop development, mid-season and late season crop development. Total days in each stage represent the growing season length.

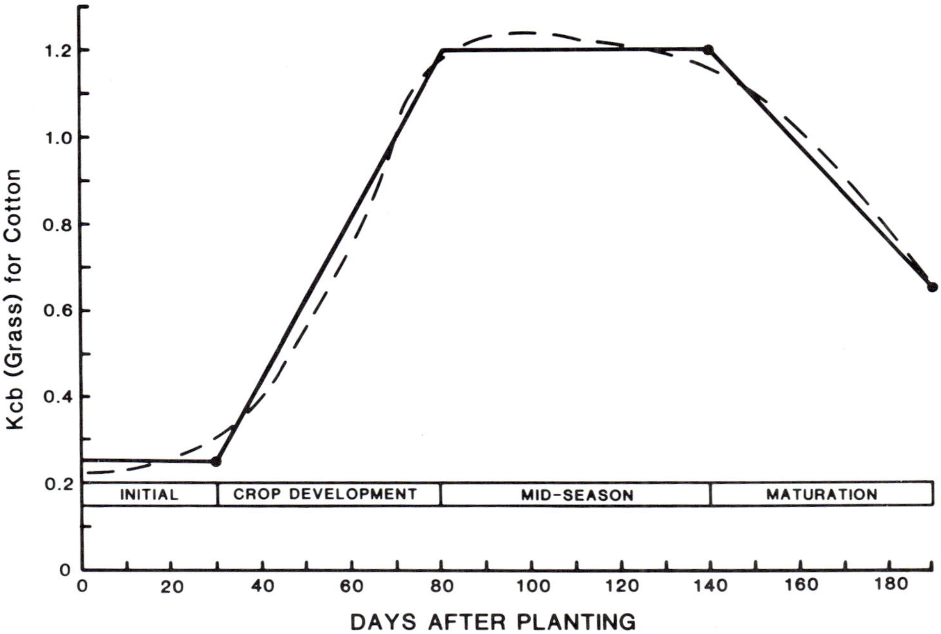

Fig. 4.1.3 Example crop coefficient curve developed for cotton using table 4.1.9.

limited soil water are complex and in many cases crop and soil specific, the relationship of transpiration to soil water has been widely debated; however a unifying theory has not been universally accepted. Howell et al. (1979) reviewed many of the relationships. Jensen et al. (1971) proposed that the relationship could be represented by

$$K_{cs} = [\ln(A_w + 1)]/[\ln(101)] \qquad (4.1.22)$$

where A_w is the percentage of available soil water with 100 when the soil is at field capacity and 0 when completely dry. This function indicates little transpirational reduction until the available soil water percentage decreases below 75% and no major reduction until about 50% of the soil water has been depleted. The prediction of available soil water is complex and difficult. In general, most trickle irrigation systems are operated so that A_w exceeds about 75% through most of the active crop growing stages.

The soil evaporation coefficient K_{so} can be estimated by

$$K_{so} = A_f(1 - K_{cb})[(1/N)^t] \quad (4.1.23)$$

where t is time in days since a rain or irrigation, A_f is the fraction of the soil surface wetted, and N is a soil textural factor which varies approximately from 1.0 to 1.5 for clays and clay loams, 2.0 to 2.5 for loams, and 3.0 to 3.5 for sands.

Pan coefficients (K_p) can be used to estimate ET_r from measurements of evaporation from various types of pans as discussed in 4.1.1.a. However, pan siting, maintenance, and climate may influence the pan coefficient. Evaporation from both open Class A pans and "screened" Class A pans in a semiarid climate is indicative of the ET_r as estimated by the combination equation calculation using the Doorenbos and Pruitt (1977) method, particularly if averaging periods of 5 to 7 days are used (Howell et al., 1983). The K_p for an "open" Class A pan is 0.81 and 0.91 for the "screened" Class A pan. Similar relations are shown in a more humid environment (Campbell and Phene, 1976). The chicken wire, 50 mm mesh screen cover used in both instances reduces the evaporation by about 10% from the standard Class A pans. Table 4.1.10 presents some pan coefficients (K_p) for Class A pans referenced to grass for two types of pan groundcover, mean relative humidity, and mean daily wind speed.

Example 4.1.4

Problem: Determine the water use of trickle irrigated cotton growing near Frenso, California, in early June (about 60 days after planting) when the average ET_r (grass) is 8 mm/day. Figure 4.1.3 can be used to estimate the K_{cb} factor. Assume that the available water percentage is 80% and that N is 1.5. How much would evapotranspiration increase over 3 days if a significant rain (>30 mm) occurred? How much would evapotranspiration increase over 3 days if a 30 mm trickle irrigation application was applied which wet 50% of the soil surface?

Solution: K_{cb} = 0.8 [figure 4.1.3]

K_{cs} = [ln(80 + 1)]/ln(101)] = 0.95

ET = {[(0.8) (0.95) + 0] (8.0)} = 6.1 mm/day

Rainfall: K_{so} (day 1) = $[(1.0)\ (1.0 - 0.8)\ (1/1.5)^1]$ = 0.13

K_{so} (day 2) = $[(1.0)\ (1.0 - 0.8)\ (1/1.5)^2]$ = 0.09

K_{so} (day 3) = $[(1.0)\ (1.0 - 0.8)\ (1/1.5)^3]$ = 0.06

ET (day 1) = {[(0.8) (0.95) + 0.13] (8.0)} = 7.1 mm/day

ET (day 2) = {[(0.8) (0.95) + 0.09] (8.0)} = 6.8 mm/day

ET (day 3) = {[(0.8) (0.95) + 0.06] (8.0)} = 6.6 mm/day

The rain increased the ET by 2.2 mm for the three days.

Irrigation: K_{so} (day 1) = [(0.5) (1.0 - 0.8) $(1/1.5)^1$] = 0.07
K_{so} (day 2) = [(0.5) (1.0 - 0.8) $(1/1.5)^2$] = 0.05
K_{so} (day 3) = [(0.5) (1.0 - 0.8) $(1/1.5)^3$] = 0.03
ET (day 1) = {[(0.8) (0.95) + 0.07] (8.0)} = 6.6 mm/day
ET (day 2) = {[(0.8) (0.95) + 0.05] (8.0)} = 6.5 mm/day
ET (day 3) = {[(0.8) (0.95) + 0.03] (8.0)} = 6.3 mm/day

The irrigation increased the evapotranspiration by 1.1 mm for the 3 days.

TABLE 4.1.10

Pan coefficients for Class A pans referenced to grass ET_r (adapted from Doorenbos and Pruitt, 1977).

Daily mean wind	Windward side distance of green crop or dry fallow	Case A: Pan placed in short green cropped area			Case B: Pan placed in dry fallow area		
		RH_{mean} %			RH_{mean} %		
(m/sec)	(m)	< 40	40-70	> 70	< 40	40-70	> 70
Light	10	0.65	0.75	0.85	0.60	0.70	0.80
< 2	100	0.70	0.80	0.85	0.55	0.65	0.75
	1000	0.75	0.85	0.85	0.50	0.60	0.70
Moderate	10	0.60	0.70	0.75	0.55	0.65	0.70
2-5	100	0.65	0.75	0.80	0.50	0.60	0.65
	1000	0.70	0.80	0.80	0.45	0.55	0.60
Strong	10	0.55	0.60	0.65	0.50	0.55	0.65
5-8	100	0.60	0.65	0.70	0.45	0.50	0.60
	1000	0.65	0.70	0.75	0.40	0.45	0.55
Very strong	10	0.45	0.55	0.60	0.45	0.50	0.55
> 8	100	0.50	0.60	0.65	0.40	0.45	0.55
	1000	0.55	0.60	0.65	0.35	0.40	0.45

4.1.1.c Soil water balance parameters

The soil water balance represents the integrated amount of water in the soil at a particular time. The integration of the field soil water content is complex under trickle irrigation culture because the water application pattern and water extraction pattern by the crop are multidimensional. Usually, the infiltration and extraction patterns under sprinkler, flood, and furrow irrigation can be closely approximated in one dimension. However, trickle irrigation infiltration and extraction patterns will be at least two dimensional and probably three dimensional. The integrated one-dimensional soil water balance can be given by

$$P + I - Q - D - E - T \pm S = 0 \tag{4.1.24}$$

where P is precipitation, I is irrigation, Q is runoff (normally considered to be a loss but could be a gain in a water-harvesting system), D is deep percolation or drainage (normally considered to be a loss but could be a gain from upward flow from groundwater), E is direct evaporation from soil and plant surfaces (dew or condensation would represent a gain if it came from the atmosphere), T is transpiration, and S is the change in the soil water content (positive for water depletion and negative for increases) with all terms expressed as volumes per unit land area (depths). The difficulty normally encountered in the application of equation 4.1.24 to trickle irrigation is that I, E, T and S are difficult to measure due to their multidimensional characteristics. As an example, Ben-Asher (1979) explored just the problems in measuring soil water content under trickle irrigation and found that many soil samples must be measured to estimate accurately S. The drainage (D) should be nearly one-dimensional near to the bottom of the root zone. Since runoff (Q) is difficult to predict, net precipitation (infiltration) or "effective precipitation" is often estimated from rainfall measurements considering such factors as soil water content, rainfall intensity and duration, land cover, land slope, etc. The drainage component must include the "leaching requirement" where salinity control is required.

4.1.1.d Irrigation water requirement

The irrigation water requirement is estimated from equation 4.1.24 by determining the effective precipitation, net drainage which includes groundwater contribution and leaching requirement and total crop water use. The irrigation requirement can be expressed as

$$I_r = D + ET - (P - Q) \pm S \tag{4.1.25}$$

where I_r is irrigation requirement in mm. If no leaching is required, D should be zero. Normally for permanent crops, S will be small (either positive or negative) since the annual water balance must be maintained near zero for the total of all components. For annual crops, S represents the amount of water in the soil at planting, and P represents the amount of rainfall received only during the growing season. In this case, S should be near full depletion by the end of the season allowing time for recharge during winter or off-season. In many arid climatic regimes, the irrigation requirement is simply the seasonal evapotranspiration less effective rain and any leaching. However, in more humid to semiarid climatic regimes the irrigation requirement is highly dependent on the seasonal rainfall and is more difficult to predict.

The irrigation requirement has two important components: (1) the peak irrigation water requirement, and (2) the seasonal total irrigation water requirement. The peak irrigation water requirement represents the irrigation rate required to meet the crop evapotranspiration rate over a time period usually selected as several days (3 or 4 days to 1 or 2 weeks). The largest expected crop evapotranspiration rate for a daily period would not exceed about 13 mm/day even under extreme advection. Consequently, the mean evapotranspiration rate for a longer time period of several days would be somewhat less and seldom exceed about 10 mm/day. In addition, the soil water reserves, under most conditions, should buffer these peak daily evapotranspiration rates such that the peak irrigation water requirement rate will seldom need to be larger than 8 to 10 mm/day (0.3 to 0.4 in/day). This peak irrigation water requirement rate largely determines the necessary water supply rate (actually the net irrigation supply rate). The total seasonal irrigation water requirement determines the depth of water needed to meet the seasonal crop water needs. The product of the irrigation requirement, I_r, and the area of irrigation represents the volume of irrigation water required. Since in many situations the irrigation supply rate and irrigation volume can be either fixed or constrained, the timely delivery of irrigation water can become complex. In some cases, the seasonal irrigation volume is constrained institutionally by water rights or other legal limitations, whereas the peak irrigation supply rate can be limited by either irrigation well-yield or by canal flow rate.

Example 4.1.5

Problem: The peak evapotranspiration rate for a crop is 10 mm/day and the total seasonal evapotranspiration for the crop is 800 mm with a 180 day growing season length. Determine the irrigation supply rate necessary to meet the daily evapotranspiration rate if the irrigation time is 18 hours/day and the irrigation efficiency is assumed to be 100% (i.e., no water losses). What is the mean daily and seasonal irrigation volume?

Solution: The irrigation supply rate would be 10 liters/m^2 per day [(10 mm/day) = (10 mm^3/mm^2 per day) = (0.01 m^3/m^2 per day) = 10 liters/m^2 per day]. This rate is equivalent to 0.56 liters/m^2 per hour[(10 liters/m^2 per day) / (18 hours/day)]. The mean daily irrigation need is 4.44 mm/day [(800 mm) / (180 days)] which is less than one-half of the peak irrigation need. The required seasonal irrigation volume would be 0.8 m^3/m^2 or 8 Megaliters per ha.

4.1.2 *Irrigation scheduling concepts*

Irrigation scheduling involves two primary decisions: (1) when to irrigate (timing), and (2) how much to apply (amount). These decisions are critical to the management of any irrigation system, are not independent of each other, and require experience to integrate fully irrigation decision with other crop cultural management decisions. The concept of trickle irrigation implies a rather high irrigation frequency compared to conventional irrigation methods. The key principle of trickle irrigation is to maintain a moist segment of the root zone with relatively small applications of water applied continuously or intermittently. The trickle irrigation cycle then becomes an infiltration-dominated process as opposed to an extraction-dominated process (Rawlins, 1973).

4.1.2.a System capacity

The irrigation system supply rate is critical to irrigation scheduling flexibility. The minimum irrigation supply rate is determined from

$$Q_s = (2.77)(ET_p)(A_i)/[(T_i)(E_a)] \qquad (4.1.26)$$

where Q_s is the supply rate in liters/sec, ET_p is the "design" peak irrigation water requirement in mm/day, A_i is the area being irrigated in ha, T_i is the net irrigation operational time in hours/day, and E_a is the irrigation application efficiency expressed as a fraction. The net irrigation operational time, T_i, can be partitioned into irrigation sets (note that A_i would then be the area per set), but also should include downtime for irrigation system repair and maintenance. In most cases, T_i should not exceed 20 to 22 hours/day. The irrigation application efficiency is usually between 0.8 and 0.95 for most trickle irrigation systems; however, E_a is difficult to determine precisely for trickle irrigation systems.

Flexibility in irrigation scheduling to meet irrigation demands greater than normal, catchup from system breakdowns, and inflexible irrigation timing (fixed interval irrigation deliveries) require that the actual irrigation supply rate exceed the minimum supply rate, Q_s. In practice, irrigation supply rates 1.5 to 2 times the magnitude of Q_s would be desirable, though often economically unfeasible. However, in many cases where groundwater is the sole source, minimum supplies seldom even meet Q_s. Surface irrigation supply causes scheduling limitations when the interval between water availability is either too long or the time of availability too short. Thus, any type or combination of limited irrigation supply affects irrigation scheduling flexibility (Burt and Lord, 1981).

4.1.2.b System uniformity

Adequate irrigation application uniformity is necessary to allow uniform crop growth and development. Nonuniformity in trickle irrigation applications is due to variations in emitter flow resulting from system pressure variations from hydraulic and elevation, emitter plugging, and emitter manufacturing variations [see chapter 3.4]. In addition, variation in soil hydraulic properties can result in nonuniform infiltration and water availability to the crop. For full cover crops, the ideal or perfect irrigation uniformity would result if each land unit received exactly the same amount of water from the irrigation system and that water infiltrated uniformly. However, for deep-rooted, widely spaced crops, the necessity of uniform wetting the soil surface is replaced by the requirement that only the water necessary for each plant be supplied uniformly and that the water infiltrates uniformly, giving each plant equal access to the water. Hence, trickle irrigation requires that the "classical" definition of irrigation uniformity must be modified to imply only uniform infiltration for each plant; however, when trickle irrigation is used on closely spaced crops like row crops then the two definitions are essentially identical.

The problem of soil spatial variability is difficult to quantify in terms of soil infiltration uniformity. Large scale soil variability can be handled through design by dividing the various soil differences into irrigation zones (sets). However, small scale, more random soil variations are difficult to incorporate into design.

The degree of irrigation uniformity has been characterized by the statistical parameters of the irrigation application (Warrick, 1983). Two leading indices of irrigation uniformity are the (1) uniformity coefficient (UC), and the (2) low-quarter distribution uniformity (DU). The UC can be expressed as follows

$$UC = 100\ [1 - (MD/X)] \tag{4.1.27}$$

where UC is in %, MD is the average absolute deviation from the mean for water infiltrated, and X is the average depth of water infiltrated. The DU can be expressed as follows

$$DU = 100\ [1 - (LD/X)] \tag{4.1.28}$$

where DU is in %, LD is the average of the lowest quarter of infiltrated water depth, and X is the average depth of water infiltrated. Warrick (1983) summarized the relationships between the UC and DU in terms of the statistical coefficient of variation (CV), where CV = s/X s is the standard deviation of the sample, and X is the sample mean for normal, log-normal, uniform, specialized power, beta and gamma distributions. Figure 4.1.4, parts (A) and (B) illustrate the relationships between UC and CV, and DU and CV for several distributions, respectively. The relationships can be approximated by

$$UC = 100\ [1 - (0.8\ CV)] \tag{4.1.29}$$

$$DU = 100\ [1 - (1.3\ CV)] \tag{4.1.30}$$

$$DU = [-60 + (1.6\ UC)] \tag{4.1.31}$$

Thus, within practical limits, the irrigation uniformity can be determined from the CV of the infiltrated depths.

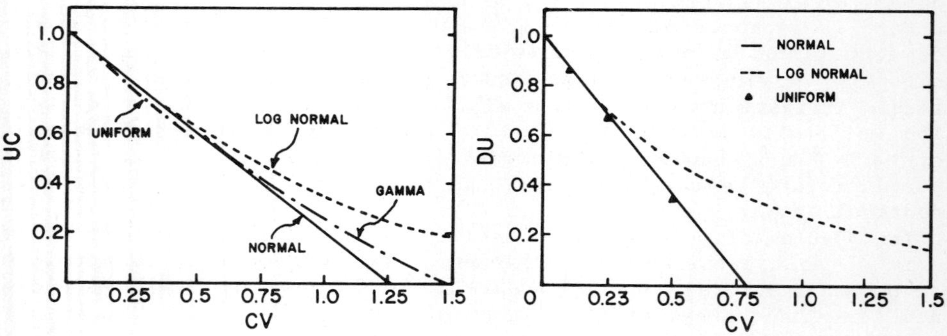

Fig. 4.1.4 Relationship between the uniformity coefficient (UC) and the coefficient of variation (CV) for several distributions (A) and the distribution uniformity of the low-quarter (DU) and the coefficient of variation (CV) for several distributions (B) (adapted from Warrick, 1983).

The emission uniformity (EU) could be used to describe the hydraulic and manufacturing variability of trickle irrigation system discharges (Karmeli and Keller, 1975). The emission uniformity is represented by

$$EU = 100 \{1 - [(1.27 \, CV)/(e^{0.5})]\}\{q_{min}/q_{avg}\} \quad (4.1.32)$$

where EU is in %, CV is the manufacturers' coefficient of variation (standard deviation per unit mean emitter flow rate at the standard operating conditions for that particular emission device, i.e., pressure and temperature, of a random sample of emission devices), e is the number of emitters per plant, q_{min} the nominal minimum emitter discharge computed at the minimum lateral pressure, and q_{avg} the average emitter discharge of the emitters under consideration. The EU is highly related to the DU since manufacturing variation in equation 4.1.32 is based on the discharge of the lowest one-quarter of the emitter sample, which is simply used as a practical minimum value. For many purposes EU and DU can be used almost interchangeably. One important difference is that EU allows for correct interpretation of multiple emitters per plant. Hence, it is likely that the uniformity of the discharge per plant would be better if four emitters with random manufacturing variation serviced each plant rather than one emitter with a large flow rate, i.e., four times the single emitter flow rate, but with the same manufacturing variation serviced each plant.

Example 4.1.6

Problem: The manufacturing coefficient of variation of a particular emitter is 0.07, and based on the hydraulic design of the system the minimum emitter flow should be 3.1 liters/hour. Four emitters serve each tree, and the mean emitter flow rate is 3.5 liters/hour. What is the EU for the trickle irrigation system?

Solution: The EU would be 85% $\{100\ \{1-[(1.27)(0.07)/(4^{0.5})]\}\{3.1/3.5\}\}$ The maximum EU considering only the hydraulic variation (q_{min}/q_{avg}) would be 89%. Using emitters with a lower manufacturing coefficient of variability of 0.03 would only increase EU from 85 to 87%. However, if only one emitter served each plant assuming similar hydraulic characteristics, the EU would be 81%. Hence, in many cases the EU is easier to optimize by minimizing hydraulic variation and using several emitters per plant than by selecting emitters with "almost perfect" manufacturing characteristics.

The uniformity of irrigation application influences irrigation scheduling through the effect of the application distribution on the mean (average) application and the extremes of high and low applications. Since the goal of irrigation is to supply each plant with like amounts of water without either over-irrigation or under-irrigation, the distribution of irrigation amounts is important. With almost equal amounts of available water, the crop growth and yield should be uniform across the field. Most crops can tolerate a modest range in applications and produce similar yields. However, "gross" under-irrigation will result in severe "water stress" in the crop manifested in reduced growth, reduced yield, and even reduced quality in many cases. On the other extreme, "gross" over-irrigation will result in severe "water logging or oxygen stress" in the crop which can reduce growth, yield, and quality; however, usually most crops can withstand substantially greater over-irrigation than under-irrigation (for example see Bielorai, 1982). But over-irrigation results in large water losses to drainage including losses of nutrients, particularly nitrates. Therefore, when water is available and/or inexpensive most farmers have chosen either by practice or design to be on the high side of irrigation applications. This is evident in the use of the low-quarter DU in which the irrigation uniformity is judged by the portion of the field receiving the least amount of irrigation. In many cases, the irrigation amounts have been adjusted to "adequately" irrigate the "driest" portion of the field allowing excess application and drainage on the "wettest" portions.

4.1.2.d System efficiency

The irrigation efficiency is generally defined as the portion of applied irrigation water "beneficially" used in relation to the total amount of irrigation water (gross) applied. Losses of water from trickle irrigation systems are generally accounted for in either runoff, evaporation from the soil, and drainage. These water balance parameters have been discussed previously. Since runoff is not serious under trickle irrigation, except when very steep slopes or where severe infiltration problems exist, and evaporation losses are usually minimal, in general, only drainage limits the irrigation efficiency. In this case, the irrigation efficiency can be estimated from the distribution of the applied water or more precisely by the distribution of the infiltrated water. The relationship has been empirically determined for sprinkler irrigation based on a normal distribution for irrigation amounts (Walker, 1979). From figure 4.1.4 and with $CV < 0.4$, the distribution type should not critically affect the UC. However, the distribution tails of most irrigation application distributions are not accurately estimated by the normal distribution. Using the normal distribution and assuming drainage losses only, the following relationship for irrigation application efficiency can be described by

$$E_a = 100\ \{1 - CV[3.634 - 1.123(A_d^{0.3}) + 0.003(A_d^{1.233})]\} \qquad (4.1.33)$$

where E_a is irrigation application efficiency in % and A_d is the fractional (dimensionless) deficiently irrigated area (Walker, 1979). In this sense E_a only accounts for the excess water applied to the field during irrigation and does not account for the adequacy of the irrigation. Furthermore, the water requirement efficiency can be determined as

$$E_r = 100\ \{E_a/[1 - CV(3.364 - 1.123(A_d^{0.3}))]\} \qquad (4.1.34)$$

where E_r is the water requirement efficiency in % (Walker, 1979). These equations have less than 4% error until the water-deficient irrigated area (A_d) becomes less than 10%. Based on these equations, the irrigation application efficiency increases for the same coefficient of variation of application as the area intentionally deficit irrigated increases. However, crop yield or production will decline with excessive deficit irrigation. Thus, the irrigation schedule should allow some nominal deficiently irrigated areas of from 5 to 15% while potentially reducing yield. A conservative estimate of the yield loss is $A_d/2$. In some cases where the excessive deep percolation reaches a shallow water table which supplies water for irrigation or where the irrigation water supply is both inexpensive and inexhaustible, then the need for high irrigation application may not be necessary. The preceding discussion illustrates the importance of the application distribution to both irrigation uniformity and irrigation efficiency. Irrigation scheduling decisions should consider both uniformity and efficiency.

4.1.2.d Frequency-duration concepts

The interval between trickle irrigations can be short. Irrigations need be only as often as necessary to prevent depleting the soil water reservoir. However, the amount of water depletion in many cases is not as critical as the energy against which the plant must work to extract the water. This energy is called the water potential. The total soil water potential is comprised of the matric and osmotic potentials. The osmotic potential is discussed in chapter 4.4. Plants function optimally when the soil water potential against which they extract water is in the range of less than -0.02 to -0.04 MPa. Trickle irrigation systems are ideal for allowing frequent, small applications of water so that the soil water potential in the crop root zone can be maintained within these narrow limits. In general, soil hydraulic conductivity and water retention characteristics dictate the optimal range in soil water potential. In heavy clay to clay loam soils, the soil water potential can be allowed to decrease close to -0.08 MPa before any serious crop water deficits develop, while in some sandy loam to sandy soils the soil water potential can not be allowed to drop below -0.04 MPa without affecting crop growth. Hence, the irrigation frequency is largely dictated by the soil physical properties. Irrigation intervals up to one week can be practical on heavier textured soils; however, in some cases on lighter textured soils daily irrigations would be recommended. Obviously, the irrigation interval should be a variable depending on the crop evapotranspiration rate. If information exists on the soil water holding characteristics and the size of the subsurface wetted volume, then the irrigation interval/period can be estimated by the amount of water "available" to the crop held above some critical soil water potential divided by the evapotranspiration rate (mm/day). In general, irrigation interval is decreased to minimize any effects of irrigation scheduling errors on crop growth.

The irrigation interval determines approximately what the irrigation amount will be in conjunction with the estimated water use rate. The irrigation amount can be estimated by

$$I_a = 100\ (F_i ET)/E_a \qquad (4.1.35)$$

where I_a is irrigation amount in mm, F_i is irrigation interval in days between irrigations, and E_a is irrigation application efficiency in %. The time of the irrigation (T_i in hr) is the irrigation amount divided by the irrigation system application rate in mm/hr. In general, the irrigation frequency should be compatible to the specific soil limitations of infiltration, water holding capacity, etc.

4.1.2.e Soil and plant measurements

The irrigation timing (frequency) and irrigation amount (quantity) are largely determined by the soil characteristics and by the crop water use rate. Since both of these parameters are difficult to estimate with precision, field measurements of soil conditions and plant conditions are necessary parts of the irrigation scheduling procedure. When properly coordinated, these soil and plant measurements can complement the field monitoring program to check the operation of the irrigation system. The soil and plant measurements serve two purposes: (1) to determine specific operating limits in which crop performance is acceptable, and (2) to verify irrigation schedules and system performance.

Measurements of soil water content or soil water potential can be used to determine accurately the exact state of the soil water profile. Gravimetric methods can be used and the volume of water determined by the product of the soil bulk density [(Mg of dry soil) / (m^3 of soil volume)] and the gravimetric water content [(Mg of water) / (Mg of dry soil)]. Field measurements using the "feel" method are also widely used although exact quantification is not possible. However, volumetric methods of soil water content [(m^3 of water) / (m^3 of dry soil)] determined by neutron probe measurements (Campbell and Campbell, 1982) can be more directly utilized in irrigation scheduling. By evaluating soil water profiles in the root zone, increases or decreases in the irrigation amounts may be easily determined. Based on this feed-back information, irrigation schedules can be adjusted to match precisely crop water demands. Many other types of soil water measuring devices including resistance blocks of gypsum, nylon, etc., thermal dissipation sensors, and tensiometers can be readily used to measure the status of the water in the soil. Placement of sites at different depths under an emitter can be particularly useful for determining whether irrigation amounts are deficient by observing deep soil water status. In addition, radial placement with respect to the emission source can indicate expansion of the wetted volume which is indicative of over- or under-irrigation. Although direct measurement of soil water should be a part of a good irrigation scheduling program, most of the methods of measuring soil water status have some type of measurement limitation on convenience, accuracy, durability, cost, etc. (see chapter 3.3).

Plant measurements include growth parameters such as ground cover, height, stem diameter, yield, etc., and water status parameters such as leaf water content, leaf water potential, crop temperature, etc. Documentation of crop performance from visual appearance as well as quantitative growth measurements can be useful in determining desirable irrigation system operating ranges. Consistent observations of relevant crop growth parameter changes, such as stem diameter, height, etc., in trees can determine whether "gross" deficiencies exist in the irrigation scheduling program. However, in general, visual or even quantified crop growth differences only appear following "moderate" crop water stress development. Plants sensitive to water stress undergo growth and yield reduction before the onset of moderate stress. Plant water status monitoring should permit rapid measurement and evaluation of crop performance and indicate crop water stress before "visual" symptoms are present. Leaf water potential measured by hydraulic press, pressure chamber or thermocouple psychrometer, and crop temperature measured by infrared thermometry are various methods for assessing plant stress (Stegman, 1983; Jackson, 1982). The interpretation of plant water status measurements is particularly difficult since the plant integrates its response to both the soil environment and the atmospheric environment. Hence, direct comparison between soil and plant measurements is very complex. The ambient vapor pressure deficit [(e_s - e_a), see equation 4.1.10, 11, etc.] at the time of the plant water status measurement has been found to be a good normalizing index of atmospheric evaporative demand. Normalization of plant water status measurements is necessary since plant water status has such a large diurnal response to the evaporative conditions, particularly between days with variable evaporative demand. Predawn plant measurements have been used to measure plant water status parameters before any evaporative demand develops. These predawn measurements should reflect the "equilibrium" soil water status; however, predawn measurements are inconvenient to take.

Soil and plant water status measurements can be used to characterize the crop response to irrigation and in some cases to determine the irrigation uniformity. If the crop response indicates water stress symptoms, then the irrigation interval can be shortened or the amount can be increased. If the the soil measurements indicate a gradual increase in soil water storage or potential, then either the irrigation amount can be decreased or the irrigation interval can be lengthened. Interpretation of soil and plant measurements and selection of sampling stations and numbers of samples necessary are related to the irrigation and soil uniformities. Hence, measurements can be easily biased by the selection of sampling location. Crop indications of water stress or excessive water applications may not be indicative of improper irrigation scheduling, but of poor system design and/or performance. In this case, some corrective action like emitter cleaning, replacement, or system alteration may be necessary. Measurement of water application by metering the rate or volume is very important for verification of the irrigation application amount which is necessary to detect changes in irrigation system performance.

4.1.2.f Computer irrigation scheduling

Computers can be used to schedule irrigations either in a control mode [see chapter 3.3] or in a predictive mode. In the predictive mode, the computer is used to calculate the crop water use, soil moisture status, and evapotranspiration rates and to calculate the predicted date of the next irrigation and the approximate irrigation amount. Irrigation scheduling models can be simple evapotranspiration calculations, complex water budgets in several dimensions, or even crop growth models. The models use various types of crop information (crop coefficient curves, dates of full cover, etc.), soil information (water holding capacities, root zone depths, allowable depletions, drainage rates, etc.) and climate information (long-term normals of weather variables, such as temperature, radiation or precipitation, various constants used in the evapotranspiration equations, etc.). The model then incorporates current weather information to calculate the evapotranspiration rates and to adjust the soil water balance to account for evapotranspiration, drainage, runoff, etc. as water losses, and the precipitation, irrigation, etc., as water gains. The evapotranspiration model requires that the irrigator set some "threshold" criterion based on allowable depletion or desired irrigation interval. In the first case, the current predicted soil water status is compared to the allowable depletion and the next irrigation date is predicted based on the expected or current evapotranspiration rate. Actual field information normally taken after an irrigation event to help account for infiltration and immediate field drainage can be used to update or correct the estimated soil water. In general, this method is not very practical for trickle irrigation because the allowable depletion is difficult to predict and the trickle irrigation amounts are usually much smaller than those used for surface or sprinkler irrigation. In the second case, the computed evapotranspiration rate can be used to indicate the necessary irrigation amount or to specify the time clock settings for a certain irrigation interval. The field measurements can be used to indicate adjustments in the irrigation amounts by either over-irrigation soil water storage increases or under-irrigation as field soil moisture depletes. This method has more practical application for trickle than the other irrigation methods.

Computer irrigation scheduling allows a convenient way to maintain accurate field records and notes in addition to the actual computing power. Computer scheduling programs should allow various report formats for end of year reconciliation, field schedules, field summaries, etc. Several versions of the original and modified irrigation scheduling programs of USDA-ARS (Jensen et al., 1970) have been adapted to various versions of microcomputers. In some cases where the record keeping is not critical, pocket calculators can be used to run simple evapotranspiration and water balance models.

4.1.2.g Economic parameters of irrigation scheduling

Modern scientific irrigation scheduling can offer many potential advantages to growers including (1) lower water use, (2) lower energy use, (3) better yields, (4) improved farm records, (5) reduced production costs for fertilizer, and (6) better farm operating efficiency due to improvement in planning and scheduling of other farm operations. With a properly designed and maintained irrigation system, the attainment of these advantages depends on the skill and expertise of the irrigation manager. Furthermore, irrigation scheduling on large-scale, regional farming areas offer additional potential public advantages such as (1) reduced return flow, (2) reduced pollution from sediment and salinity, and (3) improved water quality for downstream users. Irrigation scheduling on a regional basis should be encouraged since an economic return can be normally derived from scientific irrigation scheduling (English et al., 1980).

TABLE 4.1.11

Estimated costs for irrigation scheduling (adapted from Day and Brase, 1983).

	Type of irrigation scheduling information[a]				
	Gypsum block	Tensio-meter	Infrared thermo-meter	Neutron probe	Computer ET modeling
Initial costs					
Equipment...........	$ 250 (meter)	$ 100 (service kit)	$2,500 (remote sensor)	$3,500 (probe)	$9,500 (computer and software)
Calibration......... and licensing	0	0	0	700	0
Installation[b]....... cost/field	50	125	0	50	0
Operation costs					
Collection[c]......... fields/hr	8	8	6	4	6
Cost/season/field...	50	50	67	100	33
Data analyses costs					
Data................ fields/hr	15	20	10	10	20
Cost/season/field...	27	20	40	40	20
Cost per season[d]					
Total...........	$4,770	$5,620	$6,180	$9,700	$5,820
Per field........	95.40	112.40	124.00	194.00	116.00
Per hectare......	1.62	2.77	3.06	4.79	2.87

[a] All costs are based on 1983 U.S.$ and do not include personnel training time.

[b] Installation includes cost of in-field equipment and labor. Estimates include one site with sensors at three depths.

[c] Cost estimates are based on $20/hr labor including overhead and 6 hr/day applied time, with 20 weekly readings for direct soil measurement methods and 10 every 2 week evaluations for the computerized water budget.

[d] Seasonal cost estimates are used on amortizing equipment costs over 3 years. Installation costs are amortized over 3 years also. Totals for per-field and per-hectare costs are based on a 2,000 ha farm and 50 fields of 40 ha each.

Irrigation scheduling requires investments in labor and capital equipment. Moreover, scientific irrigation scheduling represents a fundamental change in the traditional grower attitude toward water management. Realistically, this change will occur only by economic advantages or possibly institutional requirements. Table 4.1.11 presents an estimated cost schedule for commercial irrigation scheduling services (Day and Brase, 1983). The extent of irrigation scheduling activity will depend on many factors, and there is no one absolutely perfect method for determining this. Irrigation scheduling along with the other farming decisions should be approached on a "rational" decision making basis. In this sense, each piece of data becomes another "information" resource. This total information resource can then be utilized to make "rational" decisions about irrigation scheduling. Commercial irrigation services may cost between \$2.00 to \$15.00 per ha depending on the level of consulting on the irrigation system performance (engineering), crop cultural recommendations (agronomy, viticulture, horticulture) and soil management (agronomy, soil science). Although irrigation scheduling seems complex, the process of analytically evaluating decision options is important in making optimum resource use.

PROBLEMS:

1. A fully mature orchard with full cover and 100 ha in size (note that 1 ha = 10,000 m^2) has used 250 mm of water for evapotranspiration during the month of July. How much total water volume in liters and in cubic meters was used? What was the average daily evapotranspiration flux in mm/day and liters per square meter?

2. July 25 had the following weather parameters near Fresno, California: (1) daily solar radiation of 22 MJ/m^2; (2) maximum air temperature of 35°C; (3) minimum air temperature of 20°C; (4) dew point temperature of 12°C; (5) daily average wind speed at 2 m elevation of 3.5 m/sec. Using the modified combination equation determine the water use for grass and grapes assuming that K_{cb} = 0.6 for the grapes and that the soil surface is dry for the grapes.

3. For the month of August, the warmest of the year, at Amarillo, Texas, the average maximum temperature is 35°C and the average minimum is 18°C, and the elevation is 1,200 m. Determine how much water (in mm) mature peaches would use based on the modified Jensen-Haise equation. The average solar radiation at Amarillo during August is 24 MJ/m^2. Consider Amarillo to have cold winters with killing frosts and dry, strong winds.

4. The following emitter flow rates in liters/hr were measured along a short trickle irrigation lateral:

3.9	3.7	3.8	3.5	3.8	3.4	3.3	3.2	4.2	3.1	3.4
3.0	3.3	3.0	3.6	2.9	3.1	3.3	3.1	4.3	3.0	3.5

The maximum pressure in the lateral is 86 kPa and the minimum pressure is 62 kPa. The particular emitter flows according to the equation $q_e = kH^x$ (chapter 2.1) where $q_e = 0.5\ H^{0.5}$. If only one emitter is used per plant, what is the emission uniformity, and what is the expected irrigation application efficiency if 15% of the irrigated area is deficiently irrigated?

5. Locate sufficient weather data for your location and calculate the irrigation requirement for 3-year-old grapes using the modified Blaney-Criddle equation.

REFERENCES AND SUPPLEMENTAL READING

Ben-Asher, J., 1979. Errors in determination of the water content of a trickle-irrigated soil volume. Soil Sci. Soc. Amer. J. 43: 665-668.

Bielorai, H., 1982. The effect of partial wetting of the root zone on yield and water use efficiency in a drip-irrigated and sprinkler-irrigated mature grapefruit grove. Irrig. Sci. 3: 89-100.

Blaney, H.F. and Criddle, W.D., 1950. Determining water requirements in irrigated areas from climatological and irrigation data. U.S. Dept. Agric. (Soil Conservation Service) Tech. Publ. TP-96, 96 pp.

Burman, R.D., Nixon, P.R., Wright, J.L., and Pruitt, W.O., 1980. Water requirements. In Jensen, M.E. (Ed.), Design and Management of Farm Irrigation Systems, Amer. Soc. Agric. Eng. Monograph 3, pp. 189-232.

Burt, C.M. and Lord, Jr., J.M., 1981. Demand theory and application in irrigation district operation. In Irrigation Scheduling for Water & Energy Conservation in the 80's. Proc. Irrigation Scheduling Conference, Amer. Soc. Agr. Eng., St. Joseph, Michigan, pp. 150-158.

Campbell, G.S. and Campbell, M.D., 1982. Irrigation scheduling using soil moisture measurements: theory and practice. In Hillel, D., (Ed.), Advances in Irrigation 1: 25-42, Academic Press, New York. pp. 25-42.

Campbell, R.B. and Phene, C.J., 1976. Estimating potential evapotranspiration from screened pan evaporation. Agric. Meteorol. 16: 343-352.

Day, K. and Brase, R., 1983. A Comprehensive Guide to Irrigation Scheduling. Crop Care Associates, Fresno, California, 12 pp.

Doorenbos, J. and Pruitt, W.O., 1977. Crop Water Requirements. Irrigation and Drainage Paper 24, Food and Agric. Organization, United Nations, Rome, Italy, 144 pp.

English, M.J., Horner, G.L., Orlob, G.T., Erpenbeck, J., Moehlman, M., Cuenca, R.H., and Dudek, D.J., 1980. A regional Assessment of the Economical and Environmental Benefits of an Irrigation Scheduling Service. U. S. Environmental Protection Agency Report EPA-600/ 2-80-063, 114 pp.

Hill, R.W., Johns, E.L., and Frevert, D.K., 1983. Comparison of Equations Used for Estimating Agricultural Crop Evapotranspiration with Field Research. U.S. Dept. of Interior-Bureau of Reclamation, 242 pp.

Howell, T.A., Jordan, W.R., and Hiler, E.A., 1979. Evaporative demand as a plant stress. In Barfield, B.J. and Gerber, J.F. (Eds.), Modification of the Aerial Environment of Crops, Amer. Soc. of Agric. Eng. Monograph, pp. 97-113.

Howell, T.A., Meek, D.W., and Hatfield, J.L., 1983a. Relationship of photosynthetically active radiation to shortwave radiation in the Joaquin Valley. Agric. Meteorol. 28: 157-175.

Howell, T.A., Miller, R.J., Phene, C.J., and D. W. Meek, D.W., 1983b. Evaporation from screened Class A pans in a semiarid climate. Agric. Meterol. 29: 111-124.

Howell, T.A., Meek, D.W., Phene, C.J., Davis, K.R., and McCormick, R.L., 1984. Automated weather data collection for research on irrigation scheduling. Trans. Amer. Soc. Agric. Eng. 27(2): 386-391, 396.

Idso, S.B. and Jackson. R.D., 1969. Thermal radiation from the atmosphere. J. Geophys. Res. 74: 5397-5403.

Jackson, R.D., 1982. Canopy temperature and crop water stress. In Hillel, D. (Ed.), Advances in Irrigation 1: 43-85. Academic Press, New York.

Jensen, M.E. (Ed.), 1974. Consumptive use of water and irrigation water requirements. Irrig. and Drain. Div. Report, Amer. Soc. Civil Eng., 215 pp.

Jensen, M.E. and Haise, H.R., 1963. Estimating evapotranspiration from solar radiation. J. Irrig. and Drain. Div., Amer. Soc. Civil Eng. 89(IR4): 15-41.

Jensen, M.E., Robb, D.C.N., and Franzoy, C.E., 1970. Scheduling irrigations using climate-crop-soil data. J. Irrig. and Drain. Div., Amer. Soc. Civil Eng. 96(IR1): 25-38.

Jensen, M.E., Wright, J.L., and Pratt, B.J., 1971. Estimating soil moisture depletion from climate, crop and soil data. Trans. Amer. Soc. Agr. Eng. 14(3): 954-959.

Karmeli, D. and Keller, J., 1975. Trickle Irrigation Design. Rain Bird Sprinkler Manufacturing Corp., Glendora, California, 133 pp.

Kincaid, D.L. and Heermann, D.L., 1974. Scheduling Irrigations Using a Programmable Calculator. U. S. Dept. Agric. Report ARS-NC-12, 55 pp.

Pair, C.H., Hinz, W.W., Sneed, R.E., Frost, K.R., and Schiltz, T.J., 1983. Irrigation (formerly Sprinkler Irrigation), Fifth Edition, Irrigation Association, Silver Springs, Maryland, 686 pp.

Penman, H.L., 1948. Natural evaporation from open water, bare soil and grass. Proc. Roy. Soc. (London) A193: 120-145.

Penman, H.L., 1963. Vegetation and Hydrology. Tech. Comm. 53. Commonwealth Bur. of Soils, Harpenden, England, 125 pp.

Rawlins, S.L., 1973. Principles of managing high frequency irrigation. Soil Sci. Soc. Amer. Proc. 37: 626-629.

Stegman, E.C., 1983. Irrigation scheduling: applied timing criteria. In Hillel, D. (Ed.), Advances in Irrigation 1: 1-30, Academic Press, New York.

Stigter, C.J., 1980. Assessment of the quality of generalized wind functions in Penman's equation. J. Hydrol. 45: 321-331.
Tanner, C.B., 1981. Transpiration efficiency of potato. Agron. J. 73: 59-64.
Van Bavel, C.H.M., 1966. Potential evaporation: the combination concept and its experimental verification. Water Resources Res. 2: 455-467.
Walker, W.R., 1979. Explicit sprinkler irrigation uniformity: efficiency model. J. Irrig. and Drain. Div., Amer. Soc. Civil. Eng. 105(IR2): 129-136.
Warrick, A.W., 1983. Interrelationships of irrigation unifomity terms. J. Irrig. and Drain., Amer. Soc. Civil Eng. 109(IR2): 317-332.
Wright, J.L., 1981. New evapotranspiration crop coefficients. J. Irrig. and Drain. Div., Amer Soc. Civil Eng. 108(IR1): 57-74.
Wright, J.L. and Jensen, M.E., 1972. Peak water requirements in Southern Idaho. J. Irrig. and Drain. Div., Amer. Soc. Civil Eng. 98(IR2): 193-201.
Wright, J.L. and Jensen, M.E., 1978. Development and evaluation of evapotranspiration models for irrigation scheduling. Trans. Amer. Soc. Agric. Eng. 21(1): 88-96.

Chapter 4

MANAGEMENT PRINCIPLES

4.2 FLOW MEASUREMENT AND SYSTEM MAINTENANCE

L. G. JAMES AND W. M. SHANNON

Flow measurement is extremely useful in assessing the need for trickle irrigation system maintenance as well as being an invaluable aid in system management. Pipeline leaks and breaks, emission device and filter clogging, inadequate pressure regulation, and several other problems such as those listed in table 4.2.1 may be detected by monitoring system flow. Basic flow measurement principles and practices and the preventive maintenance requirements of trickle irrigation system components are presented in the sections which follow.

TABLE 4.2.1

Possible causes of changes in trickle system flow.

<u>Increased flow</u>
- (a) improperly adjusted/open valves
- (b) pipeline leaks/breaks
- (c) pressure downstream of pressure regulators is too high
- (d) worn/oversized emission devices
- (e) system on too long (as indicated by higher than expected volumes of flow)

<u>Decreased flow</u>
- (a) improperly adjusted valves
- (b) clogged emission devices, filters, and other components
- (c) pump wear
- (d) pressure downstream of pressure regulators too low
- (e) existence of entrapped air in system
- (f) system not on long enough (as indicated by lower than expected volumes of flow)

4.2.1 *Flow measurement*

There are several types of flowmeters used to measure the volume and/or volumetric flow rate (i.e., volume per unit time) of trickle systems. Flow volume and flow rate are related by

$$V = Q\Delta t \qquad (4.2.1)$$

where V = volume of flow, Q = average volumetric flow rate, and Δt = time interval.

4.2.1.a Flowmeter performance

The performance of a flowmeter is described by its characteristic (calibration) curve. This curve relates flowmeter response to volume of flow and/or volumetric flow rate. Tests to determine flowmeter response over a range of expected flows are normally used to define the characteristic curve of a flowmeter. Typical linear and nonlinear characteristic curves are shown in figure 4.2.1.

A linear flowmeter has a characteristic curve with an essentially constant slope over the range of operation (line A in figure 4.2.1). For most nonlinear flowmeters, flow is proportional to the square root of flowmeter response and the shape of the characteristic curve is similar to curve B in figure 4.2.1. The percent change in flowmeter response resulting from a 1 percent change in flow, i.e., flowmeter sensitivity, is constant over the range of operation of linear flowmeters. For nonlinear flowmeters, sensitivity increases with increasing flow.

Because flowmeter characteristic curves are fitted to calibration tests and tend to "smooth" the data, they do not show deviations of indicated from true flow. Indicated flow is that given by the flowmeter or calculated from its readings while true flow is determined by a high-accuracy measuring device used in the calibration test. Plots of such parameters as the coefficient of discharge, meter correction term, meter factor, and K-factor against flow or quantities such as the Reynolds number do, however, describe deviations from true flow. Figure 4.2.2 illustrates such a plot.

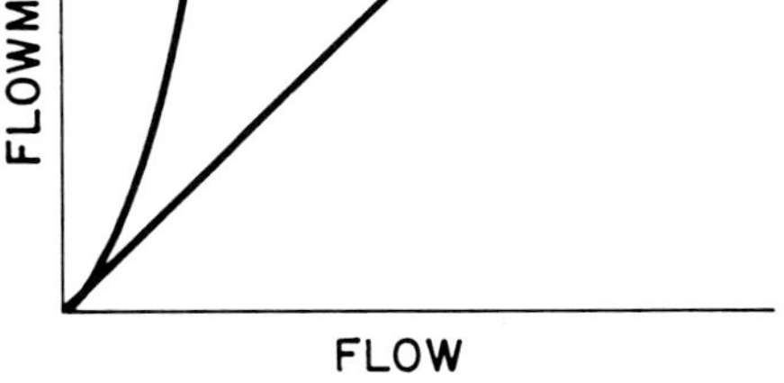

Fig. 4.2.1 Characteristic (calibration) curves for linear (line A) and nonlinear (curve B) flowmeters.

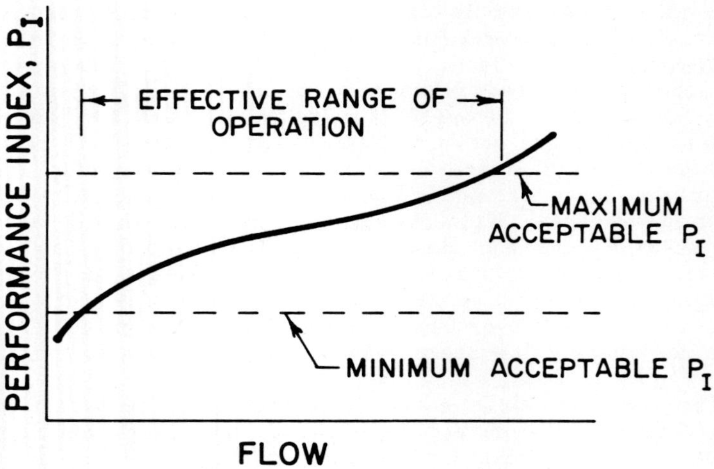

Fig. 4.2.2 Effective range of flowmeter operation.

Once an acceptable range of accuracy has been specified, plots similar to figure 4.2.2 can be used to establish an effective range for the flowmeter. The dashed lines in figure 4.2.2 are the upper and lower limits of acceptable values for the particular flowmeter performance index, P_I, being used. Flows for which the flowmeter performance equals or exceeds the minimum acceptable P_I, but does not exceed the maximum acceptable P_I are within the flowmeters' effective range. Acceptable accuracy can not be expected outside the flowmeters' effective range.

The coefficient of discharge is defined as the ratio of indicated to true flow and is used extensively with differential pressure meters such as venturi meters, nozzles, and orifices. It is generally plotted against Reynolds number calculated for the throat of the meter.

The meter correction term is normally used with volume meters and is defined by equation 4.2.2. This equation is:

$$\Delta = \frac{V_T - V_I}{V_I} \tag{4.2.2}$$

where, Δ is the meter correction term, and V_T and V_I are the true and indicated flow volume, respectively.

The meter factor is also normally used with volume meters and especially with turbine and positive displacement meters. The meter factor is a ratio of the true to indicated volume of flow and equals the meter correction term plus 1.

The K-factor describes the performance of meters such as turbine meters whose output is a series of electrical pulses. The K-factor is a ratio of the number of pulses to the volume of flow. It is assumed that the number of pulses is proportional to the volume passed.

Discrimination, repeatablilty, and accuracy are terms which also describe flowmeter performance. Discrimination describes how finely a flowmeter can measure, i.e., the number of significant figures which can be read. The repeatability of a flowmeter is an indication of its ability to give the same result when the same flow is measured several times in succession. Flowmeter accuracy is often defined by equation 4.2.3.

$$ACC = \frac{100\ (F_I - F_T)}{F_T} \qquad (4.2.3)$$

where ACC = flow meter accuracy in percent, and F_I and F_T = indicated and true flow (may be either flow volume or volumetric flow rate).

Accuracy is sometimes defined as a percentage of full-scale flowmeter reading (i.e., full-scale flow is used in the denominator of equation 4.2.3 rather than F_T).

4.2.1.b Flowmeter types

Several types of devices can be used to measure the volume of flow and/or volumetric flow rate in trickle irrigation systems. These devices may be classified as differential pressure, rotating mechanical, ultrasonic, or insertion meters.

(i) Differential pressure flowmeters. Differential pressure flowmeters create a pressure difference which is proportional to the square of the volumetric flow rate. One popular way of creating the pressure difference is to cause water to flow through a contraction. The water gains kinetic energy and losses potential (i.e., pressure) energy as it accelerates through the contraction. Manometers, bourdon gages, and pressure transducers are normally utilized to measure the pressure difference.

Venturi and orifice plates are the main contraction type differential pressure flowmeters used with trickle irrigation systems. A typical venturi meter is diagrammed in figure 4.2.3. As water passes through the throat of the venturi its velocity increases, producing a pressure drop between the inlet and throat. In the section downstream of the throat the gradual increase in cross-sectional area causes the velocity to decrease and the pressure to increase. The pressure drop between the venturi's inlet and throat is related to the volumetric flow rate. This relationship is given by equation 4.2.4

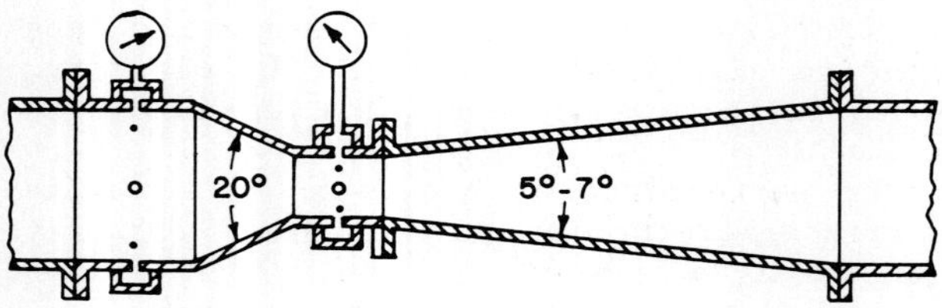

Fig. 4.2.3 Venturi tube flowmeter.

$$Q = \frac{C\ d^2K(P_1 - P_2)^{1/2}}{[1 - (d/D)^2]^{1/2}} \tag{4.2.4}$$

where Q = volumetric flow rate (L/min, gal/min)
C = flow coefficient
D = diameter of upstream section (cm, inches)
d = diameter of contraction (cm, inches)
P_1 = pressure in upstream section (kPa, lb/in^2)
P_2 = pressure in contraction (kPa, lb/in^2)
K = unit constant ($K = 6.66$ for Q in L/min, d and D in cm, and P_1 and P_2 in kPa. $K = 29.86$ for Q in gal/min), D and d in inches, and P_1 and P_2 in lb/in^2).

The flow coefficient C varies with Reynolds number. C for a venturi meter equals 0.98 for Reynolds numbers for the throat which exceed 200,000. Venturi tubes for chemical injection can sometimes be used to measure flow.

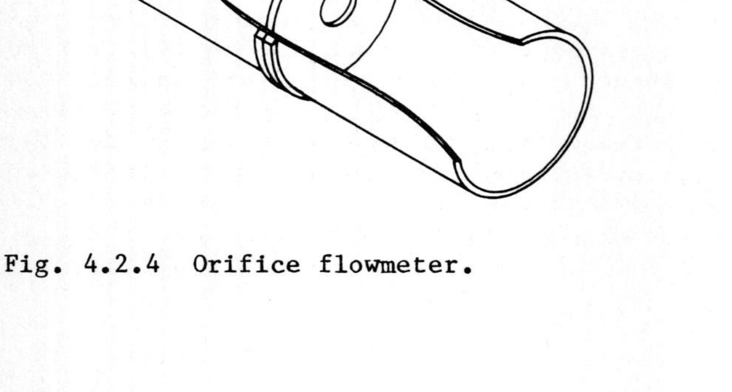

Fig. 4.2.4 Orifice flowmeter.

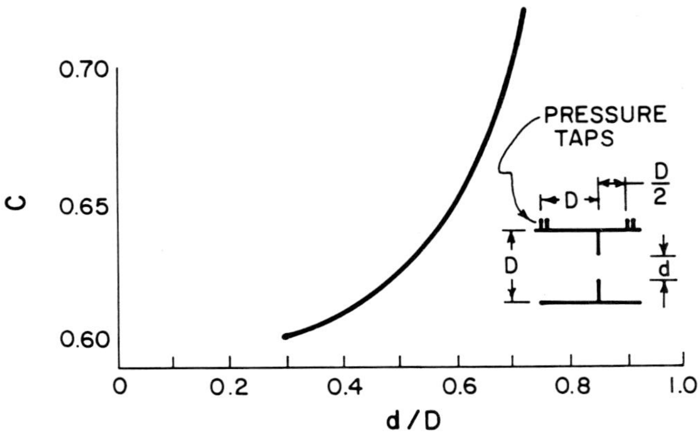

Fig. 4.2.5 Flow coefficient (C in equation 4.2.4) for square edged circular shaped orifice of different ratios of orifice opening to inside pipe diameter for Reynolds numbers greater than 100,000.

An orifice consists of a thin plate with a square-edged hole which is clamped between flanges in a pipe. Most orifice plates have a circular shaped hole concentric with the pipe as in figure 4.2.4. The principle of operation of an orifice is similar to a venturi meter. Increased velocity in the orifice creates a pressure drop between the up and down stream sides of the orifice plate. The pressure drop across an orifice normally exceeds that for a venturi meter with the same d to D ratio. Orifices are, however, less expensive than venturi meters.

Equation 4.2.4 can be used to compute the volumetric flow rate through an orifice plate for various pressure drops. The flow coefficient for square-edged, circular-shaped, concentric orifices varies with the ratio of d to D and Reynolds number. Figure 4.2.5 can be used to obtain values of C when the Reynolds number computed using the orifice diameter exceeds 100,000 and pressure taps are located 1.0 and 0.5 pipe diameter up and downstream, respectively, of the orifice plate.

Eccentric and chord orifice plates like those in figure 4.2.6 are recommended for use with sediment laden waters. These orifices prevent sediment accumulation in the pipe, but are less accurate then concentric orifices.

Choosing the diameter of the contraction in venturi and orifice meters is extremely important. It is desirable that the diameter of the contraction, d, be large enough to minimize head loss. However, if d is too large relative to D, accurate measurement of the pressure difference may be impossible.

Another important type of differential pressure flowmeter is an elbow meter. Pressure differences between the inside and outside

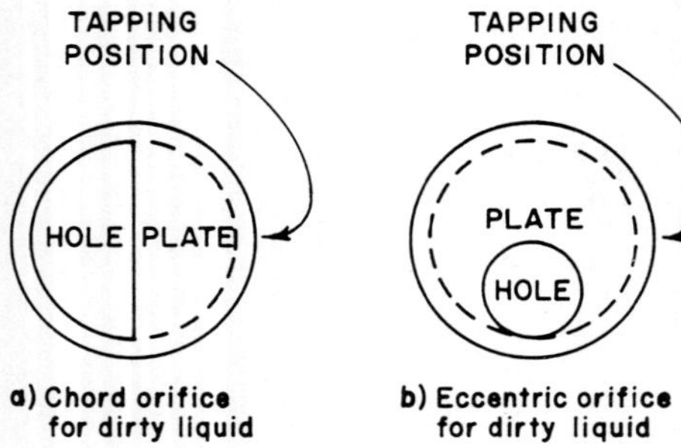

Fig. 4.2.6 Chord and eccentric orifice plates.

walls of an elbow are related to volumetric flow rate. Equation 4.2.5 is used to compute volumetric flow rate when the pressure differences and cross-sectional area of the elbow are known.

$$Q = C_e K A (P_o - P_i)^{1/2} \tag{4.2.5}$$

where Q = volumetric flow rate (L/min, gal/min)
C_e = elbow meter flow coefficient
A = cross-sectional area of elbow (cm^2, in^2)
P_o = pressure on outside of elbow (kPa, lb/in^2)
P_i = pressure on inside of elbow (kPa, lb/in^2)
K = unit constant ($K = 8.49$ for Q in L/min, A in cm^2, and P_o and P_i in kPa. $K = 38.02$ for Q in gal/min, A in $inches^2$, and P_o and P_i in lb/in^2)

As shown in figure 4.2.7 the elbow meter flow coefficient, C_e, ranges between 0.63 and 0.83 depending on size, shape and type of elbow.

(ii) Rotating Mechanical Flowmeters. There are many types of rotating mechanical flowmeters used in trickle systems. These devices normally have a rotor which revolves at a speed roughly proportional to the flow rate, and a mechanical system for recording and displaying the total volume of flow and/or the volumetric flow rate. The rotor may be a propeller or axial-flow turbine, or a vane wheel with the flow impinging tangentially at one or more points.

Calibration tests are usually needed to accurately relate rotor revolutions to flow. The lowest flow rate that can be accurately measured by a rotating mechanical flowmeter (i.e., the lower limit of the flowmeter's effective range) depends on the amount of bearing friction that can be tolerated while the occurrence of cavitation often establishes the upper limit. Head loss through most rotating mechanical flowmeters is moderate.

(iii) Bypass flowmeters. A bypass or shunt meter is another type of flowmeter used with trickle irrigation systems. As shown in figure 4.2.8 a bypass meter is an orifice or other differential pressure

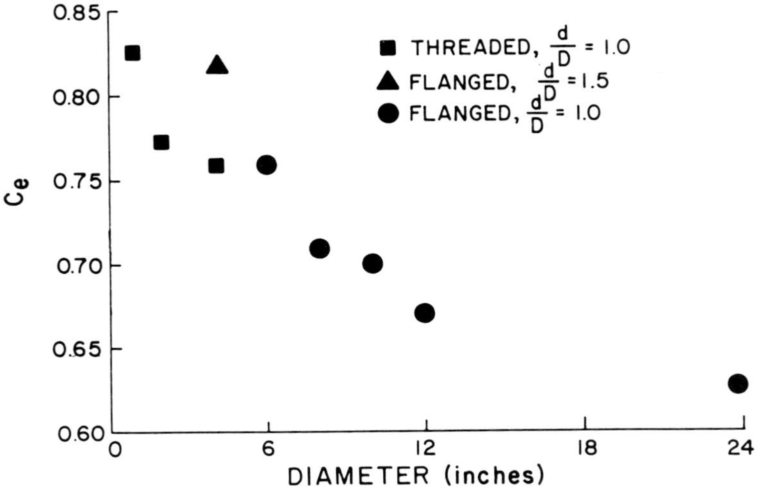

Fig. 4.2.7 Flow coefficient (C_e in equation 4.2.5) for different diameter elbow meters for Reynolds numbers greater than 100,000.

device with a small mechanical flowmeter across the pressure taps rather than a pressure measuring device. The relationship between volumetric flow rate in the main pipe and flow in the bypass line is essentially linear for properly designed bypass flowmeters. Both volume of flow and volumetric flow rate can usually be obtained.

(iv) Ultrasonic flowmeters. Various types of devices that use beams of ultrasound to measure flow velocity and hence volumetric flow rate are called ultrasonic flowmeters. The electronic circuitry required by ultrasonic flowmeters usually also provide the volume of flow. Some of these devices require the presence of suspended particles, air bubbles and/or fluid turbulence to measure flow while others do not. Ultrasonic meters do not obstruct flow and, thus, cause no loss of pressure and do not have mechanical parts to wear.

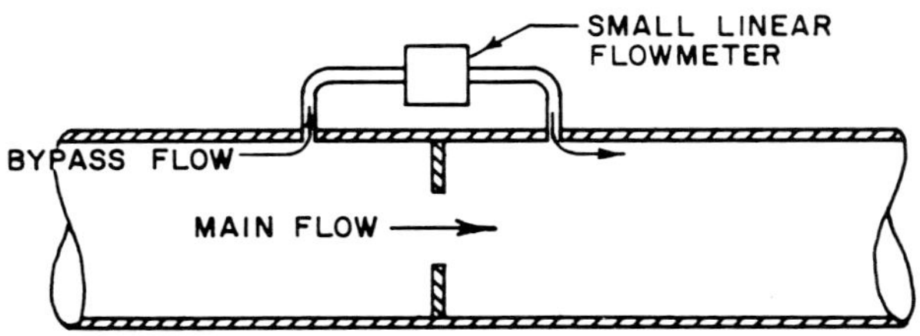

Fig. 4.2.8 Schematic of a bypass meter.

Because ultrasonic beams will travel through the wall of a pipe, ultrasonic flowmeters can be either portable or built-in. Portable models have the transmitter and receiver mounted in a housing which is clamped onto the outside of the pipe. This eliminates the trouble and expense of breaking open the pipe for flowmeter installation. With built-in ultrasonic flowmeters the transmitter and receiver are factory mounted in a short section of pipe. Built-in meters are installed in the pipeline and cannot be conveniently moved to another location. Built-in ultrasonic flowmeters are more accurate than clamp-on units because the relative positions of the transmitter and receiver is fixed and the cross-sectional area of the flowmeter sections is precisely known.

The single-path, diagonal-beam meter is one of the earliest and most widely used types of ultrasonic flowmeters. This type of device transmits two ultrasonic signals diagonally across the pipe as shown in figure 4.2.9. One of the signals travels downstream and the other upstream. The difference in travel times of the two beams is related to the flow velocity in the pipe. Because this relationship is normally obtained by calibration and since the flow velocity along only a single line across the pipe is used, flowmeter accuracy is highest when the actual velocity profile within the pipe is similar to the one which existed when the flowmeter was calibrated. Both built-in and clamp-on models of single-path, diagonal-beam, ultrasonic flowmeters are available.

The accuracy of single-path, diagonal-beam meters can be improved by using a multi-chordal, diagonal-beam arrangement similar to that in figure 4.2.10. Flowmeter accuracy is less sensitive to the velocity profile within the pipe since several beams of ultrasound travel diagonally across the pipe rather than just one. These meters, however, require more complicated circuitry for signal processing. Because of the close tolerances needed for high accuracy, multi-chordal meters are available only as built-in units.

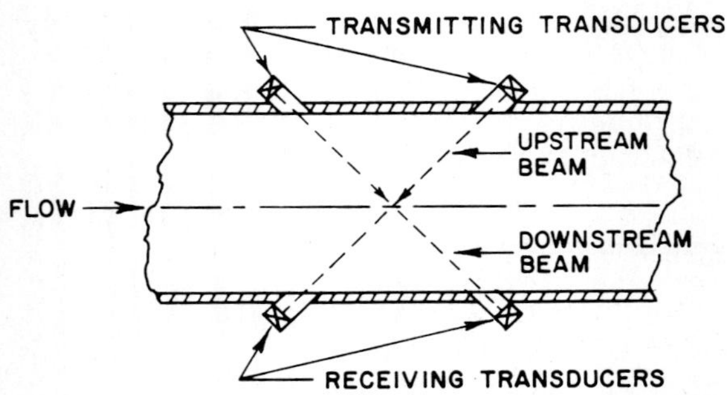

Fig. 4.2.9 Principle of the diagonal-beam, ultrasonic flowmeter.

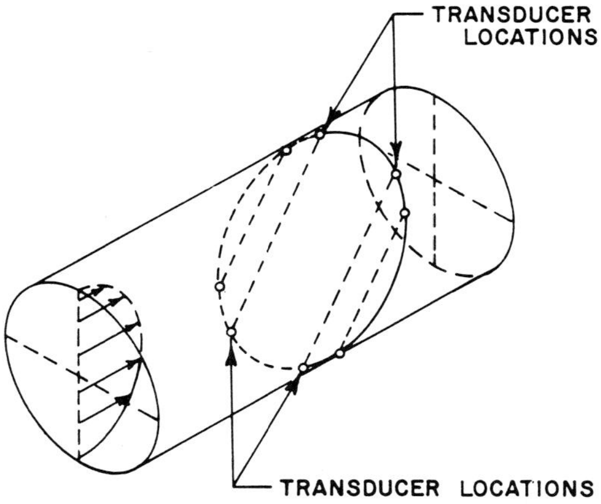

Fig. 4.2.10 A four-chordal, diagonal-beam, ultrasonic flowmeter.

Cross-correlation, ultrasonic flowmeters employ two transverse beams of ultrasound, one located a short distance upstream of the other, as in figure 4.2.11, to measure flow velocity within a pipe. The volumetric flow rate is calculated from the time required for flow discontinuities such as aggregations of suspended particles, air bubbles, or fluid turbulence (eddies) to pass from the up- to downstream beams. Like single-path, diagonal-beam meters, the accuracy of cross-correlation meters is extremely sensitive to deviations of the actual

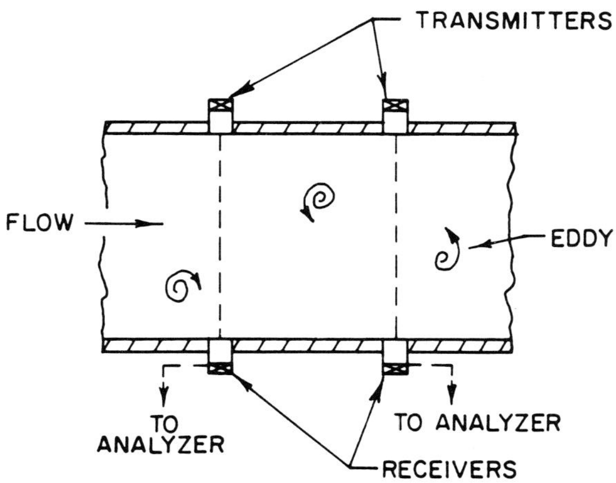

Fig. 4.2.11 Principle of the cross-correlation, ultrasonic flowmeter.

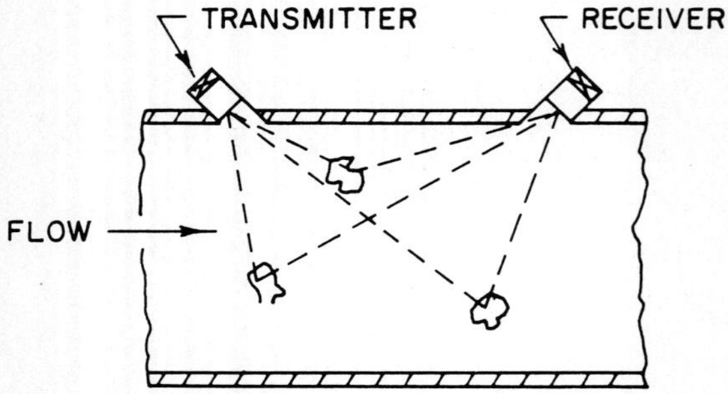

Fig. 4.2.12 Principle of the Doppler-effect, ultrasonic flowmeter.

velocity profile from the one which existed during calibration. Cross-correlation meters require the presence of flow discontinuities and are well suited to clamp-on operation.

A Doppler-effect, ultrasonic meter (figure 4.2.12) measures the velocity of suspended particles or small air bubbles being carried in the flowing water. Some of the signals which are transmitted into the pipe are reflected to the receiver. Transmitted and reflected signals are compared and the volumetric flow rate determined. In addition to being sensitive to the velocity profile within the pipe, Doppler-effect flowmeter readings are affected by changes in the velocity of sound in the water caused by temperature and density variations. Like cross-correlation meters, Doppler-effect meters are well suited to clamp-on operation.

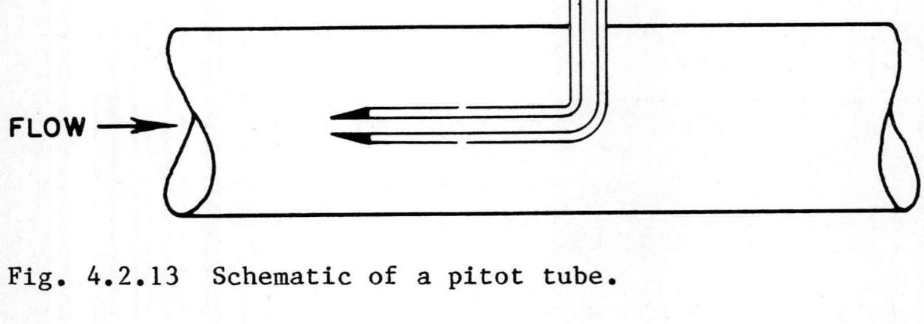

Fig. 4.2.13 Schematic of a pitot tube.

(v) Insertion flowmeters. The pitot tube is probably the best known insertion meter. The pitot tube in figure 4.2.13 consists of two, small diameter concentric tubes pointing directly upstream. The inner tube measures the total flow energy (kinetic plus potential energy) while the outer tube senses only potential (pressure) energy. Because water in the mouth of the inner tube is brought to rest and its kinetic energy converted to potential (pressure) energy the difference in pressure between the inner and outer tubes equals the kinetic energy of flow. The velocity of flow can be computed from this pressure difference using equation 4.2.6, which is

$$V = CK(\Delta P)^{1/2} \tag{4.2.6}$$

where V = velocity of flow (m/sec, ft/sec)
ΔP = pressure difference (kPa, lb/in^2)
K = unit constant ($K = 1.41$ for V in m/sec and P in kPa $K = 12.19$ for V in ft/sec and P in lb/in^2)
C = flow coefficient (C for well designed pitot tubes is approximately 1.0)

The volumetric flow rate can be calculated when the pipe diameter and the average velocity of flow within the pipe are known using equation 4.2.7.

$$Q = KD^2V \tag{4.2.7}$$

where Q = volumetric flow rate (L/min, gpm)
D = pipe diameter (mm, inches)
V = average velocity of flow within the pipe (m/sec, ft/sec)
K = unit constant ($K = 4.71\times10^2$ when Q is in L/min, D is in mm, and V is in m/sec. $K = 2.45$ when Q is in gal/min, D is in inches, and V is in ft/sec).

Because the pitot tube like the one in figure 4.2.13 measures the velocity at a single point and since the velocity varies across a pipe, it is usually necessary to measure velocity at several locations within a pipe to accurately determine V. When lower accuracy is acceptable and the velocity profile is symmetric V can be obtained from a single velocity measurement at a point located 3/4 the pipe radius from the pipe center.

Several insertion type flowmeters similar to the one in figure 4.2.14 are available commercially. These meters typically have tubes with several strategically located, upstream facing holes to "average" the total energy (kinetic plus potential energy) across the pipe. They also have another tube for sensing the potential energy (static pressure). The differential pressure (total-static pressure) obtained with these type of meters can be used in equations 4.2.6 and 4.2.7 to determine the volumetric flow rate. The flow coefficient, C, in equation 4.2.6 must, however, be determined by calibration since the total energy is usually not the true average.

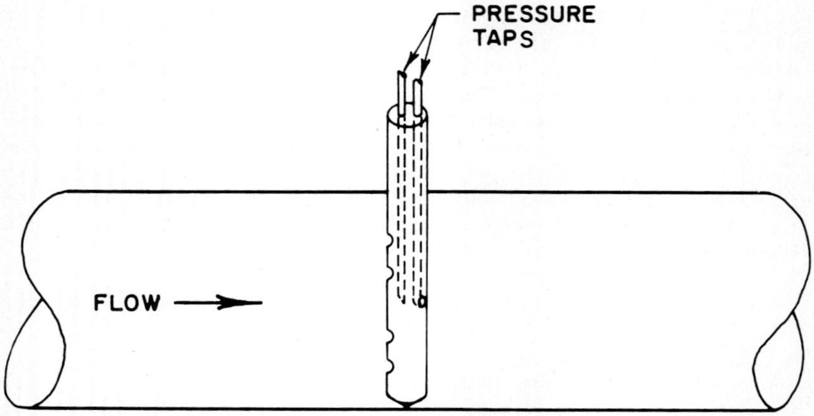

Fig. 4.2.14 A commercially available insertion type flowmeter.

4.2.1.c Flowmeter installation

Flowmeters perform best when velocity profiles are symmetric and flow does not rotate. Asymmetrical velocity profiles and flow rotation are caused by bends, valves and other fittings which significantly disturb flow. It is usually recommended, therefore, that flowmeters be installed in long, straight sections of pipe free of fittings which distort flow. There must be enough straight pipe on each side of the flowmeter to prevent upstream and/or downstream disturbances from affecting flowmeter performance.

Some flowmeters are more sensitive to flow disturbances than others. The length of straight pipe required upstream of flowmeters varies from 5 to 50 pipe diameters. The minimum length of straight pipe required downstream of flowmeters is 5 to 10 pipe diameters. Thus, flowmeters require anywhere from 12 to 60 pipe diameters of straight pipe free of fittings which distort flow for best performance.

When an adequate length of straight pipe is not available, flowmeter accuracy can frequently be improved by in-place calibration and/or by installing a flow straightener. In-place calibration involves determining the characteristic curve for the flowmeter after it has been installed in the pipe. This procedure allows the effect of flow distortions to be included in the characteristic curve.

Four different flow straightener designs with varying abilities to correct velocity profile distortions and reduce flow rotation are shown in figure 4.2.15. Tube bundle straighteners effectively reduce flow rotation, adequately correct asymmetric velocity profiles, but create substantial pressure loss. The AMCA and etiole straighteners remove flow rotation and have negligible head loss across them but do not correct velocity profile distortions. Perforated plate straighteners are easily installed between two flanges and effectively correct

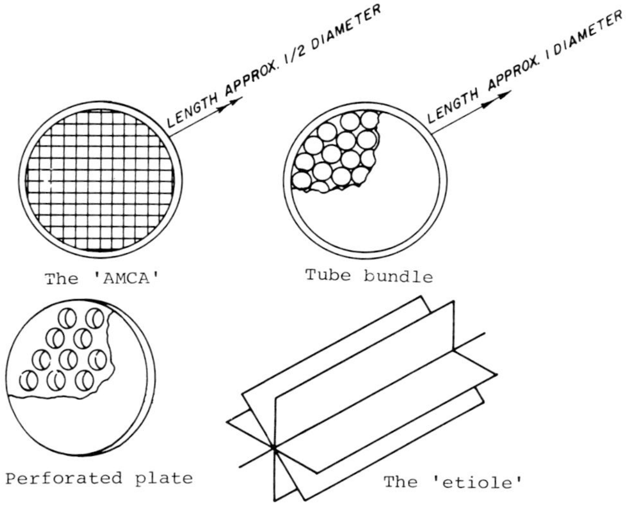

Fig. 4.2.15 Some typical flow straighteners.

distorted velocity profiles, but do not reduce flow rotation. Pressure loss across a perforated plate straightener can be excessive.

4.2.2 ***System maintenance***

Complete and timely system maintenance is a prerequisite for successful trickle irrigation. Proper maintenance will normally extend the life, improve the performance, and reduce the operating costs of trickle irrigation systems. A carefully developed and diligently implemented maintenance program will also reduce unplanned system shutdowns which interrupt often critical irrigation schedules. This is especially important with trickle irrigated crops since they commonly have high market values and/or limited root zones for storing water. The preventive maintenance requirements of the major components of a trickle irrigation system are discussed in the following sections.

4.2.2.a Pumping plants

Many trickle irrigation systems have pumping plants to lift water from ponds, reservoirs, lakes, streams, canals, and wells, and to provide pressure for trickle system operation. Pumping plants normally have either horizontal or vertical centrifugal pumps powered by either electric motors or internal combustion engines.

(i) Centrifugal pumps. Horizontal centrifugal pumps are used with surface sources and normally have mixed or radial flow impellers and volute type casings. The pump and power unit can be positioned above the water surface or in a dry pit below the water surface. This facilitates access to the pump for maintenance but makes it necessary to prime the pump (i.e., fill the suction line and pump casing with water) before pumps positioned above the water surface can be started.

Vertical centrifugal pumps can be used with either surface or ground water sources. These pumps often have several mixed or radial flow impellers and bowls (turbine-type casings) connected in series. The bowls guide water from the outlet of one impeller into the eye of the next impeller. The power unit may be located above the water surface or submerged beneath the pump (i.e., a submersible pump). When the power unit is above the water surface, a long drive shaft is required. Submergence eliminates the need for priming but makes maintenance difficult and expensive.

Table 4.2.2 lists preventive maintenance recommendations for centrifugal pumps. A flowmeter and pressure gauge should be installed downstream of centrifugal pumps and the flow rate and pressure closely monitored. In addition, a vacuum gage should be installed on the suction side of horizontal centrifugal pumps and watched closely to detect suction side problems. Any deviations in flow rate or pressure from desired levels should be promptly and thoroughly investigated.

During the irrigation season, pumps should be inspected at least every other day for irregularities in noise, bearing temperatures (pumps should feel "cool" to the touch), and leakage. Excessive noise may indicate problems including the presence of foreign material in the pump or piping, worn bearings, impeller or shaft damage, various suction line difficulties (with horizontal centrifugals) as well as a variety of other problems. Pump owners manuals often contain "trouble shooting" sections which are extemely useful in diagnosing and correcting many common pump problems associated with changes in flow rate and pressure, noise, vibration, leakage, and bearing temperature. The owners manual, dealer and/or manufacturer of the pump should be consulted whenever a problem is detected.

(ii) Power Units. Table 4.2.3 lists preventive maintenance recommendations for electric motors and control panels. Manufacturer's literature should be consulted for specific installation and maintenance instructions.

The amount of power being used by the electric motor(s) serving a pumping plant should be checked periodically. The kilowatt hour meter used by the power supply company for billing purposes and equation 4.2.8 (or other equation provided by the power supply company) can be used to determine power use.

$$HP = \frac{(R)(K)(M)}{(0.2072)(t)} \tag{4.2.8}$$

where HP = horsepower

R = number of revolutions of power meter disk during time t

K = disk constant (do not confuse with dial constant)

M = multiplier (ratio of current transformers used to rating of meter)

t = time for R revolutions of meter disk

TABLE 4.2.2

Preventive maintenance for centrifugal pumps.

Preparation for off-season

(a) drain all water from pump and piping (it may not be practical to remove submerged vertical pumps from wells, lakes, reservoirs, sumps, etc. to drain them)
(b) if possible, remove suction lines from water source and store
(c) make sure all oil- or grease-lubricated bearings are well covered with lubricant
(d) cover shaft and other exposed metal with protective lubricant to prevent corrosion
(e) remove the suction cover or volute of horizontal centrifugal pumps to check wear ring and impeller wear, and to clean debris from the impeller and volute
(f) remove packing gland and packing to check wear on shaft sleeve
(g) repack pump (do not tighten)
(h) make sure all passages for liquid are not obstructed
(i) loosen all belt drives and insert a piece of grease-proof paper between belts and pulley

Preparation for use before the irrigation season

(a) if there is a trash screen, make sure that it is clean and properly installed
(b) make sure foot valve on suction line of horizontal centrifugal pumps operates properly
(c) install and/or check suction line (must be air tight)
(d) make sure all passages for liquid are not obstructed
(e) tighten packing gland to proper setting
(f) change bearing oil and/or lubricate bearings with grease gun
(g) check impeller adjustment of vertical turbine pumps
(h) check if vertical turbine pumps or suction lines of horizontal centrifugal pumps are adequately submerged (by checking static water level)
(i) start pump and check noise, vibration, leakage, flow rate and pressure, and bearing temperature (after one hour of operation)
(j) check drawdown in wells (to determine if there is proper submergence during pumping)

During the irrigation season

(a) at least every 2 days check
- noise
- vibration
- leakage
- bearing temperatures
- flow rate and pressure
- trash screen (must be clean and properly placed)
- pumping level (to determine if there is proper submergence of pump or suction line)

(b) lubricate pump as per manufacturer's recommendations

TABLE 4.2.3

Preventive maintenance for electric motors.

Preparation for the off-season
- (a) clean dust, debris, and caked-on dirt and oil from motor
- (b) visually check motor winding insulation
- (c) lubricate all bearings
- (d) cover motor to protect against rodents, insects and dust being sure to provide ventilation between motor and cover to prevent condensation
- (e) lock service cabinet in "off" position
- (f) cover exposed service cabinets to protect against moisture and dust

Preparation for use before irrigation season
- (a) change motor bearing oil and/or use grease gun to lubricate bearings
- (b) change oil in reduced voltage starters
- (c) clean all debris, vegetation and rodent or insect nests from motor
- (d) check that motor ventilation vents are open and unobstructed
- (e) make sure safety shields are attached and functioning
- (f) check that overhead service lines are free of tree branches and other physical obstructions
- (g) replace conductors having frayed, cracked, or worn insulation
- (h) make sure all conduit or shielded cables are in good condition
- (i) test all coils and heaters for continuity and shorts and clean all magnet surfaces
- (j) make sure the interior of service cabinet is free of moisture, corrosion, insects, rodents, snakes, etc.
- (k) check and tighten all electrical connections
- (l) make sure all contact points are free of corrosion and pitting (clean copper contacts with fine sandpaper or file; replace replace all contacts that are severely pitted)
- (m) operate all moving parts by hand before applying power

During the irrigation season
- (a) at least every 2 days check
  - noise
  - vibration
  - temperature of both windings and bearings
  - that ventilation screens are clean
- (b) lubricate as per manufacturer's recommendations
- (c) check electrical demand of motor (using the demand meter and equation 4.2.8)

TABLE 4.2.4

Preventive maintenance for internal combustion engines (after Pair et al., 1983).

Preparation for the off-season

(a) run engine to thoroughly warm up oil in the crankcase
(b) stop engine and drain crankcase oil
(c) replace drain plug and refill crankcase with high-grade engine oil
(d) start engine and run slowly for two minutes to complete oil distribution on all surfaces
(e) stop engine and remove all spark plugs
(f) pour 60 mL (2 oz) of engine oil into each spark plug hole
(g) with ignition switch off, crank engine for several revolutions to distribute this oil over the cylinder walls and valve mechanism
(h) replace spark plugs
(i) drain oil from crankcase
(j) drain cooling system and close drain cocks (including block, water pump, heat exchanger, oil cooler and radiator)
(k) drain all fuel from tank, lines and carburetor bowl, replace all plugs and close drain cock (if liquified propane gas is used, drain vaporizer-regulator)
(l) lubricate all accessories and seal all openings airtight with weatherproof masking tape including air cleaner inlet, exhaust outlet, and crankcase breather tube
(m) check oil filler cap, gas tank and radiator cap
(n) spray all accessories and electrical equipment with suitable insulating compound
(o) insert a strip of grease-proof paper under the "V" belt pulley to prevent fan belt from bonding to pulley
(p) remove battery and store fully charged
(q) if engine is outside, cover with a waterproof covering

Preparation for use before the irrigation season

(a) remove all tape from openings that have been sealed
(b) open fuel tank valve
(c) shut water drain cocks and add coolant
(d) check oil drain plug - be sure it is tight, replace oil filter and add correct amount of oil to engine
(e) remove spark plugs and spray cylinder walls with a light engine oil
(f) replace spark plugs and crank engine several revolutions by hand to spread oil on cylinder walls
(g) fill fuel tank
(h) lubricate all engine accessories
(i) if a distributor is used, clean inside and outside of cap
(j) inspect cap and rotor for cracks, lubricate distributor sparingly with suitable lubricant (if magneto is used, inspect breaker points for wear and gap, and lubricate rotor)

TABLE 4.2.4 (continued)

(k) check all terminals and electrical connections
(l) if oil bath air cleaner is used, clean and fill with correct grade of oil
(m) start engine, run slowly for a few minutes, watch oil pressure and if it fails to come up to correct reading, stop engine at once and investigate the cause
(n) check oil level in crankcase, bring oil level up to proper mark on dipstick

During the irrigation season

(a) at least every 2 days check engine
- temperature
- oil pressure
- fuel consumption
- vibration
- noise

(b) change oil and lubricate as per manufacturer's recommendation

---

Electric motor operating temperature is an extremely important maintenance parameter. When a properly selected electric motor is operating at rated load it should feel "warm" to the touch. Excessive operating temperatures drastically shorten expected motor life and indicate the presence of a variety of pumping plant problems. The "trouble shooting" section of the owners manuals for the motor and/or pump will provide useful information for identifying the cause of motor overheating. If the problem can not be identified or corrected, dealers and/or manufacturers should be consulted.

Gasoline, diesel, liquified petroleum gas, and natural gas internal combustion engines are used to drive trickle irrigation pumps. Manufacturer's literature will provide instructions for operating and maintaining these engines. Care should be taken not to overload internal combustion engines if long-life and minimum repair costs are to be obtained.

Table 4.2.4 lists recommended procedures for preparing internal combustion engines for storage and use as well as recommendations for preventive maintenance during the irrigation season. Table 4.2.4 will provide guidance when specific manufacturer's instructions are not available.

4.2.2.b Filtration equipment

Filtration requirements of trickle irrigation systems vary widely depending upon emission device design and construction, and irrigation water quality. This section covers the inspection and maintenance of centrifugal separators, screen filters, and media filters. It does not include prefiltration devices such as trash racks or screens, or sediment tanks or settling basins. Table 4.2.5 summarizes preventive maintenance measures for filtration equipment.

TABLE 4.2.5

Preventive maintenance for filtration equipment

<u>Preparation for the off-season</u>
- (a) flush and drain filtration equipment
- (b) inspect interior components of media filters, centrifugal separators, and screen filters for wear, damage, corrosion and other signs of deterioration
- (c) check condition of seals, gaskets, and valve seats
- (d) service valves (see table 4.2.10)
- (e) disconnect electrical conductors from power source
- (f) check that electrical conductors are not frayed, worn or broken

<u>Preparation for use before the irrigation season</u>
- (a) make sure all electrical connections are clean and tight
- (b) check that all electrical contacts are free of corrosion, dirt, and wear
- (c) verify that filtration equipment, including the automatic control system, operates properly
- (d) remove media filter tank covers and inspect the media for proper level and cleanliness
- (e) check backwash restrictor valve for proper adjustment

<u>During the irrigation season</u>
- (a) at least every other day
  - be sure that filtration equipment and associated automatic controls are operating properly
  - check the need for manual flushing
- (b) at least monthly
  - remove media filter tank covers and inspect the media for proper level and cleanliness
  - check backwash restrictor valve for proper adjustment
  - inspect all couplings and connections for leaks
  - inspect all components of automatic control systems including hydraulic tubing, screens, and pressure regulators
- (c) service valves as per manufacturer's recommendations

(i) <u>Centrifugal separators</u>. Centrifugal separators can efficiently remove suspended mineral particles above approximately 100 micrometers. Their efficiency is proportional to the amount of head loss through the separator, with greater head loss producing a greater removal efficiency. They are most commonly used when pumping out of a well which produces a large amount of sand, or in particularly dirty waters which would overload screen or media filters.

System flow rates should be checked occasionally to assure that the separator is operating in the flow range recommended by the manufacturer. Excessive flow rates will cause higher head losses and higher power requirements. Flow rates below those recommended by the manufacturer will substantially decrease sediment removal efficiency. In cases where the flow rate varies beyond acceptable limits, it is preferable to manifold smaller units together than to use a single larger unit. For example, if farm development plans call for the initial development of 100 acres (40.5 ha) having a flow requirement of 500 gpm (1900 L/min) and the development of 100 additional acres (40.5 ha) in 5 years, then a single 500 gal/min (1900 L/min) unit should be installed initially with a second 500 gal/min (1900 L/min) unit installed later in parallel, rather than one 1000 gal/min (3800 L/min) unit installed initially. Day to day variations in the flow requirement can be handled in the same way through the use of isolation valves on the separators.

In general, separators require little maintenance. They should be checked frequently for proper operation and sequencing of automatic purge controls, if so equipped. Manual purge valves should be operated frequently to prevent excessive solids build up in the collection chamber. Proper purging frequencies and durations are determined through experimentation. Figure 4.2.16 shows approximate weights and volumes of sediment collected for differing concentrations and flushing frequencies per 100 gal/min (78 L/min) of system flow, assuming 100% removal efficiency. Separators, particularly those of lighter weight construction, are susceptible to wear from suspended sediments. The result will be a gradual loss of efficiency and will show up in more frequent media and screen filter flushing cycles, and in increased emitter clogging rates.

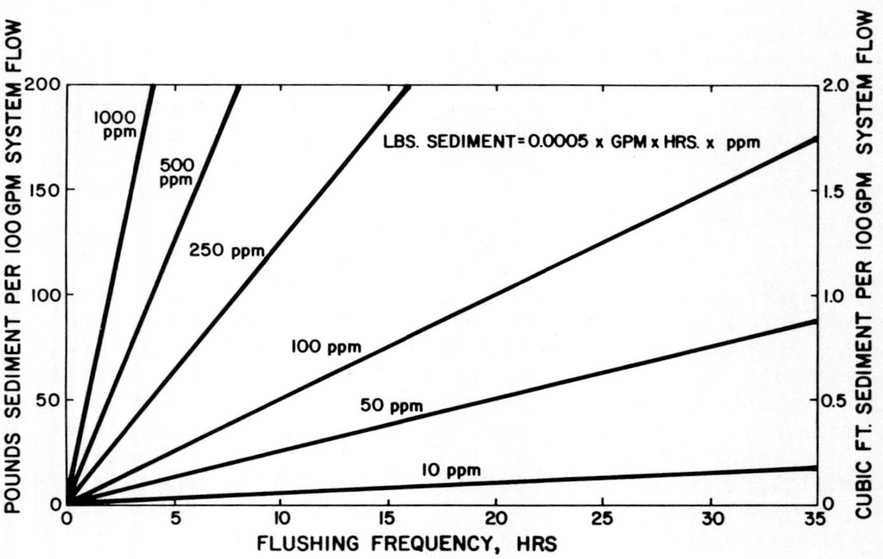

Fig. 4.2.16 Sediment accumulation in centrifugal separator per 100 gpm (378 L/min) of system flow.

(ii) Media filters. Media filters, or sand filters as they are commonly known, are widely used in trickle irrigation. They are particularly effective in removing suspended materials not removed by centrifugal separators, i.e., finer sediments and suspended organics, and are generally very reliable and require little maintenance. Proper filter selection and proper set up and adjustment, however, are crucial to satisfactory operation. This includes matching the filter media with the selected emission device in consideration of existing water quality and the effects of other filtration devices, i.e., screens or separators. It is generally recommended that media material be fine enough to retain all particles larger than one-sixth the size of the smallest passage in the trickle system.

Flow rates through each filter should not exceed 25 gpm per square foot (0.10 L/cm^2) of filter area, and desirably should be between 15 (0.06 L/cm^2) and 20 gal/min per square foot (0.08 L/cm^2). The lower values are recommended for particularly dirty source waters. Media filters are cleaned by reversing the flow sequentially through each tank so that the filter bed or media is lifted and expanded allowing trapped sediment to be released through turbulent mixing. Figure 4.2.17 is a schematic of the normal filtering process and the backwash cycle. Backwashing can be accomplished manually, but automatic systems based upon timers or pressure differential switches are becoming more common.

Setting the backwash flow rate is perhaps the most important adjustment for media filters. Excessive backwash flow rates will expand the media to the point that the media itself is expelled from the tank. Insufficient backwash flow will not expand the media enough to purge all the entrapped sediment. The backwash flow rate must be properly adjusted to achieve maximum filter performance.

Backwash flow rates vary with the type of media. Some general guidelines are presented in table 4.2.6 for different media types. The exact flow rate should be determined experimentally in the field.

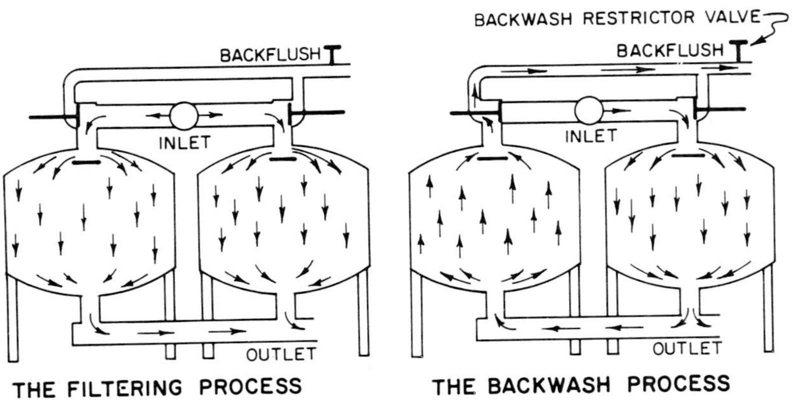

Fig. 4.2.17 Schematic of the normal filtering and flushing cycle of a media filter.

TABLE 4.2.6

Approximate backwash flow rates for different filter sizes and media types (courtesy Yardney Corp.).

Media type	Minimum backwash flow (gal/min) by filter size[a] 18"	24"	30"	36"	48"
No. 8 Crushed granite	51	91	141	201	360
No. 11 Crushed granite	26	48	74	105	188
No. 16 Crushed silica sand	32	57	89	126	225
No. 20 Crushed silica sand	26	48	74	105	188

[a] Rates are for tanks of indicated diameter oriented vertically.

The following procedure shows how to adjust the backwash flow control in a newly installed system:

(1) Turn on the system and operate until design pressure and flow rate are attained.
(2) Open the backwash restrictor valve a small amount.
(3) Release all entrapped air from each tank by partially opening and closing each tank valve.
(4) Close the valve on one tank. This changes that tank from filtering mode to backwash mode. Run backwash water through a screen or other sampling device.
(5) Gradually open the backwash restrictor valve until a small amount of media from the backwash water appears on the screen.
(6) When media begins to show in the backwash water, close the backwash restrictor valve until the water is essentially clear of media. A trace of media is acceptable since it is desirable that the lighter granules (fines) in the bed be allowed to wash out. After completing the above adjustments, all tanks should be backwashed extensively to remove contaminants and fine material usually found in newly installed media.

Filters should be backwashed when the pressure differential across them reaches approximately 35 kPa (5 lb/in^2). If the source water is particularly clean, the filters should be flushed at least daily to prevent migration of sediment through the filter bed or consolidation of the sediment in a layer above the bed. Automatic systems can be equipped to flush on any fixed time interval and usually have an automatic pressure differential override.

Water hammer can develop in the backwash discharge line if the flushing control valves close too rapidly. This can be corrected by adding a vacuum relief valve on the backwash line at the upstream filter.

Air can become entrapped in the top of the filters causing poor filtration and excess headloss. This is normally released during each flushing cycle but it is preferable to provide a continuous type air relief valve at the head of the filter manifold to prevent this occurrence.

(iii) Screen filters. Screen filters are generally used as a backup for media filters, but may be used alone if the source water is not particularly contaminated or if the filtration requirement is not as great. Filter screen material is usually woven synthetic or stainless steel mesh, but perforated or slotted sheets or rolls may be used as well as wire screens similar to well screen. Selection of the proper screen size should be made in consideration of the same factors affecting selection of other components of the filtration/irrigation system.

Screens are cleaned either manually or automatically in a variety of ways. In the simple single cartridge systems, the filter screen must be removed and washed by hand. Others are cleaned in a similar manner to media filters, by reversing the flow and "backwashing" them. Other systems are cleaned by "through flushing" or opening a valve on the end of the filter barrel to accelerate the flow through the filter and wash them clean (figure 4.2.18). Systems must be designed to assure that adequate pressure is available to achieve the minimum flushing flow rate or velocity recommended by the manufacturer.

Flushing flow rates recommended by the screen manufacturer should be carefully followed. Synthetic screen materials are susceptible to tearing or distortion when flushing velocities or pressures are excessive. Inadequate cleaning will occur if flushing rates are lower than recommended. Flushing is usually initiated when the differential pressure across the screen reaches approximately 5 lb/in^2 (35 kPa).

Filters should be disassembled seasonally and inspected for damage. If screens become clogged they may be cleaned with a brush. Use a nylon brush for the synthetic and lighter weight stainless steel meshes and a wire brush for the heavier wire meshes.

Concentrated chlorine or fertilizer solution may corrode filter components, including stainless steel, and should not be left in the filters unless specifically warranteed by the manufacturer to withstand attack by the solution in question. (See chapter 4.3, table 4.3.1 on corrosion resistance to chemicals.) It may be necessary to inject chemicals downstream of filters if this becomes a problem.

4.2.2.c Chemical injection equipment

Fertilizers and other chemicals are normally injected into trickle systems using either the pressure difference technique, an eductor, or a metering pump. These three principal injection methods are diagrammed in figure 3.2.6.

The injection system in 3.2.6.a uses a venturi or other eductor to create an area of low pressure within the trickle system where chemicals may be added. The regulator valve is adjusted to meter in the desired amount of chemical.

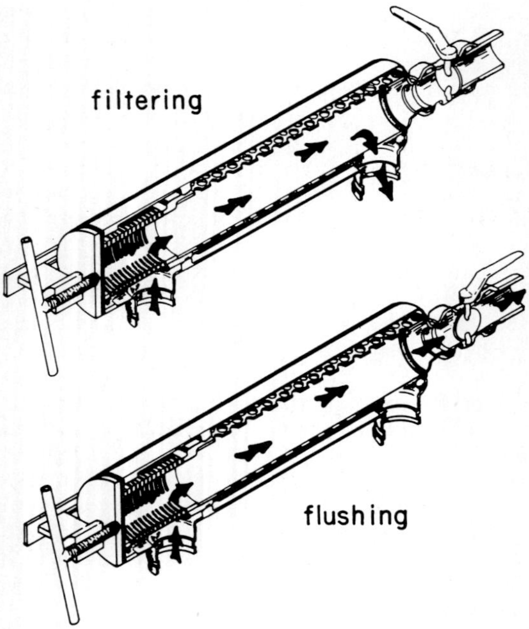

Fig. 4.2.18 Schematic of the normal filtering and flushing processes for a screen filter.

In a pressure difference type system (figure 3.2.6.b,c) chemical injection occurs when there is a pressure difference across the pressurized chemical holding tank and the regulating valves are open. The pressure difference is created by a valve, pressure regulator, or other device which causes a sufficient pressure change. A disadvantage of differential injection is that the concentration in the tank is diluted as injection continues.

The use of a metering pump is the most precise chemical injection technique. Packed-plunger and diaphragm (both mechanically and hydraulically activated) pumps similar to the ones in figure 4.2.19 are the most common types of metering pumps used. Electric and water motors, gasoline engines, or belts driven by the pumping plant power unit power the metering pumps. Discharge and suction side check valves require regular and careful maintenance since dirty and/or worn valves and/or seats limit pump capacity. Packed-plunger pumps require periodic adjustments to compensate for packing and plunger wear. Diaphragms must be checked and the oil in hydraulically activated diaphragm pumps changed periodically.

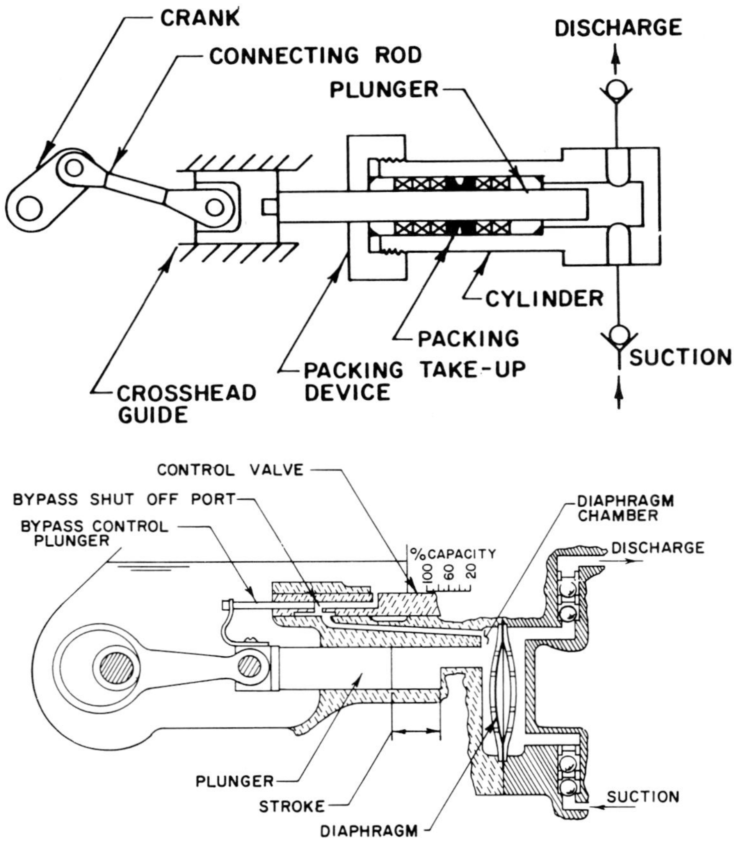

Fig. 4.2.19 Schematic of packed-plunger and mechanically-activated, diaphram pumps typically used for chemical injection.

Table 4.2.8 lists several preventive maintenance measures for chemical injection equipment. Maintenance of regulating, check and vacuum release valves is especially important. Each of these valves must be clean and operating properly to accurately meter chemicals into the trickle system and to prevent chemical contamination of the water source during water or power failure. Screens and strainers for removing foreign materials must be clean and in good condition. Metering pumps, motors, and gear reducers must be clean, lubricated, and properly adjusted. The corrosive properties of many injected fertilizers and chemicals require that all injection system parts directly contacting chemicals be corrosion resistant and that injection systems be thoroughly flushed with water after each use. Antifreeze is sometimes used to flush metering pumps.

TABLE 4.2.8

Preventive maintenance for chemical injection equipment.

Preparation for the off season

(a) thoroughly flush and drain injection equipment
(b) clean all exterior injection equipment surfaces
(c) clean interior surfaces of corrosion, etc.
(d) lubricate pumps, motors and gear reducers
(e) change oil in hydraulically activated diaphragm pumps
(f) check condition of diaphragm
(g) examine condition of check valves and backflow prevention equipment
(h) check condition of pump packing
(i) clean all filters/strainers/screens
(j) visually inspect all electrical components
(k) loosen all belt drives and insert a piece of grease-proof paper between belts and pulley
(l) cover shaft and other exposed metal with protective lubricant to prevent corrosion
(m) check condition of gaskets and seals

Preparation for use before irrigation season

(a) make sure all liquid passages are free of obstructions
(b) check oil levels in motors, pumps and gear reducers
(c) check motor/pump/gear reducer noise, vibration, bearing temperature
(d) check for leaks
(e) verify injection rate(s)

During the irrigation season

(a) every 24 to 48 hours of operation
- visually inspect hoses, valves, pumps, motor, tank, and other injection equipment
- check for leaks
- check oil levels in motors, pumps, gear reducers
- check motor/pump noise, vibration, bearing temperature
(b) lubricate motor, pump, and gear reducers as per manufacturers recommendations
(c) flush pump and tank after every use
(d) periodically clean filters/strainers/screens

4.2.2.d Flowmeters

The maintenance requirements of flowmeters vary with meter type. Preventive maintenance measures for flowmeters are listed in table 4.2.9.

Rotating mechanical flowmeters must be cleaned, lubricated, and checked for wear and damage each year. Electrical conductors, connections and contacts in ultrasonic flowmeters and electronic pressure sensing devices used with pressure difference and insertion meters need to be checked regularly for wear, dirt and corrosion. Manometers and pressure gauges must be free of leaks and other signs of deterioration. Insertion meters, orifice plates, venturi tubes, and elbow meters need to be examined for wear and corrosion annually. The accuracy of all flowmeters needs to be verified at least once a year. This may be accomplished by comparing the volume and/or volumetric flow rate determined with the flowmeter being verified to that measured with a test meter. Test meters can sometimes be obtained from an electric power supplier or flowmeter company. Tanks of known volume can be used in lieu of test meters. The volumetric flow rate equals the volume of the tank divided by the time required to fill the tank (equation 4.2.1).

TABLE 4.2.9

Preventive maintenance for flowmeters.

<u>Preparation for the off-season</u>
- (a) clean all dirt, grease, moisture, and other foreign material from the exterior of the meter
- (b) remove accumulations of organic matter, dirt, corrosion, etc. from interior surfaces of meter
- (c) check for bent impeller blades, worn blade and orfice edges, pitted or eroded meter sections
- (d) check bearing and shaft wear in rotating meters
- (e) check for damaged flow straighteners
- (f) remove batteries from ultrasonic meters and store fully charged
- (g) if possible, store meter in clean, dry place
- (h) lubricate flowmeters

<u>Preparation for use before the irrigation season</u>
- (a) clean all debris, insect and rodent nests, and other foreign material from meter
- (b) repair leaks in gaskets and fittings of meter and pressure sensing equipment
- (c) make sure that electrical conductors are not frayed, worn or broken
- (d) check that all electrical connections are clean and free of corrosion
- (f) verify flowmeter accuracy

<u>During the irrigation season</u>
- (a) every time the flowmeter is read inspect the meter for
  - missing hardware
  - loose screws
  - fogged or broken register lens
  - leakage
  - other signs of wear or deterioration
- (b) lubricate rotating meters as per manufacturers recommendations
- (c) monitor pressure drop across meter and clean meter filters/strainers/screens when needed

4.2.2.e Valves

Valves form an integral part of trickle irrigation systems. The nature of the valving for a given installation will depend on the level of automation, degree of pressure regulation, and number of sets required. Several types of automatic, manual, check and air release valves are used in trickle systems.

Of these, the most trouble free and reliable are manual and check valves, followed by air release and automatic valves. Due to the relative complexity of automatic valves and the contaminated (sediment laden) environment in which many operate, they are most subject to failure and most in need of preventive maintenance. Simpler valves require less maintenance and may require only a periodic inspection. Table 4.2.10 summarizes the preventive maintenance requirements of trickle system valves.

TABLE 4.2.10

Preventive maintenance for valves.

<u>Preparation for the off season</u>

(a) completely drain all valves
(b) lubricate valves
(c) clean corrosion, dirt and other foreign material from valves
(d) make sure valves are open

<u>Preparation for use before the irrigation season</u>

(a) after disassembling, thoroughly clean and inspect automatic diaphragm valves
(b) inspect valve packing
(c) verify that valves operate properly

<u>During the irrigation season</u>

(a) inspect valves weekly for leakage and proper operation
(b) lubricate valves as per manufacturers recommendations

(i) <u>Automatic valves</u>. Automatic (or diaphragm actuated) valves are typically found at pump and filter stations to regulate mainline pressure, control backwash cycles in the filters, or control flow through branching mainlines. Solenoid control valves may be found on submains or laterals to control the flow of water to individual blocks. Their primary function is on-off service; however, they may be equipped for pressure regulation and back flow prevention. They are typically electrically operated from the control center. It is possible to perform the functions of some automatic control valves with manual valves equipped with remote controlled operators. Although this may be desirable with some large valves, it is normally not justified by the expense involved and is not generally found in trickle irrigation systems.

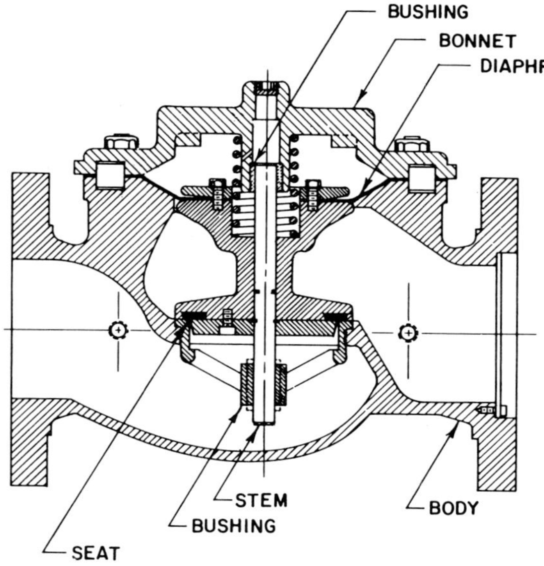

Fig. 4.2.20 Schematic of an automatic-diaphragm valve.

Automatic valves require periodic maintenance to assure satisfactory operation. Maintenance schedules depend on the use of the valve and the cleanliness of the water. At a minimum, it is recommended that all diaphragm valves be disassembled and thoroughly cleaned and inspected at least once a year. Most valve manufacturers provide for easy servicing or replacement of major valve components. This can usually be accomplished without removing the valve from the line. Of particular importance in servicing diaphragm valves is cleaning deposits from the valve stem. The valve stem rides in bushings normally provided in the valve seat and bonnet (figure 4.2.20). Deposits on the stem may cause the valve to bind and prevent normal operation. Cleaning requires polishing with No. 400 wet or dry sandpaper in water. Dipping in a 5 percent muriatic (hydrochloric) acid solution may remove encrustations consisting principally of lime deposits. The polished valve stem should be clean and free of deposits, scratches, abrasion or corrosion. Similarly, the bushings should be clean and permit free operation of the valve stem. Valve stems and bushings are subject to wear from suspended sediments and should be replaced if either too loose or too tight. Valve seats (usually integral with the lower stem bushing) should be carefully examined and replaced if worn or defective. Resilient parts including diaphragm, seating disc, and O-ring seals should be replaced if any signs of wear or chemical attack are apparent.

Diaphragm valves may be fitted with a number of auxiliary controls and features to provide operating flexibility and convenience. Some of these are listed below:

(1) Pressure reducing control. This is a small valve which senses pressure changes in the outlet of the main valve and adjusts the pressure in the bonnet or valve cover to compensate for any changes. Improper operation may result from contamination, obstructions, improper assembly, or damaged or worn parts.

(2) Isolation valves. These are small valves in the pilot control system which are used to isolate the system from mainline pressure. They must be open during normal operation.

(3) Opening and closing speed controls. These are small adjustable valves in the pilot control system which regulate the opening and closing speed of the main valve by throttling the fluid entering or exiting the cover. They are subject to clogging by fine sediments if adjusted too tightly.

(4) Wye strainer. Wye strainers are installed in the pilot control system to protect its components from contamination. It must be disassembled and cleaned periodically. Cleaning may be facilitated, however, by installing a small valve or hose bibb in place of the plug in the strainer to wash out the screen.

After reassembly, it is important to release all air from the main valve cover and at all high points in the pilot control system. This is accomplished by loosening, but not removing, the cover plug and pilot control fittings while the valve is under pressure. After all the air has been released, the fittings may be retightened.

(ii) Manual valves. Manual valves are found throughout trickle systems and are used for control purposes, isolation, and flushing. They include gate, globe, butterfly and ball valves. Butterfly valves are commonly used in the larger sizes [greater than 6 inches (15.2 cm)] and gate valves are used for the smaller sizes. They may be used above or below ground depending upon the design philosophy. Hose ends may be equipped with gate valves for flushing purposes; however, a more common practice is to merely pinch the hose closed and secure it with a "figure 8".

Manual valves are generally trouble free and require little maintenance or service throughout the life of the system. Packing should be inspected around all valve stems and should be replaced or glands adjusted when leaks occur. Plastic ball valves are sometimes used on hose risers or around chemical injection equipment. Plastic components may be susceptible to cracks or other physical damage, particularly freeze damage. All parts of the system must be drained prior to the onset of freezing temperatures to prevent damage.

Gate and butterfly valves are perhaps the most commonly used manual valves in a trickle system. They are used for on-off service and require little maintenance with the possible exception of the occasional replacement of the rubber seat on butterfly valves. Globe valves may be used for throttling. Rubber seats can generally be easily replaced in these valves.

(iii) Other valves. Other valve types found in drip systems include check valves and air release valves. Check valves are normally used only at the pump station and particularly when pumping out of a sump or deep well. Air release and vacuum relief valves are located at high points on mains, submains, and laterals.

Air release valves are generally placed at high points in mainlines, submains, and pump stations. They release entrapped air on system startup, and allow air to enter the pipeline under conditions of negative pressure. Air release valves are of two main types: the large orfice type which seals completely when the pipeline is pressurized, and the combination large and small orifice type which releases entrapped air while the system is operating. The second type is generally found in the field. Air release valves should be periodically checked for proper operation. They should not leak (except when releasing air) and floats should not stick or bind. It is occasionally necessary to replace the rubber seats which are subject to cracking, particularly in hot dry weather.

Check valves and pressure relief valves require little routine maintenance and if damaged or defective should be sent to the factory for repair. No repairs or adjustments should be attempted on pressure relief valves and should be left to qualified personnel.

4.2.2.f Controllers

Various electro-mechanical and electronic controllers are used with automated trickle systems. Controllers range from mechanical clocks that open/close a single valve on a preset time schedule to microcomputers, which are programmed to interrogate a series of soil and/or climatic sensors, decide when to begin and end irrigation, start-stop pumps and open-close valves to accomplish the irrigation, and remember how much water and fertilizer were applied to each block within the field. Many controllers are also able to diagnose system malfunctions and take corrective action. Some even turn the system off during rain storms and then restart the system when the storm ends.

A timer-type of controller uses a clock (either solid state or motor driven electric) as the means for programming the starting and sequence of irrigation. The controller supplies electrical or hydraulic power to activate remote on-off valves located strategically throughout pipe network. From one to thirty or more valves may be controlled by a controller. Communication between the controller and valves is via wires, hydraulic or pneumatic conduit or radio telemetry.

In addition to controlling on-off valves, microprocessor and microcomputer based controllers can be programmed to control pumps, injection equipment, and filters etc., using data from tensiometers, pyranometers, evaporation pans, thermocouples, humidity meters, anemometers, flowmeters, pressure transducers and other sensors. These controllers poll soil and/or climatic sensors according to an irrigator specified schedule. The controller uses a program to compute the need for irrigation of various crops and blocks within the field using these data. It then operates the pumps, filters, injection equipment, and valves needed to accomplish the irrigation. Data from flowmeters and pressure sensors are used to determine the need for such things as flushing and to detect system malfunctions. Because of their complexity, controller and sensor repairs are usually best handled by trained service personnel. There are, however, preventive maintenance measures which the irrigator can perform. These measures are summarized in table 4.2.11.

TABLE 4.2.11

Preventive maintenance for controllers and sensors.

Preparation for the off-season
- (a) clean controller and sensors
- (b) check condition of controller panel seals
- (c) remove and store batteries
- (d) flush and drain hydraulic control conduits
- (e) disconnect field wires
- (f) check for frayed, worn or broken electrical conductors

Preparation for use before the irrigation season
- (a) make sure all electrical connections are clean and tight
- (b) check that electrical contacts are free of corrosion, dirt, and wear
- (c) inspect all hydraulic and pneumatic control conduits for leaks
- (d) verify that all accessory equipment and sensors operate properly

During the irrigation season
- (a) visually check all external components weekly
- (b) disconnect field wires during electrical storms
- (c) disconnect batteries when controller is to be out of service for one week or more

Because dirt and moisture are the primary enemies of controllers, protective seals in the controller panels must be diligently checked and maintained. Batteries should be removed whenever the controller is to be out-of-service for periods of one week or more. Electrical conductors, connections, and contacts must be free of dirt, corrosion and wear. Hydraulic and pneumatic conduits should be inspected for leaks and the proper operation of sensors and accessories should be verified regularly. It is also recommended that field wires be disconnected during electrical storms.

4.2.2.g Pipelines

Pipe networks for trickle irrigation systems typically consist of main, submain or manifold, and lateral lines. Main lines convey water from the water source and distribute it to the submains. The submains provide water to the laterals which supply point source emitters or microsprinklers. For line source emitters, porous and single- or multi-chambered tubes which discharge water along their entire length are used in lieu of laterals. Preventive maintenance measures for pipelines are summarized in table 4.2.12.

TABLE 4.2.12

Preventive maintenance for pipelines.

<u>Preparation for the off-season</u>
- (a) while system is operating, check for leaks (mark all leak locations for subsequent repair)
- (b) flush and drain mains, submains, and laterals
- (c) open all valves
- (d) inspect pipes for corrosion (consult pipe supplier for protective measures)

<u>Preparation for use before the irrigation season</u>
- (a) remove bird and animal nests from pipelines
- (b) flush mains, submains and laterals
- (c) inspect pipes for leaks

<u>During the irrigation season</u>
- (a) inspect pipelines for leaks
- (b) flush mains, submains and laterals as required

Polyvinyl chloride (PVC), steel and asbestos cement (AC) are the most common materials used for mains and submains while laterals are usually made of black polyethylene. PVC and polyethylene pipes resist corrosion for most water conditions and have a relatively long life when protected from surge pressures. Steel pipes are strong and durable, but subject to corrosion on both the inside and outside of the pipe. Asphalt and asbestos paper are often used to protect the outside of steel pipes, whereas cement and asphalt are the most commonly used inner coatings. AC pipe has a high resistance to corrosion and long life expectancy, but is subject to cracking when pipeline support is unstable. Use of AC pipe is not recommended with soils and/or waters high in water soluble sulfate.

PVC, polyethylene and steel pipelines may be either buried or laid on the ground surface. AC mains and submains are buried. Leaks in above ground pipes are more easily detected and repaired than those in buried pipe. Surface pipelines are, however, subject to damage by farm machinery, animals, vandals, etc. Above ground pipelines should be inspected regularly for leaks during the irrigation season.

Provisions for removing water from pipelines (either by gravity or pumping) should be provided to facilitate repair and to protect pipelines in locations when subfreezing temperatures occur.

Pipelines should be flushed regularly to remove deposits of materials carried into the system in the irrigation water. This is especially important in submains and laterals where diminishing flow along the pipe causes velocities to steadily decline in the downstream direction. Significant deposits of silt and clay sized materials can even occur in trickle systems which have well designed and maintained filtration equipment. Provisions for flushing each main and submain pipe and every lateral should be provided. When mains and submains are being flushed field or block valves should be closed during flushing to prevent the suspended sediment from being carried into the field laterals and clogging emission devices.

The frequency of flushing varies from daily to twice per season (at the beginning and end of the season) depending on the rate of deposit accumulation. Experimentation is usually required to establish the proper frequency of flushing. It may be necessary to adjust the frequency of flushing during the irrigation season as the concentration and/or particle size distribution of suspended material entering the system changes.

Most problems occurring with the water transmission and distribution system can be prevented through proper design. A correct hydraulic design is especially important for trickle systems to achieve satisfactory pressure distribution and application efficiency. This includes selection of appropriate pipe material, pressure class and size, and proper location of blowoffs and air releases. Problems occurring through improper design should be referred to a qualified irrigation engineer for solution.

Air releases are crucial to satisfactory performance of the transmission and distribution system, particularly in varying topography. Maintenance and inspection of air release valves were previously covered in the section entitled "Valves". Air pockets which form in high spots on the pipeline cause a restriction to the flow and must be released either manually or automatically. Air locks are hard to locate and correct in an existing system.

4.2.2.h Emission devices

Emission devices include point and line source emitters which operate either above or below the ground surface and microsprinklers that spray water over the land surface. Emitters generally have smaller passages for discharging water and are more prone to physical, chemical and biological induced clogging than are microsprinklers. Chemical treatment and filtration of the source water to control emission device clogging is discussed in section 3.2, Water Treatment. Pipeline flushing to reduce clogging rates is discussed in section 4.2.2.g. Table 4.2.13 lists preventive maintenance measures for emission devices.

TABLE 4.2.13

Preventive maintenance for emission devices.

Preparation for the off-season
(a) spot check emission device discharge and pressure (compare to catalog to evaluate wear)
Preparation for use before irrigation season
(a) visually check emission devices for clogging, damage, and other signs of deterioration
During the irrigation season
(a) check discharge and pressure of critical emission devices monthly (low spots and ends of laterals)
(b) visually check emission devices for clogging, damage, and other signs of deterioration at least once during the irrigation season

Above ground emission devices can be damaged by machinery, animals, etc., while buried emitters are subject to attack by burrowing rodents. Emission device passageways often enlarge over time as they are eroded by flowing water. Emission device wear is especially pronounced when the water being discharged contains high concentrations of abrasive materials and is often indicated by decreased uniformity of application and increased system flow rates. The discharge and corresponding pressure of individual emitters can be measured and compared to catalog information to evaluate emission device wear.

It is impractical to make frequent checks of all emitters for clogging, particularly in large fields. The first obvious sign of clogged emitters may be a decrease in a total system flow rate or signs of moisture stress in the crop. Clogging can be detected early by periodically sampling a set of emitters. A systematic procedure is recommended where the most critical emitters in each field (low spots and ends of laterals) are checked monthly for pressure and flow. Any clogging tendency can thus be detected and corrective measures implemented. All emitters should be checked visually at least twice per year for clogging.

In-line emitters are less subject to damage than on-line emitters. Clogged or damaged in-line emitters must be replaced, usually by cutting the line, removing the emitter, and replacing it with a new one. Some online emitters may be returned to service by flushing them or otherwise removing the blockage. However, in a large operation it is usually more economical to replace the emitter than to clean it.

New trickle irrigation technologies, including new component design and water management techniques, have advanced the state of the art to drastically reduce emitter clogging and greatly improve overall system reliability and efficiency. Despite the claims of some manufacturers, however, there are no clog-free emitters and no panacea for a maintenance free system. In general, the most reliable and trouble-free systems have been those of simple construction with few moving parts.

REFERENCES AND SUPPLEMENTAL READING

Finkel, H.J., 1982. Handbook of Irrigation Technology. Volume I, CRC Press, Boca Raton, Florida, 369 pp.

Finkel, H.J., 1982. Handbook of Irrigation Technology. Volume II, CRC Press, Boca Raton, Florida, 223 pp.

Glanville, R.H., 1976. Metering. In Karassik, I.J., Krutzsch, W.C., Fraser, W.H. and Messina, J.P. (Eds.), Pump Handbook, McGraw-Hill. New York, pp. 10-231 to 10-234.

Hayward, A.T.J., 1979. Flowmeters: A Basic Guide and Source-book for Users. John Wiley and Sons, New York, 197 pp.

Howell, T.A., Stevenson, D.S., Aljibury, F.K, Gitlin, H.M., Wu, I.P., Warrick, A.W. and Raats, P.A.C., 1980. Design and Operation of Trickle (Drip) Systems. In Jensen, M.E. (Ed.), Design and Operation of Farm Irrigation Systems, Monograph 3, Amer. Soc. Agric. Eng., St. Joseph, Michigan, pp. 663-717.

Karassik, I.J., 1976. Centrifugal Pump Construction. In Karassik, I.J., Krutzsch, W.C., Fraser, W.H. and Messina, J.P. (Eds.) Pump Handbook. McGraw-Hill. New York, pp. 2-31 to 2-124.

Lansford, W.M., 1936. The Use of an Elbow in a Pipe Line for Determining the Flow in the Pipe. Eng. Exp. Sta. Bull. 289, Univ. of Illinois, Urbana, Illinois, 36 pp.

Pair, C.H., Hinz, W.W., Frost, K.R., Sneed, R.E. and Schiltz, T.J., 1983. Irrigation (formerly Sprinkler Irrigation). Fifth Edition, Irrigation Association, Silver Springs, Maryland, 686 pp.

Roberson, J.A. and Crowe, C.T., 1980. Engineering Fluid Mechanics. Houghton Mifflin Co., Boston, Massachusetts, 661 pp.

Streeter, W.L., 1966. Fluid Mechanics. McGraw-Hill, New York, 705 pp.

Schweitzer, P.A., 1972. Handbook of Valves. Industrial Press, New York, 180 pp.

Vennard, J.K. and Street, R.L., 1974. Elementary Fluid Mechanics. John Wiley and Sons, New York, 740 pp.

CHAPTER 4

MANAGEMENT PRINCIPLES

4.3 FERTILIZATION

D. E. ROLSTON, R. J. MILLER and H. SCHULBACH

4.3.1 *Introduction*

Trickle irrigation offers the opportunity for precise application of fertilizers to the soil. Because roots develop extensively in a restricted volume of soil wetted by drip irrigation, application of fertilizers through the irrigation system can efficiently place plant nutrients in that zone in which roots are of highest concentration and in which the water is also being removed from the soil system. At the same time, fertilizer application prevents nutrient deficiencies that can develop because of the limited soil volumes explored by roots. However, several basic principles must be followed in applying nutrients through trickle irrigation systems in order to place the fertilizer correctly, decrease potential nutrient losses, avoid excessive fertilizer application, and prevent clogging of the system by precipitated compounds.

4.3.2 *Criteria for applying fertilizers through trickle systems*

All chemicals applied through irrigation systems must meet the following criteria (Bucks and Nakayama, 1980). They must (1) avoid corrosion, softening of plastic pipe and tubing, or clogging of any component of the system, (2) be safe for field use, (3) increase or at least not decrease crop yield, (4) be soluble or emulsifiable in water, and (5) not react adversely with salts or other chemicals in the irrigation water. In addition, the chemicals or fertilizers must be distributed uniformly throughout the field. Achieving such uniformity of distribution requires efficient mixing, uniform water application, and knowledge of the flow characteristics of water and fertilizers in the distribution lines.

Distribution of fertilizers to a field depends on flow characteristics of water in the system and the uniformity of water application by emitters or orifices. Most systems are designed so that the flow rate of water and chemicals in the system is high enough to ensure uniform distribution of the chemicals throughout the distribution lines. Generally, nutrients are distributed uniformly if they are injected after the water distribution system is full of water and if nutrient injection is stopped before the irrigation period ends (Bester et al., 1974). Variability of water discharge at each emitter may be quite large and is dependent on quality control by the manufacturer. The standard deviation varies between 1 and 16% of the mean discharge of various kinds of emitters operated at various pressures (Davis and Pugh, 1974). A trickle irrigation system should be designed with no more than a 5% discharge variation, 94% emission uniformity (Karmeli and Keller, 1975). As a system clogs, discharge variation will increase and result in more variability of fertilizer application.

To avoid clogging, chemicals applied through trickle systems must meet certain requirements. The chemicals must be completely soluble. If more than one material is used in preparing a concentrated stock solution for subsequent injection into the trickle lines, the chemicals must not react with each other to form a precipitate. The chemicals must also be compatible with the salts contained in the irrigation water.

The particles of some solid fertilizers that meet solution requirements are coated with a clay or wax to prevent caking in storage. The coatings can cause scum to form on the surface or cause sludge to deposit on the bottom of stock solutions. Several precautionary measures can prevent such residues from reaching the emitters, such as locating the discharge tube some distance above the bottom of a stock solution tank and periodically removing any scum or sludge. Wetting agents can be helpful in emulsifying wax coatings and preventing scum formation. In some situations, it may be important to locate the injection point before the filter in the control head; this will minimize emitter clogging, because any contaminants will be filtered out.

The effect of various chemicals on the system's life expectancy depends greatly on the particular characteristics of the chemicals applied. For example, chemicals of low or high pH may corrode metal parts of the irrigation system, such as copper, iron, zinc, aluminum, brass, and bronze alloys. Therefore, the components of the system that come in contact with corrosive solutions should consist of stainless steel, plastic, or other noncorrodible materials. However, most irrigation systems are constructed entirely of plastic, and applied chemicals will not corrode pipes or emitters unless materials are applied that contain solvents. Precipitation of applied chemicals is a critical problem and must be controlled carefully to prevent clogging of the system. If in doubt about the mixing compatibility of chemicals, one should flush the lines thoroughly before applying a different chemical through the system.

Another clogging problem associated with fertilizer applications is the increase in algae or microbial populations due to increased amounts of nutrients in the water. Algae and other microorganisms produce slimes that can quickly clog filters and other parts of the irrigation system. To control the formation of slimes, bactericides of various types have been used with varying degrees of success, depending on conditions of their particular use. For situations in which microbes grow in sand filters, fertilizer should be injected downstream from the sand filter.

Applying nutrients through a trickle irrigation system can result in (1) increased fertilizer use efficiency, because the material is applied to that part of the soil in which the water and the roots are located (2) decreased labor and energy cost by making use of the water distribution system for distributing the nutrients, and (3) the ability to apply nutrients at various times of crop growth and not be dependent on the ability to get into the field with machines.

Possible hazards of applying fertilizers through trickle irrigation systems are those associated with potential clogging of the system if the pH and various mixes of nutrients are not managed carefully. Also, injection of various nutrients and chemicals into the irrigation system can result in possible contamination of the water supply if devices are not used to prevent backflow of nutrients into the well or other water source. Some of the chemicals applied through trickle irrigation systems are quite corrosive to metal parts and can also cause skin burning, if safety devices are not provided to protect workers against unplanned discharge or spillage of chemicals. Protective goggles, face shields, and clothing should be worn when making chemical dilutions, especially for materials like phosphoric acid.

4.3.3 *Equipment and methods for fertilizer injection*

Fertilizers can be injected into trickle irrigation systems by selecting appropriate equipment from a wide assortment of available pumps, valves, timers, computers, power sources, tanks, venturis, meters, and aspirators. Most injection systems are designed to accommodate other agricultural chemicals such as herbicides, nematicides, insecticides, fungicides, algicides, chlorine, and acids. The selection of the equipment will be based on the specific irrigation system. The primary considerations depend on the type of fertilizer to be applied. Most fertilizers are very corrosive to the metal components of pumps, fittings, valves, and filters. The goal should be to keep the system as simple as possible. Table 4.3.1 shows the severity of corrosion damage of fertilizers to common metals.

Acid should not be added to cement or asbestos cement lines. Plastic lines can be damaged when materials added to the system contain solvents or carriers that settle out in the lines under low velocities, interact with the plastic, and weaken the walls so that blowouts can occur.

Fertilizer solutions are preferably stored in chemically inert plastic containers. Metal tanks can corrode and can be the source of rust-like fragments which can plug emitters.

Two injection points should be provided, one before and one after the filter. This arrangement can be used to by-pass the filter if filtering is not required, and thus avoid corrosion damage to the valves, filters, and filter screens or to the sand media of sand filters. Furthermore, the discharge line from the fertilizer tank should have a filter, and similarly, the injection hose line should be equipped with an in-line hose filter or screen. The intake or suction side of the injector should be equipped with a filter or strainer. Injection points must be installed so that the injected fertilizers are properly mixed before the flow divides in several directions. Injection pumps must not be allowed to run dry. Low water cut-off or pressure-sensitive switches must be installed to protect pumps.

TABLE 4.3.1

Severity of corrosion damage of fertilizers to common metals (adapted from Martin, 1953).

Kind of metal	Calcium nitrate	Ammonium nitrate	Ammonium sulfate	Urea	Phosphoric acid	Di-ammonium phosphate	Complete fertilizer solution 17-17-10[a]
Galvanized iron	2	4	3	1	4	1	2
Sheet aluminum	0	1	1	0	2	2	1
Stainless steel	0	0	0	0	1	0	0
Phospho-bronze	1	3	3	0	2	4	4
Yellow brass	1	3	2	0	2	4	4
pH of fertilizer solution	5.6	5.9	5.0	7.6	0.4	8.0	7.3

0, none; 1, slight; 2, moderate; 3, considerable; 4, severe.

- Metal sheets stood in fertilizer solutions 4 days.
- Solutions made by dissolving 5 kg material in 400 L water.

[a] Commercial mixture made by mixing ammonium sulphate, di-ammonium phosphate, and sulfate of potash.

The size or capacity of the injection system depends on the concentration, rate, and frequency of application. Naturally, less fertilizer solution and more frequent applications require smaller, less costly units.

Water application rates and application times vary considerably depending on crop, emitter spacing, and greenhouse or container-plant applications. Vegetable crops grown in the field and orchards and vineyards may have moderate, to long irrigation periods, which allow for later starting of the injection and earlier shutdown to flush out the system. Flushing of the system after injection is necessary not only to reduce corrosion hazards, but also to reduce microbial growth. This also offers the option of applying higher concentration of nutrients for short durations with additional time allowed for flushing of lines.

When potted plants are irrigated by trickle irrigation in nursery operations, the actual application rates are usually very high, and no filling or flushing time is available. For example, a 4 L/hr emitter running into a 0.3-m-diameter container would have an application rate equal to about 50 mm/hr, and that into a 0.15-m container would have an application rate of 200 mm/hr. In irrigation of container plants, the applied water must contain the necessary plant nutrients that may not be adequately supplied in the containers because of limited soil volume.

4.3.3.a Methods of injection

Suction of fertilizer through the intake of the pump is a common method of application. However, corrosive fertilizer materials will cause the pump to deteriorate. In some cases, the water pressure on the suction side is such that it does not allow the fertilizer solution to flow into the pump. This may happen when the water sources have a pressure of 103 kPa (15 lb/in^2) or greater.

Pumping is the most common method of injecting fertilizer into a trickle irrigation system. Injector energy is available from electrical motors, internal combustion engines, water driven hydraulic motors and pumps, and impeller driven power units.

The positive injection pumps include the single or multiple piston pumps, diaphragm pumps, gear pumps, and roller pumps. Where two or more different types of fertilizers are required, multiple pump units can be used to avoid or reduce precipitation problems. All of the injection pumps can be regulated to achieve the desired or required rate, usually by adjusting the length of stroke of the piston pump or by selecting the appropriate pulley diameter. Another means of adjusting pumping speed is with variable-speed motors. Pumps may be adapted to receive signals from a water meter, which then can be used to apply precise amounts of fertilizer. Where a specific concentration is desired, information on the irrigation flow rate is needed in order to inject the necessary amount of fertilizer. No standard concentration has been established for application with trickle systems. The nutrient concentration of the modified half-strength Hoagland's solution can be used as a guide (table 4.3.2) in setting concentrations. It is desirable to monitor the nutrient status of the plant tissue, the soil nutrient concentration, and soil pH to see that nutrient levels are adequate.

Water-driven pumps usually are more complex, more difficult to maintain and more expensive. However, they are useful where electricity is unavailable or where gasoline-driven units can be operated for

TABLE 4.3.2

Modified half-strength Hoagland's solution.

Nutrient element	Concentration me/L	Concentration mg/L
NO_3-N	7.5	103
H_3PO_4-P	3	30
K	3.5	140
Ca	5.5	110
Mg	2	24
SO_4-S	2	32
Fe	—	2.5
B	—	.25
Mn	—	.25
Zn	—	.025
Cu	—	.01
Mo	—	.005

only short periods. Water-driven pumps also allow the proportioning of the stock fertilizer solution to maintain the desired concentration of nutrients in the irrigation water. Furthermore, the injection system does not require an expensive pressurized fertilizer tank.

Pressure differential (PD) units are another method of injecting fertilizer into trickle irrigation systems. A schematic diagram of a PD unit is given in figure 4.3.1. The PD units take advantage of the system's pressure-head differences.

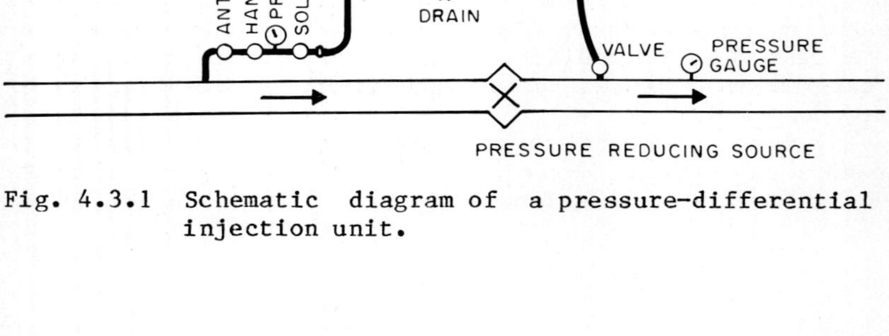

Fig. 4.3.1 Schematic diagram of a pressure-differential fertilizer injection unit.

Pressure differences can be developed by valves, venturi, elbows, or pipe friction. Most PD systems use closed tanks so that the tanks must withstand the pressure of the irrigation system.

The main advantage of the PD applicators is the absence of moving parts. They are simple in operation and require no electric-, gasoline-, or water-powered pumps. They can operate whenever water is flowing and where a pressure drop is present. The primary disadvantage of the PD units is that the rate of application is not constant and changes continuously with time; thus, a uniform concentration of a nutrient cannot be maintained. This system is difficult to use when the unit must be cycled through various sections of the entire field. Consequently, when one fertilizer tank is to serve more than one section in a rotation, most of the fertilizer within the tank must be discharged before it is refilled and moved to the next section.

The amount of mixing that occurs in the tank will depend on the solubility of the fertilizer, size and shape of the tank, specific gravity of the fertilizer, rate of flow through the tank, and temperature.

When it is necessary to know the percentage of the material that remains in the tank or that has been discharged from the tank, the following equation can be used, assuming perfect mixing within the tank:

$$n = 100 \exp(-xt/100) \qquad (4.3.1)$$

in which n is % mixture remaining in the tank, exp is the exponential function (2.718), x is the flow rate through the tank, and t is time.

Knowledge of the dilution rate allows one to judge adequately when to drain the fertilizer container and to recharge without discharging excessive fertilizer waste. Figure 4.3.2 gives curves developed from the preceding equation for three values of percent-of-tank volume flowing through the tanks per unit of time. For example, if a 300-L container has a flow of 15 L/hr, then the 5% curve is used. If only half of the material is to be applied, then the unit should be turned off after 18 hr. The curve also indicates that at least 100 hr are required to reduce the concentration to less than 1%. When the same 300-L container has a flow of 15 L/min, one-half of the material should have been applied in 18 min.

Flow rate through the tank can be controlled by valves. High flow rates are readily determined by several types of meters. However, for very low flow rates, accurate meters may not be available and an alternative method must be used. Disk orifices in the hose line can control flow. The operation is based on the pressure difference developed between both sides of the orifice.

Orifice size (metric) can be calculated from the equation

$$D = \left(\frac{15.13Q}{C\sqrt{P}}\right)^{1/2} \qquad (4.3.2)$$

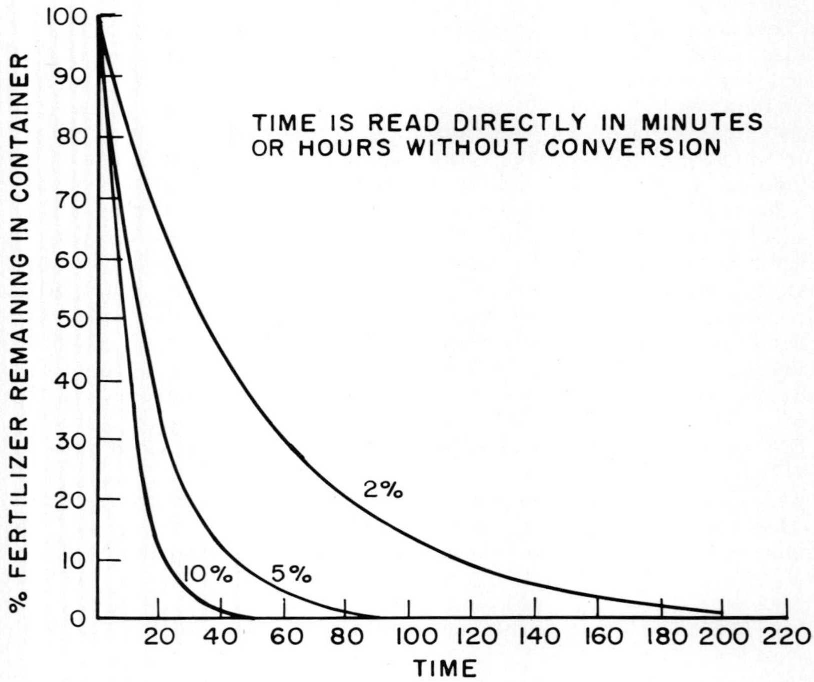

Fig. 4.3.2 Relationships for % fertilizer remaining in a fertilizer container as a function of time using the pressure-differential injection unit.

in which D is the diameter in mm, Q is the flow rate in L/min, C is the orifice coefficient (0.62) and P is the pressure in kPa. The orifice size can also be determined from the nomogram given in figure 4.3.3. Conversion factors for the various units are also listed in the Appendix section.

Fertilizer injection should not begin until all lines are filled with water and the emitters are running. For many systems, it is preferable for chemical injection to begin one hour after the system has been operating and for injection to cease one hour before the system is to be turned off. This allows adequate time for most systems to fill and be operating fully. Some small or moderate-sized systems may take 30 minutes or less to fill. Applying chemicals into a partially filled system will result in poor fertilizer distribution. There also needs to be adequate time to flush the fertilizer from the system to avoid corrosion and to reduce microbial growth.

A modification of the pressure differential system is a tank that contains a collapsible plastic bag into which the fertilizer is added. Water is admitted to the area between the tank and the bag, which

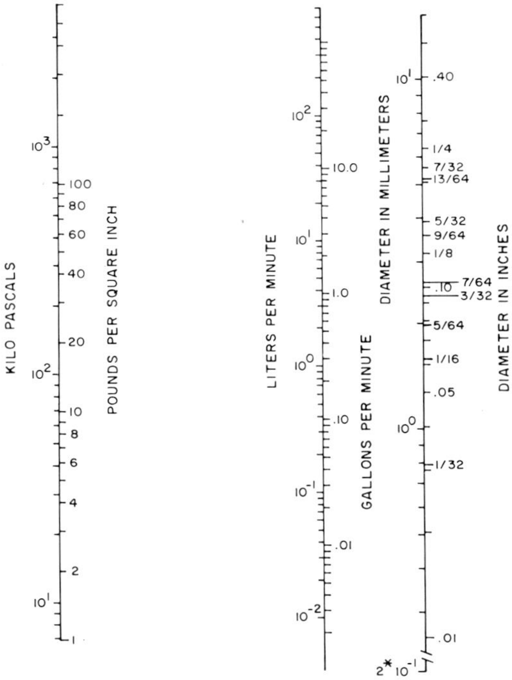

Fig. 4.3.3 Nomogram for determining orifice sizes for desired flow rates when the coefficient of discharge equals 0.62.

forces the fertilizer from the bag into the system. The plastic bags must be replaced frequently and can cause some inconvenience.

Another modification of the tank approach is the replacement of the plastic bag with a rubber diaphragm. The fertilizer solution is added to one side of the diaphragm. Water is added at a controlled rate to the opposite side to force the fertilizer from the tank.

Some venturi injection systems allow fertilizer to be added directly into the system from open tanks without being diluted. A portion of the irrigation water is bypassed through a venturi, which functions as an aspirator to pull the solution into the system. Because of high pressure losses, the larger venturis may require booster pumps. Solution injection rates are regulated by flow meters and valves.

Surface application of fertilizer is a practical alternative to applying fertilizer through the trickle system. The material may be weighed or measured for each plant and may also be divided to give equal portions to each of several emitters of a plant. This is a good method of applying slowly soluble materials that may cause excessive wear on pumps or that may produce a precipitate and plug the emitters.

TABLE 4.3.3

Solubility and composition of some commercial fertilizer materials (adapted from Hawkes et al., 1980).

Material	Approximate parts solubility in 100 parts cold water	Average percent nutrient composition of materials			
		N	P	K	Others
Major nutrients					
Ammonium nitrate	118	33.5	—	—	—
Ammonium sulfate	71	21	—	—	—
Calcium nitrate	102	15.5	—	—	21Ca
Diammonium phosphate	43	21	11.5	—	—
Monoammonium phosphate	23	11	10.5	—	—
Orthophosphoric acid	550	—	49	—	—
Potassium chloride	35	—	—	52	—
Potassium nitrate	13	14	—	39	—
Potassium sulfate	12	—	—	45	18S
Sodium nitrate	73	16	—	—	—
Superphosphate, single	2	—	4-5	—	20Ca 12S
Superphosphate, double	4	—	9-10	—	13Ca 10S
Urea	78	45-46	—	—	—
Micronutrients					
Copper sulfate	22	—	—	—	25Cu
Ferrous sulfate	29	—	—	—	20Fe
Manganese sulfate	105	—	—	—	25Mn
Sodium borate	5	—	—	—	11B
Sodium molybdate	56	—	—	—	40Mo
Zinc sulfate	75	—	—	—	22Zn
Fe-EDDHA [a]	9	—	—	—	6Fe
Fe-DTPA [a]	22	—	—	—	10Fe

[a] Information obtained from CIBA-Geigy Corporation. The chelating compound EDDHA is ethylenediamine di-(o-hydroxphenylacetate), and DTPA is diethylenetriamine pentaacetic acid.

4.3.4 *Nitrogen fertilization*

4.3.4.a Fertilizer sources

Nitrogen (N), being one of the major plant nutrients, is often supplied in order to obtain optimum crop production. Nitrogen availability is usually limited in the soil compared with other plant nutrients because its various forms can be leached, volatilized, dentrified, or fixed in the organic fraction of the soil.

Some major sources of fertilizer-N are anhydrous ammonia, urea, urea sulfate (US-28), urea ammonium nitrate (UAN-32) (names of commercial products used for benefit of reader and do not imply endorsement by authors), ammonium sulfate, aqua ammonia, ammonium phosphate, ammonium nitrate, and calcium nitrate. Table 4.3.3 provides a list of fertilizer materials, their solubilities, and the percent elemental composition of the materials.

4.3.4.b Water-quality interactions with N

Although water quality must be considered (Fabry, 1978) when N is applied through a trickle irrigation system, it is less of a problem than other nutrients such as phosphorus. The injection of anhydrous ammonia or aqua ammonia into irrigation water will bring about an increase in pH that may be conducive to the precipitation of calcium, magnesium, and phosphorus or the formation of complex magnesium ammonium phosphates, which are insoluble. This can be especially serious if bicarbonate is also present in the irrigation water (Rolston et al., 1979). In contrast to the preceding, NH_4NO_3 causes a sharp decrease in soil pH and a sharp increase in soluble aluminum in the wetted zone (Edwards et al., 1982). A decrease in soil pH in the soil near the emitters is also found when anhydrous ammonia is applied through a subsurface system (Mitchell, 1981). Nitrate salts, such as potassium nitrate or calcium nitrate, are relatively soluble in water and cause only a slight shift in the pH of the water and soil.

Nitrogen injected in the form of ammonium phosphate can cause serious clogging of the irrigation system. If calcium and magnesium are present in the irrigation water, the phosphate can form complex precipitates. Other N salts generally play negligible roles in clogging or change in pH.

One of the favored forms of N for use in this system is urea, because it is a highly soluble nitrogen fertilizer that does not react with water to form ions unless the enzyme urease is present. The enzyme, however, is often found in water containing large amounts of algae or other microorganisms. Since urease is not removed by filtration, its presence could cause hydrolysis of urea to the ammonium ion.

4.3.4.c Distribution in soil

Reactions to N fertilizer source differ not only with the irrigation water but also with the soil. Thus local water quality and soil conditions should be considered when selecting a N fertilizer source. Nitrogen in the ammonium (cation) form and at low fertilizer application rates will adsorb onto the soil (clay) colloids, thus moving only a minimal distance from the source of application. Depending on the rate of application, concentration of ammonium ions in the vicinity of the emitter can be very high (Bacon and Davy, 1982). When the ammonium ions are present in a high concentration (high fertilizer rate), they can overcome the capacity of the exchange sites on the soil colloids and thus move to a greater soil depth. The ammonium ions in the nearly saturated zone below the emitters are not nitrified, but nitrification can occur in the unsaturated zones further away from the emitters. Thus, the ammonium can act like a time-release fertilizer (Laher and Avnimelech, 1980). Normally, most ammonium in the soil will be transformed biologically to nitrate within 2 to 3 weeks at soil temperatures of 25 to 30°C.

Application of ammonium fertilizer to the soil surface can result in some loss into the atmosphere due to ammonia volatilization, especially when the soil pH is greater than 7. This loss will be increased if the irrigation water has a pH substantially greater than 7 (as would be true when anhydrous ammonia or aqua ammonia is injected into the irrigation system). Although large losses by ammonia volatilization are possible when ammonia is applied at a point immediately below an emitter, only a small part of the total ammonium ions are adsorbed on the exchange sites of the surface soil. Thus, when the wetted soil surface below each emitter is no greater than 200 to 300 mm in diameter, ammonia volatilization losses can be expected to be relatively small.

Urea, which is relatively soluble in irrigation water and is not strongly adsorbed by soil, will move deeper below an emitter than the ammonium salts. Urea has another advantage in that its concentration near the soil surface will often be small, thereby minimizing ammonia volatilization loss. After hydrolysis of urea to ammonium, the ammonium ions react with and are fixed by the soil. Because urea is only slightly adsorbed on the soil and moves with the irrigation water, flexibility in urea placement is achieved through water management. The availability of urea solutions, which are easily metered into the irrigation system, also offers an advantage over the dry N materials.

Nitrate will move with the water along the wetting front and when too much water is applied, the N will leach below the root zone (Goldberg et al., 1971). Thus, only part of the root zone will have access to the increased N levels, and fertilization efficiency is less compared to when the N is uniformly distributed in the wetted soil and root volume.

Any non-nitrate form of N applied to the soil will eventually be converted to nitrate and will readily move with irrigation water. The movement of nitrate in irrigation water implies that a single application of N through the system may not be the most efficient fertilizer application practice. Efficiency may be improved by applying fertilizer throughout the growing season as needed to meet the plant's nutrient requirements. Indeed, growers commonly use this practice when applying N fertilizer to trickle irrigated crops. Once N fertilizer is in the form of nitrate, it is susceptible to leaching losses. The ability to manage closely the water with the trickle irrigation system should result in minimum leaching of applied nitrate. However, a trickle irrigation system is not 100 percent efficient because of nonuniform flow among the emitters so that nitrate distribution can also be uneven, and leaching can occur. When growth conditions are unfavorable, another potential loss of N is through denitrification of nitrate. When adequate soluble carbon is available and soil water content approaches saturation, oxygen will become exhausted in the soil profile and nitrate may be reduced to volatile gases such as dinitrogen and nitrous oxide and diffuse upward to the atmosphere. Thus, it becomes important to avoid saturating the soil, which is generally not a problem, except directly under the emitter. It might be expected that a trickle irrigation system, which keeps soil at a high water high water content locally, would result in increased denitrification. However, denitrification occurs over a narrow range of water content. Little denitrification takes place if soil water pressure is smaller (drier) than -10 kPa (Rolston et al., 1978). There are times when the soil water pressure immediately below an emitter may be lower (wetter) than -10 kPa, but little nitrate would normally be in this zone unless N is continuously or frequently applied with the irrigation water. As nitrate moves out of the nearly saturated regions (near emitters) of the soil profile, the potential for denitrification should be reduced.

4.3.4.d Crop response, rates, and timing

When fertilizer efficiencies are nearly equal, the required rates of fertilization should be about the same for N applied through the trickle irrigation system compared to that applied with other types of irrigation. However, N fertilizer frequently will be poorly distributed in the soil because nitrate N moves easily with water and advances with the wetting front. Thus, it might be expected that fertilizer efficiency will be increased if N is applied through the trickle irrigation system (Phene et al., 1979). Frequent N applications using trickle irrigation improves the efficiency of fertilizer use by potatoes more than twofold over that of conventional fertilization methods (figure 4.3.4). Applying fertilizer through the system also increases the timing flexibility, so that N may be applied at times tailored to fit the plant requirements.

Results of research on trickle irrigated tomatoes (Miller et al., 1976) indicate that N is used more efficiently when applied through the trickle system than when banded and furrow irrigated or banded and trickle irrigated. When fertilizer is banded beside the plant row, furrow irrigation is the better irrigation method, because the applied water moves the N toward the plants. Trickle irrigation systems, often located between the plants and the banded fertilizer, move the banded N away from the plants and, therefore, result in less efficient use of the fertilizer. However, if banded fertilizer is placed between the plant row and the trickle line or directly below the trickle emitters, the water moves the fertilizer into the plant root zone. Although different rates, materials, and timing are more flexible for fertilizer applications made through trickle irrigation systems, sufficient N must be available in the root zone early in the season. Placement of emitters in relation to plant root zones is critical in that the direction of water movement in the root zone will influence the movement of N fertilizer applied through the system or will influence the availability of residual soil N from prior cultural practices.

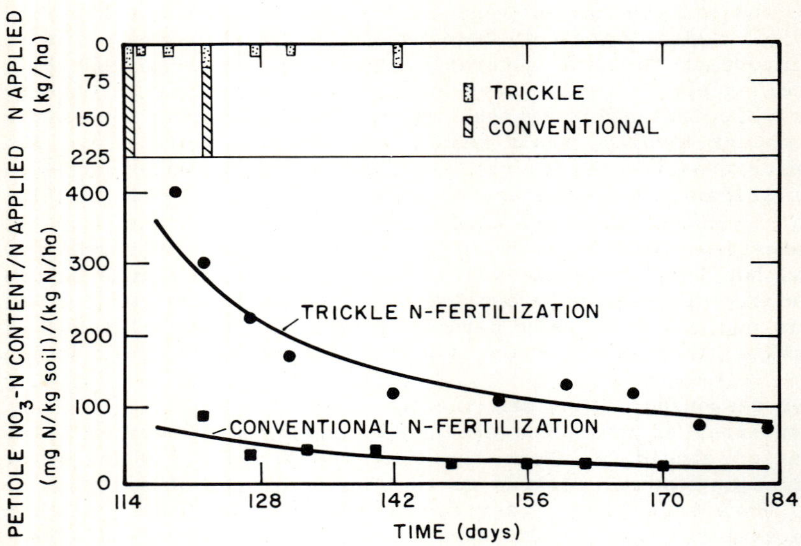

Fig. 4.3.4 Petiole nitrate content of potato leaf per unit of N applied for high frequency and conventional nitrogen fertilization methods (after Phene, 1979).

4.3.5 *Phosphorus fertilization*

4.3.5.a Fertilization sources

Generally, injection of phosphorus (P) fertilizer through a trickle irrigation system has not been recommended. Most P fertilizers have created chemical or physical precipitation problems and subsequent clogging of the trickle irrigation system. Further, the fixation rate of P by soils is high, and subsequent movement from its point of placement is limited. However, various sources of P and application methods will at least partially circumvent the problems of clogging and distribution of P in the soil. Without regard to cost, some of the inorganic sources of P for crop growth are given in table 4.3.3. These sources not only have different P contents, but also have different solubilities in irrigation water. In addition to these differences, the pH and calcium and magnesium contents of the irrigation waters play a major role in precipitation and subsequent clogging of the trickle emitters.

4.3.5.b Water quality interaction with P

The possibility of precipitation of insoluble phosphates is extremely high when phosphorus fertilizer is added to irrigation water high in calcium and magnesium. The result is the clogging of emitters or trickle lines with calcium and/or magnesium phosphates. However, if precautions are taken, phosphoric or sulfuric acid can be added to a trickle irrigation system to prevent such problems. Because phosphoric acid will form insoluble precipitates if the water has large amounts of calcium and magnesium, the pH of the irrigation water must be kept low enough for the salts to remain soluble. This can be done by injecting sufficient amounts of phosphoric or sulfuric acid into the irrigation water, which keeps the pH low enough while the pulse of phosphorus fertilizer is being applied (Rauschkolb et al., 1976). Phosphoric acid has been applied successfully through a trickle irrigation system and causes no precipitation or clogging even when the irrigation water is relatively high in bicarbonate plus calcium and magnesium. To achieve this, the phosphoric acid must be injected at a point beyond any metal connections or filters in order to avoid corrosion. To prevent precipitates from forming at the boundary between the phosphoric acid solution and irrigation water at the end of the phosphorus injection cycle, a short pulse of sulfuric acid must be added immediately after the phosphoric acid to keep the pH low until all P are removed from the system. Sulfuric acid additions may not be necessary in some cases. If the irrigation water is low in calcium and magnesium, few problems should be encountered in applying phosphoric acid through trickle irrigation systems.

4.3.5.c Distribution in soil

Inorganic P is precipitated and strongly adsorbed in most soils, and if applied uniformly on the soil surface at usual fertilization rates, P will not normally move more than 20 or 30 mm with the irrigation water. Thus, if the fertilizer is not mechanically worked deeply into the soil, most of it will remain near the soil surface, where availability will be limited to plant roots. The problem of P immobility in soil can be greatly alleviated by banding fertilizer below the seed at planting time. If an adequate amount of P is placed in the plant root zone, the plant generally can obtain the P it needs, although even banded P fertilizer is often used inefficiently by the plants. Orthophosphate movement is much greater when applied through a trickle irrigation system than with other application methods (Rauschkolb et al., 1976). When applied at a rate of 39 kg of P/ha (figure 4.3.5), orthophosphate moved 250 mm horizontally and 305 mm vertically in the soil profile from emitters placed approximately 1 m apart on beds of approximately 1.7 m. Trickle application can increase P movement 5-to 10-fold in contrast to when the same rates are applied conventionally and uniformly; movement is greater when applied in the trickle system because of the more concentrated and, thus, more effective application rate obtained by applying the P over a small surface area. Thus, the increased orthophosphate movement is the result of the saturation of soil reaction sites by P near the point of application and the subsequent mass flow of P with the soil water. The distance of P movement also can increase with an increase in the application rate (figure 4.3.5). Figure 4.3.6 shows the effective application rates corresponding to various radii of nutrient movement.

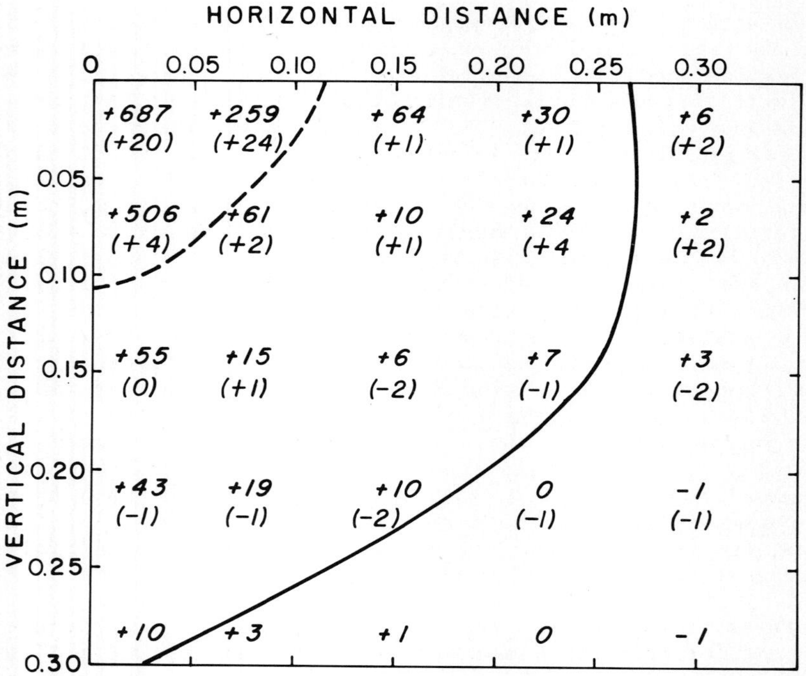

Fig. 4.3.5 Influence of application rate on phosphorus distribution beneath a field emitter on Panoche clay loam. Broken and solid curves are approximate maxima for bicarbonate-soluble phosphorous movement at 6.5 (values in parentheses) and 39 kg of P/ha (values not in parentheses) applications of orthophosphoric acid, respectively (after Rauschkolb et al., 1976).

The vertical and horizontal movement of phosphate applied in a relatively concentrated form through the trickle system at the time of planting is illustrated in figure 4.3.7. In this instance, the mono-ammonium phosphate at rates of 112, 224, and 336 kg/ha was applied with trickle tubing placed between two plant rows (Miller et al., 1980). Again the data show that P can move in a soil when applied in amounts large enough to overcome the fixing capacity of the sites of the soil. Not only can the P move farther with increased rates, but it can also be maintained at a high concentration in the root zone after 3 months, which would be available for subsequent uptake by the crop.

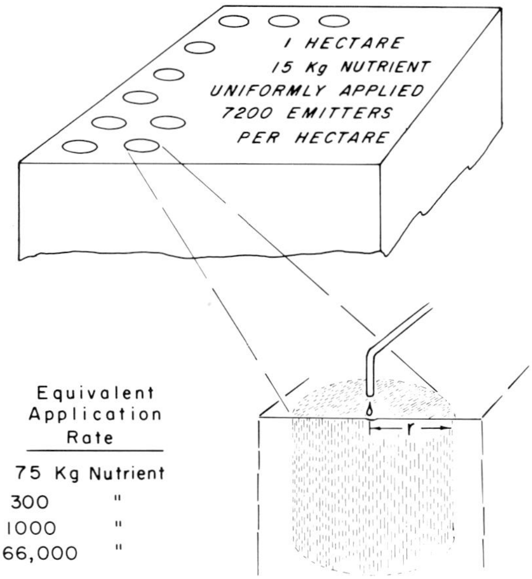

Radius r of Fertilizer Application	Equivalent Application Rate
300 mm	75 Kg Nutrient
150 "	300 "
80 "	1000 "
10 "	66,000 "

Fig. 4.3.6 Radii of fertilizer application and equivalent application rate for nutrient application through a trickle system.

4.3.5.d Crop response, rates, and timing

In growing crops under trickle irrigation, only that limited portion of the soil wetted by the emitters can support root growth. At the same time, only the plant nutrients in that limited portion of wetted soil are available for plant uptake. This situation not only presents some problems, but also provides an opportunity to "spoon feed" needed plant nutrients to a crop during the growing season. Although native soil P may be present in large amounts, its availability for rapid uptake by a crop may not be sufficient over the short time period. To avoid this possibility of limited uptake when the crop demand may be high, the use of trickle irrigation to apply needed P could be a practical solution. However, the flexibility of a trickle irrigation system to apply timely P fertilizer is less important in terms of timeliness than it is in applying needed N to a crop.

Although there are fixation reactions in the soil that tend to remove P slowly from the soil solution and make it less available to plants, there seems to be little benefit from applying P fertilizers continuously or nearly so during the growing season.

Definite benefits can be obtained by applying P fertilizer through a trickle irrigation system in terms of crop response (Miller et al., 1980). Figure 4.3.8 shows the uptake of P and yield of sugar cane obtained from a single application of monoammonium P at planting. Differences possibly may have been greater in plant response and P uptake if the crop could be allowed to mature (2 years).

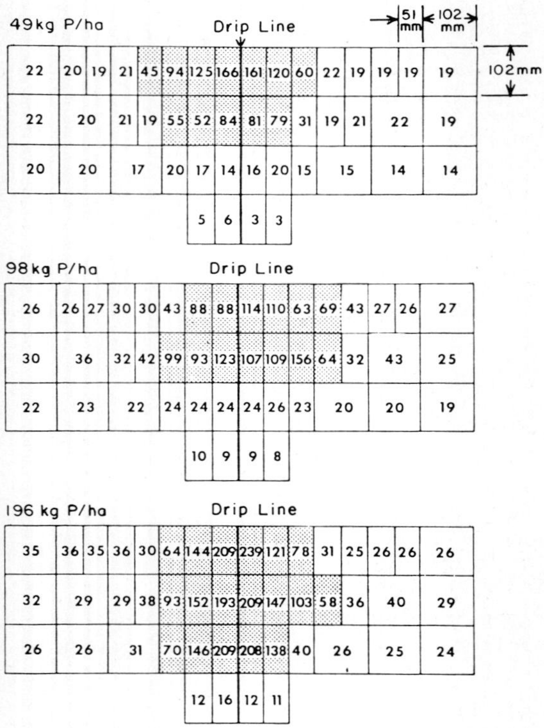

Fig. 4.3.7 Distribution of sodium bicarbonate extractable P beneath trickle lines for three rates of application of phosphorus. The shaded area indicates concentrations above background (after Miller et al., 1980).

A plant generally needs P early in its growth; therefore, it is important that P be applied before planting, at planting, or shortly after planting. There are situations, however, where applying P through a trickle irrigation system adds flexibility in correcting deficiency symptoms that might develop in mid-season or late-season for annual crops and in later years for perennial crops. All things considered, applying P fertilizer through a trickle irrigation system should make an efficient use of the nutrient, labor, and energy in producing a crop.

4.3.6 *Potassium fertilization*

4.3.6.a Fertilizer sources

Because potassium (K) is an important plant nutrient and is required in relatively large amounts to produce high-yielding crops, K must

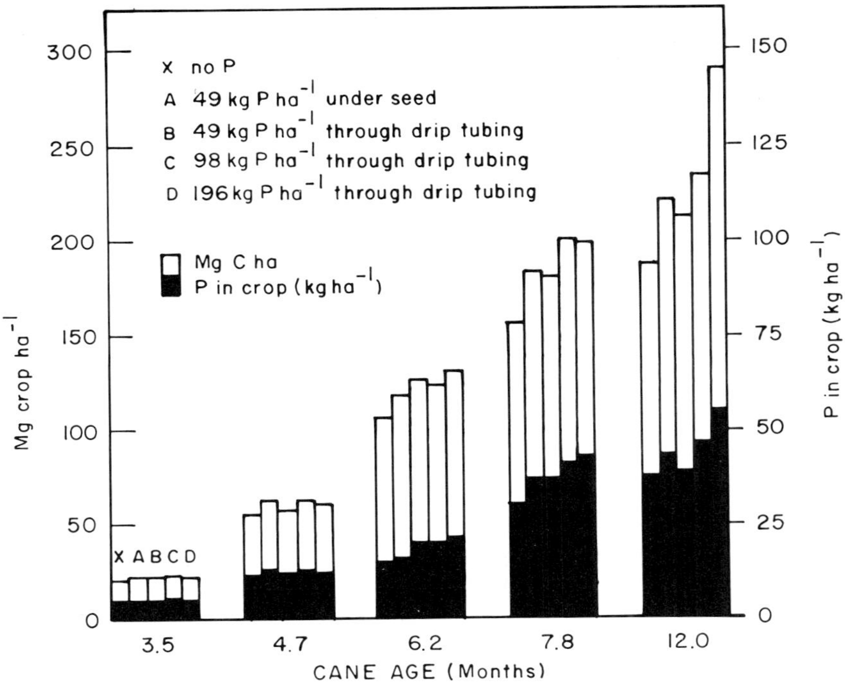

Fig. 4.3.8 Uptake of P and yield of sugarcane obtained from a single application of monoammonium phosphate at planting as a function of time and various application rates and placements (after Miller et al., 1980).

be monitored to ensure that it will not be a limiting factor in crop production.

Considerable data are available to demonstrate methods of application and the benefits of the use of K fertilizer, but most of the data have been obtained from conventional agricultural practice with only a few for trickle irrigation. The choice of K fertilizer is based on crop needs, crop tolerance, application method, other elements in the fertilizer, fertilizer availability, and cost. Table 4.3.3 lists K fertilizer materials, their solubility, and the percent K in each material. Fertilizer KCl accounts for 90% of the K fertilizers used, followed by K_2SO_4, $K_2SO_4 \cdot 2MgSO_4$, KNO_3, K_2HPO_4, and KH_2PO_4. The price is relative to the availability of the materials. Potassium chloride may be less desirable for some plants because of chloride toxicity, but for most other plants it has been used successfully where rates have been reduced or applications are split.

4.3.6.b Water quality interaction with K

No adverse chemical reactions are expected with the common K fertilizers when they are added alone to water. However, reduced solubility and/or fertilizer incompatability is possible when different fertilizer types are mixed. An example is a mixture of calcium nitrate and potassium sulfate, which will yield insoluble calcium sulfate.

4.3.6.c Distribution in soil

Although many soils generally contain large amounts of K, it is not all available. Unavailable K accounts for 90% of the total, slowly available K accounts for about 8%, and readily available K accounts for about 2%. The distribution of available K is usually greater at the surface and decreases with depth. This is because the K is part of organic matter and because the K is leached from organic residues during normal soil developmental processes. The ratio of other nutrients such as calcium or magnesium to K in the soil and subsoil may also reduce K availability. Where land has been graded or leveled, the cut areas (surface removed) can be expected to have a lower concentration of available K, and the fill areas to have a greater concentration.

Availability and uptake of K depend to a large extent on soil moisture conditions. The wetted soil volume of trickle irrigated soil depends on the age of plants, plant spacing, and management. Even under mature orchard conditions, the wetted volume may be less than 20% of the total soil volume. As the percentage of the wetted soil volume is reduced, the available K is limited to that in the moist soil. Plants obtain little, if any, K from dry soils.

4.3.6.d Crop response, rates, and timing

Potassium deficiency in crops cannot always be solved by irrigating to increase the wetted soil volume. This is because much of the increase in wetted soil volume may take place in the subsoil where K levels are lower, thus K fertilizer would be required.

Since some K deficiency has occurred under trickle irrigation and probably will become more prevalent as crops utilize the limited K available under trickle irrigation, it is desirable to monitor the K levels in the plants by leaf analysis.

Since trickle irrigation does offer the advantage of repeated small or continuous applications, the fertilizers can be partitioned to rates and frequencies that are effective and economical. This is important, since K is more readily exhausted than many of the other nutrients. Furthermore, plants will take up more K than is necessary. With lower application rates, this so-called luxury consumption can be avoided.

Potassium deficiency is likely to occur under wide plant spacings, such as exists in orchards where there are fewer emitters per unit of surface area. For annual crops, such as strawberries, tomatoes, and vegetable crops, K-containing fertilizers can be incorporated into the beds at planting time to ensure adequate K. This avoids the need for applying K through the trickle system, but it requires more equipment. For long-lived orchards, preplant K fertilizations have not been recommended.

Potassium sulfate is generally preferred where crops are sensitive to high chloride concentrations. A mixture of K_2SO_4 and KCl fertilizers has also been used to reduce costs. Stone fruits, pecans, citrus, strawberries, and avocados are among those plants considered sensitive to chloride, whereas tomatoes, cotton, sugar beets, safflower, alfalfa, and grapes are not.

Application rates for correcting K deficiency in prune orchards under conventional irrigation systems have been 1,000 to 2,000 kg of either KCl (52% K) or K_2SO_4 (45% K) per ha. At 250 trees per ha, this is 4 to 8 kg per tree. In contrast, K application is in the order of 0.5 kg per grape vine.

A trickle irrigation system is convenient and efficient to operate and provides a low-cost approach for correcting K deficiency by allowing low rates and frequent applications, thus avoiding excess use and buildup of salts while maintaining high nutrient availability. However, the rates, timing, material selection, and application method should be evaluated for each site. An example in which K application through the trickle irrigation system results in much greater K levels in leaves than other methods is shown in figure 4.3.9. Potassium content in the leaves can be almost twice as high when K is added with the trickle irrigation water.

Because water is applied to a limited soil volume with trickle irrigation systems, salts tend to accumulate in and around the periphery of the wetted volume. When highly soluble fertilizers, such as KCl and many of the other fertilizers are applied in a confined volume of water and soil, the salinity values can become very high. Soil salinity should be measured periodically to check for salinity buildup. The consequences of high salinity on plant growth and yield are discussed in Chapter 4.4 on salinity.

4.3.7 *Micronutrient fertilization*

4.3.7.a Fertilizer sources

Chelates and sulfate compounds of various micronutrients are generally used for correcting micronutrient deficiencies. The solubility and composition of some commercially available micronutrient fertilizers are listed in table 4.3.3. Iron and zinc chelates have reasonably high water solubility and may create the least problems. However, their cost is still fairly high, and sufficient research results are unavailable to evaluate adequately their effectiveness for correcting micronutrient deficiencies when applied through trickle irrigation systems.

4.3.7.b Water quality interactions with micronutrients

Micronutrients such as iron, zinc, copper, and manganese may react with salts in the irrigation water and cause precipitation and clogging. Many of the micronutrients can be applied as chelates.

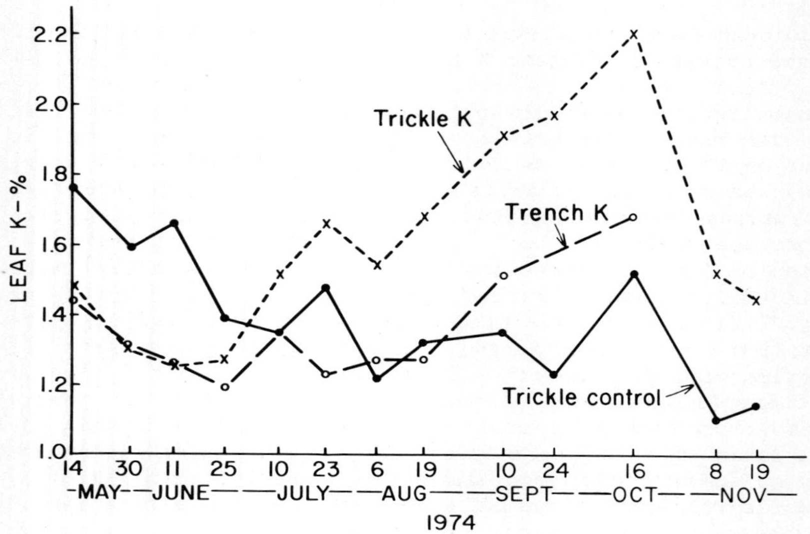

Fig. 4.3.9 Seasonal leaf K levels. Trench K received 2.1 kg K as potassium sulfate applied per tree in mid-May in parallel trenches 0.15 to 0.20 m deep and sprinkler irrigated. Trickle K received potassium sulfate applied in trickle irrigation water at 192 mg/L K concentration from June 4 to September 10 with a total of 2.1 kg of K applied per tree. Trickle control received no K (after Uriu et al., 1980).

They are fairly soluble and consequently will cause little clogging or precipitation. Growers have experienced difficulty dissolving zinc sulfate for preparing stock solutions. Some problems are encountered with precipitation and emitter clogging when zinc sulfate is put into the trickle irrigation lines. Chelates or sulfate salts of micronutrients can be predissolved and metered as a liquid into the trickle irrigation system. This permits the precise control of the quantity applied, since only a small amount is needed per application.

4.3.7.c Distribution in soil

Because of the generally greater mobility of chelated micronutrients in soils, it might be desirable to use this form for surface application of micronutrients. However, adequate data are unavailable on whether cationic micronutrients can move sufficiently through the soil under an emitter to achieve proper placement in the root zone. Because of the small quantities applied and the high affinity of micronutrients in the soil exchange site, it is expected that the cationic micronutrients would not move significantly away from the source.

4.3.7.d Crop responses, rates, and timing

Little research information is available on the efficacy of applying micronutrients through trickle irrigation systems. Although normal plant requirements for these nutrients are low, these compounds are highly adsorbed by the soil so that their availability to the plant can be limited. Application of zinc chelate through trickle irrigation systems can improve pecan tree growth at a lower cost than foliar applications (Lindsey and New, 1974). However, leaf concentrations of zinc were generally lower than when foliar applications were used. Zinc deficiency in avocados can be corrected successfully by injecting zinc sulfate into the trickle irrigation system at rates of approximately 0.15 kg/tree, at costs less than those associated with foliar applications (Francis, 1977).

Because micronutrient compounds are strongly adsorbed to the soil medium, probably little advantage would be obtained by applying the micronutrients continuously in low concentrations with each irrigation. In fact, better movement into the soil occurs when the materials are applied at a relatively high concentration at one time. Considerably more research is required to evaluate fully the amounts and timing of micronutrient applications through trickle irrigation.

4.3.8 *Rate calculations and examples*

The actual plan and fertilization schedule of trickle irrigated crops depend on site-specific conditions such as cultural practices, soil, crop, farm area, nutrients required, climate, supply, qualities, amount of water to be applied, fertilizer injector, and system design. Rates and concentrations should be calculated to avoid overfertilization.

For perennial crops with wide spacing, the applied nutrients may result in an apparent very high application rate, since only the wetted volume receives the fertilizer. Additionally, if the desired amount is applied over a short time, the high concentration of nutrients may cause plant damage. The nutrient balance will be upset, toxicity may result, soil pH may be changed rapidly, and the salinity content of the water and the soil may become excessive.

When the trickle system provides the major source of water for the crop, the plant nutrient needs can usually be met by applying the nutrient at a concentration at or below that of half-strength Hoagland's solution (table 4.3.2). This also permits nutrient adjustments to avoid mixing incompatible nutrients. Most crop nutrient needs can be met at a concentration of 100 mg/L in the irrigation water, since a 0.3-m depth of irrigation water will provide about 300 kg nutrient per ha.

The injection rate of fertilizer into the system, for a desired application rate to an area, is determined by:

$$Q_f = \frac{F_r A}{N_c T} \qquad (4.3.3)$$

in which Q_f is the quantity of fertilizer to be injected (L/hr or gal/hr), F_r is the fertilizer rate per application (kg/ha or lb/ac), A is the area to be fertilized (ha or ac), N_c is the nutrient concentration (kg/L or lb/gal) in the stock solution, and T is the time of injection (hr).

After the projected Q_f injection rate has been calculated, it must be evaluated as to concentration of nutrients in the irrigation water. This can be determined by

$$C_f = \frac{KF_r}{W} \quad (4.3.4)$$

in which C_f is the concentration of fertilizer in irrigation water in mg/L, K is a constant equal to 100 for metric units and 4.415 for English units, F_r is the fertilizer rate (kg/ha or lb/ac), and W is the net amount of irrigation water applied during the injection period (mm or in.).

When the desired concentration of nutrient in the irrigation water C_f has been selected, the rate of injection can be determined from flow rate in the system, density, and % of the nutrient in the fertilizer from

$$Q_f = \frac{KC_fQ}{\rho Y} \quad (4.3.5)$$

in which Q_f is the quantity of fertilizer to be injected (L/hr or gal/hr), K is a constant equal to 0.36 for metric units and 0.006 for English units, C_f is the desired nutrient concentration (mg/L), Q is the volume or rate of flow (L/sec or gal/min), ρ is the density of the fertilizing solution (kg/L), and Y is the percentage of fertilizer in solution as a whole number.

When dry materials are considered an alternative, the rate of application is

$$D_f = \frac{K_2CQ}{Y} \quad (4.3.6)$$

in which D_f is the dry fertilizer per hour to be injected (kg/hr), K_2 is the constant equal to 0.36 for metric units and 0.05 for English units, C is the desired concentration (mg/L), Q is the volume or rate flow (L/sec or gal/min), and Y is the percentage of nutrient as a whole number.

Example 4.3.1

Problem: A grower applies nitrogen to his orchard at a rate of 0.5 kg N (1.1 lb) per tree. The tree spacing is 6 m (20 ft) by 6 m (20 ft); assume 267 trees/ha (108 trees/ac). There are six emitters per tree; assume a wetted diameter of 0.6 m (2 ft) for each emitter. The fertilizer was applied with an irrigation of 6.35 mm (0.25 in). (a) Calculate the actual surface rate of application in kg/emitter (lb/ emitter), and (b) calculate the concentration of the fertilizer in mg/L in the water.

Solution: a. $0.5 \div 6 = 0.08$ kg/emitter

$$\pi\ (0.6/2)^2 = 0.28\ \text{m}^2 \text{ wetted area.}$$

$$\frac{(0.08)(10{,}000)}{0.28} = 2857\ \text{kg/ha}$$

$$1.1\ \text{lb} \div 6 = 0.18\ \text{lb/emitter.}$$

$$\pi\ (2/2)^2 = 3.14\ \text{ft}^2 \text{ wetted area}$$

$$\frac{(0.18)(43560)}{3.14} = 2497\ \text{lb/ac}$$

b. Using equation 4.3.4

$$\frac{(100)(133.5)}{6.35} = 2102\ \text{mg/L (ppm)}$$

$$\frac{(4.415)(118.8)}{(0.25)} = 2098\ \text{mg/L (ppm)}$$

Example 4.3.2

Problem: If the grower wishes to reduce the nitrogen content of the irrigation water to 100 mg/L, how many liters or gallons of 32% liquid nitrogen (UAN-32) with density of 1.33 kg/L, 10.5 lb/gal, should be applied per hour per acre? Flow = 0.68 L/sec or 10.8 gal/min.

Solution: Using equation 4.3.5

$$Q_f = \frac{(0.36)(100)(0.68)}{(1.33)(32)} = 0.58\ \text{L/hr}$$

$$Q_f = \frac{(0.006)(100)(10.8)}{(1.33)(32)} = 0.15 \text{ gal/hr}$$

Example 4.3.3

Problem: A grower plans to fertilize tomatoes with liquid 7-7-7 through his trickle system at a rate of 336 kg N/ha (300 lb/ac). The rows are 1.5 m wide (5 ft) and 45 m long (150 ft). He will inject the liquid fertilizer weekly for 16 weeks during irrigation. He irrigates 30 rows at a time; assume the liquid weighs 1.22 kg/L (10.2 lb/gal); and assume irrigation application = ET of 6.35 mm/day or 0.25 in/day.

Determine:
1. How many liters (or gallons) of 7-7-7 will be needed per ha (or acre)? 3934 L (420 gal)
2. How many liters (or gallons) per week for 30 rows? 50 L (13.56 gal)
3. How much nitrogen in solution? 331 mg/L (ppm)
4. How many hectares (or acres) per irrigation set? 0.20 ha (0.52 ac)

Example 4.3.4

Problem: Grape plantings on a spacing of 3.66 m (12 ft) by 2.44 m (8 ft) are fertilized with 0.5 kg (1 lb) of K_2SO_4 per vine. There are two 2-L (1/2-gal) emitters for each vine. The wetted diameter of each emitter equals 0.6 m (2 ft). ET is met daily; the current rate is 5 mm or 0.20 inches per day. The fertilizer is applied during the first 11 hours of a 12-hour set. The vine population is 1119 vine/ha (453 vines/ac), and K_2SO_4 = 45% K. (a) Calculate the actual rate applied to the wetted surface per ha and per acre, and (b) calculate the concentration of potassium (K) in the irrigation water.

Solution: a. $$\frac{(10^4)(0.5/2)}{\pi\ (0.6/2)^2} = 8842 \text{ kg } K_2SO_4\text{/ha}$$

$$\left(\frac{43560/2}{\pi\ (2/2)^2}\right) = 6933 \text{ lbs } K_2SO_4\text{/acre}$$

b. $$\frac{(100)(1119)(0.50)(0.45)}{(5)(11/12)} = 5493 \text{ mg/L (ppm)}$$

$$\frac{(4.413)(453)(1)(0.45)}{(0.20)(11/12)} = 4909 \text{ mg/L (ppm)}$$

REFERENCES AND SUPPLEMENTAL READING

Bacon, P.E. and Davey, B.G., 1982. Nutrient availability under trickle irrigation. II. Mineral nitrogen. Soil Sci. Soc. Amer. J. 46: 987-993.

Bester, D.H., Lotter, D.C. and Veldman, G.H., 1974. Drip irrigation on citrus. Proc. 2nd Int. Drip Irrig. Congr., San Diego, California, pp. 58-64

Bucks, D.A. and Nakayama, F.S., 1980. Injection of fertilizer and other chemicals for drip irrigation. Proc. Agri-Turf Irrig. Conf., Houston, Texas, Irrigation Association, Silver Spring, Maryland, pp. 166-180.

Davis, S.A. and Pugh, W.J. 1974. Dripper flow consistency. Proc. 2nd Int. Drip Irrig. Congr., San Diego, California, pp. 281-286.

Edwards, J. H., Bruce, R.R., Horton, B.D., Chesness, J.L. and Wehunt, E.J. 1982. Soil cation and water distribution as affected by NH_4NO_3 applied through a drip irrigation system. J. Amer. Soc. Hort. Sci. 107: 1142-1148.

Fabry, C. 1978. One drop at a time. Fertilizer Solutions. 22: 18-28.

Feigin, A., J. Letey, J. and Jarrell, W.M. 1982. Nitrogen utilization efficiency by drip irrigated celery receiving preplant or water applied N fertilizer. Agron. J. 74: 978-983.

Francis, L., 1977. Fertilization with drip irrigation: Concepts, practices practices and problems. Proc. Annual Technical Conf., Salt Lake City, Utah, Irrigation Association, Silver Spring, Maryland, pp. 80-87.

Goldberg, D., Gornat, B. and Bar, Y., 1971. The distribution of roots, water and minerals as a result of trickle irrigation. J. Amer. Soc. Hort. Sci. 96: 645-648.

Handbook of Chemistry and Physics, 54th Edition. 1973-74. CRC Press, Cleveland, Ohio.

Hawkes, G.R., Campell, K.B., Holgested, E. and Shaw, E.J., 1980. Sixth Edition, Western Fertilizer Handbook.

Johnson, H., Jr., 1977. Hydroponics: A Guide to Soilless Culture Systems. Univ. of Calif. Leaflet 2947, Berkeley, California.

Karmeli, D. and Keller, J., 1975. Trickle Irrigation Design. Rain Bird Sprinkler Manufacturing Corp., Glendora, California, 133 pp.

Laher, M. and Avnimelech, Y., 1980. Nitrification inhibition in drip irrigation systems. Plant Soil 55: 35-42.

Lindsey, K.E. and New, L.L., 1974. Application of fertilizer through drip irrigation systems in West Texas. Proc. 2nd Int. Drip Irrig. Congr., San Diego, California, pp. 400-404.

Martin, W.E., 1953. Do fertilizers ruin sprinkler systems? Proc. Irrig. Sprinkler Conf., Univ. of Calif., Davis, California.

Miller, R.J., How, K. and Hilton, H.W., 1980. Liquid phosphate fertilizer applied through drip irrigation tubing: Movement in soil and plant uptake. Hawaiian Sugar Planters' Assoc. Agron. Report 37, 6 pp.

Miller, R.J., Rolston, D.E., Rauschkolb, R.S. and Wolfe, D.W., 1976. Drip irrigation of nitrogen is efficient. Calif. Agric. 30: 16-18.

Mitchell, W.H., 1981. Subsurface irrigation and fertilization of field corn. Agron. J. 73: 913-916.

Phene, C.J. and Beale, O.W. 1976. High frequency irrigation for water nutrient management in humid regions. Soil Sci. Soc. Amer. J. 40: 430-436.

Phene, C.J., Fouss, J.L. and Sanders, D.C., 1979. Water nutrient herbicide management of potatoes with trickle irrigation. Amer. Potato J. 56: 51-59.

Rauschkolb, R.S., Rolston, D.E., Miller, R.J, Carlton, A.B. and Burau, R.G., 1976. Phosphorus fertilization with drip irrigation. Soil Sci. Soc. Amer. J. 40: 68-72.

Rolston, D.E., Koffman, D.L. and Toy, D.W., 1978. Field measurement of denitrification: I. Flux of N_2 and N_2O. Soil Sci. Soc. Amer. J. 42: 863-869.

Rolston, D.E., Rauschkolb, R.S., Phene, C.J., Miller, R.J., Uriu, K., Carlson, R.M. and Henderson, D.W., 1979. Applying Nutrients and Other to Trickle Irrigated Crops. Univ. of Calif. Bull. 1893, Berkeley, California, 14 pp.

Schulbach, H., 1975. Fertilizer Injection with Pressure Differential Systems. Univ. of Calif. Cooperative Extension, Soil and Water Newsletter 25, 4 pp.

Uriu, K., Carlson, R. M., Henderson, D.W., Schulbach, H. and Aldrich, T.M., 1980. Potassium fertilization of prune trees under drip irrigation. J. Amer. Soc. Hort. Sci. 105: 508-510.

Chapter 4

MANAGEMENT PRINCIPLES

4.4 SALINITY

G. J. HOFFMAN

A mixture of soluble salts is present in the root zone of all soils. If the concentration of these salts becomes excessive, losses in crop production will result. Yield reductions may result from osmotic stress caused by the total soluble salt concentration, from toxicities or nutrient imbalances created when specific solutes become excessive, or from a reduction of water penetration through the root zone caused by excess sodium inducing a deterioration of soil permeability. The key to alleviating salinity stress is leaching, the net downward movement of soil water through the root zone. Salts are leached whenever water applications exceed evapotranspiration, provided soil infiltration and drainage rates are adequate. In some regions, rainfall periodically flushes salts; in others, provisions must be made for adequate leaching.

Worldwide, over 50 million hectares are affected by salinity. Salinity is estimated to be a potential threat to about half of the 20 million hectares of irrigated land in the Western United States, with crop production being limited by salinity on about one-fourth of this land. The threat of excess salinity is particularly high with trickle irrigation because of the potential for the system to apply water uniformly and efficiently. If the amount of water applied does not meet the evapotranspiration demands of the crop plus an appropriate amount for leaching, any salts present in the irrigation water will accumulate in the root zone and a salt-affected soil will develop.

Plant response to soil salinity is considered to be independent of the irrigation method provided the space-time distribution of salinity in the root zone is integrated properly. Thus, the criteria for water quality is applicable to all irrigation systems with the exception of foliar damage from wetting crop leaves. With the appropriate determination of the average root zone salinity, one can use crop salt tolerance data and leaching requirement criteria to make irrigation, drainage, and agronomic management decisions.

4.4.1 *Plant response to salt-affected soils*

The predominant influence of salinity stress on plants is growth suppression. In most cases, growth decreases linearly as salinity increases beyond a threshold level. This effect is similar whether the salinity is increased by adding nutrients or salts like sodium chloride, sodium sulfate, or calcium chloride, which are common in saline soils. Growth suppression typically is a nonspecific salt effect, depending more on osmotic stress created by the total concentration of soluble salts than on the level of any specific solute. However, an excess concentration of an individual solute can be detrimental to some crop species in addition to that caused by its osmotic effect.

4.4.1.a Plant growth characteristics

Soil salinity distribution within a field is typically so variable that one area of a field can show distinctively different salinity symptoms from another area. Bare spots, poor spotty stands, and severely stunted plants are all signs of high salinity. However, moderate salinity levels often restrict plant growth without any obvious injury symptoms. Salt-affected plants usually appear normal although their growth is stunted. Salt-affected leaves may be smaller and may have a darker blue-green color than normal leaves. Chlorosis (the yellowing or blanching of green plant parts) is not a typical characteristic of salt-affected plants.

Wilting is not a regular characteristic of salt-affected plants because it typically occurs when water availability decreases rather abruptly, as in a drying soil. Under saline conditions, moderately low plant water potentials are always present and water potential changes are usually gradual. Plants are, therefore, hardened by the continual stress and are less apt to exhibit abrupt changes in turgor.

Acute injury symptoms, such as marginal or tip burn of leaves, occur as a rule only in woody plants in which these symptoms indicate toxic concentrations of chloride or boron. Most nonwoody plants do not exhibit such leaf injury symptoms although some accumulate as much chloride or boron as woody species that show injury. In rare instances, excess sodium may cause calcium deficiency symptoms.

4.4.1.b Crop salt tolerance

The relative salt tolerances of selected agricultural crops that are commonly trickle irrigated are given in table 4.4.1. This alphabetical crop list provides two essential parameters sufficient to evaluate salt tolerance: (1) the threshold salinity level (the maximum allowable salinity that does not reduce yield measurably below that of a nonsaline control treatment) and (2) the percent yield decrease per unit of salinity increase beyond the threshold. All salinity values are reported as σ_e (the electrical conductivity of soil saturation extracts reported in units of dS/m and corrected for temperature to 25°C) and rounded to two significant digits.

Relative yield (Y) in percent at any given soil salinity, expressed as the electrical conductivity of the soil saturation extract (σ_e), can be calculated by the equations

$$Y = 100 - B(\sigma_e - A), \qquad \sigma_e \geq A \qquad (4.4.1)$$

$$Y = 100, \qquad \sigma_e \leq A \qquad (4.4.2)$$

TABLE 4.4.1

Salt tolerance of trickle irrigated crops as a function of the electrical conductivity of the soil saturation extract (σ_e) where relative yield (Y) in percent = 100 - B(σ_e - A) where $\sigma_e \geq$ A (from Maas and Hoffman, 1977).

Crop	Salt Tolerance Threshold[a]	Percent Yield Decline[b]	Qualitative Salt Tolerance Rating[c]	Crop	Salt Tolerance Threshold[a]	Percent Yield Decline[b]	Qualitative Salt Tolerance Rating[c]
	dS/m[d]	%/(dS/m)			dS/m[d]	%/(dS/m)	
Almond	1.5	19	S	Grapefruit	1.8	16	S
Apple	---	---	S	Lemon	---	---	S
Apricot	1.6	24	S	Lettuce	1.3	13	MS
Avocado	---	---	S	Okra	---	---	S
Bean	1.0	19	S	Olive	---	---	MT
Beet, Garden	4.0	9	MT	Onion	1.2	16	S
Bentgrass	---	---	MS	Orange	1.7	16	S
Bermudagrass	6.9	6.4	T	Peach	1.7	21	S
Blackberry	1.5	22	S	Peanut	3.2	29	MS
Boysenberry	1.5	22	S	Pepper	1.5	14	MS
Broadbean	1.6	9.6	MS	Plum	1.5	18	S
Broccoli	2.8	9.2	MS	Potato	1.7	12	MS
Cabbage	1.8	9.7	MS	Radish	1.2	13	MS
Carrot	1.0	14	S	Raspberry	---	---	S
Cotton	7.7	5.2	T	Ryegrass	5.6	7.6	MT
Cowpea	1.3	14	MS	Spinach	2.0	7.6	MS
Cucumber	2.5	13	MS	Strawberry	1.0	33	S
Date Palm	4.0	3.6	T	Sugarcane	1.7	5.9	MS
Fescue, Tall	3.9	5.3	MT	Sweet Potato	1.5	11	MS
Grape	1.5	9.6	MS	Tomato	2.5	9.9	MS

[a] Salt tolerance threshold is the mean soil salinity at initial yield decline.
[b] Percent yield decline is the rate of yield reduction per unit increase in salinity beyond the threshold.
[c] Qualitative salt tolerance ratings are sensitive (S), moderately sensitive (MS), moderately tolerant (MT), and tolerant (T).
[d] dS/m = decisiemens per meter = 1 millimho per cm, referenced to 25°C.

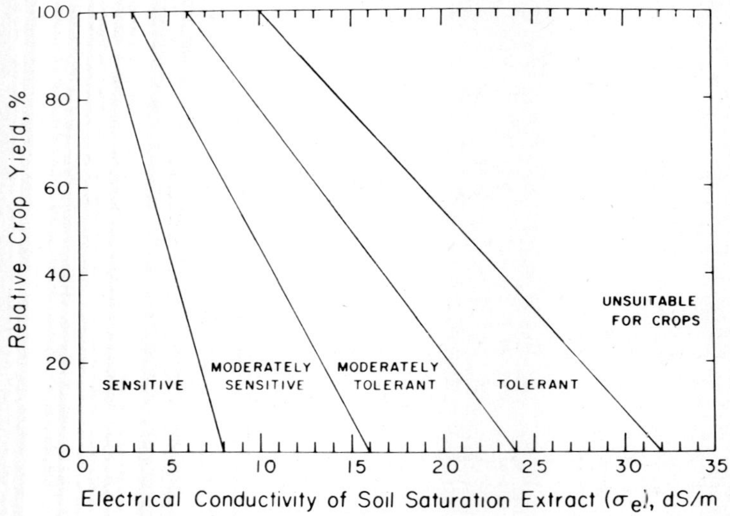

Fig. 4.4.1 Divisions for qualitative salt-tolerance ratings of agricultural crops. Symbols given compare with those in table 4.4.1 (after Maas and Hoffman, 1977).

where A is the salinity threshold value and B is the yield decrease per unit salinity increase, as given in table 4.4.1. A qualitative salt tolerance rating is also presented in table 4.4.1 for quick, relative comparisons among crops. The boundaries for the qualitative ratings are defined in figure 4.4.1. Four qualitative ratings are defined to correspond with previously published terminology ranging from sensitive to tolerant.

Example 4.4.1

Problem: The average soil salinity in the root zone of an orange grove has an σ_e of 4 dS/m. What is the anticipated loss in yield from excess salinity?

Solution: The salt tolerance equation for oranges is Y = 100 - 16(4 - 1.7). The values of 16 for B and 1.7 for A are given in table 4.4.1. The expected yield would be 63% or the anticipated loss from excess salinity would be 37%.

Example 4.4.2

Problem: If your field had an average σ_e of 3 dS/m, which crops could be grown without any loss in yield from salinity?

Solution: From the 100% relative yield value in figure 4.4.1, any moderately tolerant or tolerant crop listed in table 4.4.1 could be grown without yield loss.

4.4.1.c Factors influencing salt tolerance

Many crops seem to tolerate salinity equally well during seed germination and later growth stages. Germination failures that occur on saline soils are not normally caused by crops being especially sensitive during germination, but rather to exceptionally high concentrations of salt where the seeds are planted. These high salt concentrations near the soil surface are a consequence of upward water movement and evaporation in the absence of water applications. This can be a serious problem in fields previously irrigated with a trickle system where salinity is a hazard and provisions have not been made to leach the salts concentrated in the seed bed at the edge of the wetting front. The salt tolerance of some crops, particularly cereals, does change with growth stage.

Differences among varieties of a given crop, while not common, must also be considered in evaluating crop salt tolerance. Most known varietal differences are among grass species and legumes, but differences among varieties are expected to increase in the future as new varieties are developed.

Rootstocks are also an important factor affecting the salt tolerance of tree and vine crops. Varieties and rootstocks of avocado, grapefruit, and orange that differ in their ability to absorb and transport chloride have different salinity tolerances. Similar effects of rootstocks on salt accumulation and tolerance have been reported for stone fruit trees and grapes.

Environmental factors such as temperature, atmospheric humidity, and air pollution markedly influence crop salt tolerance. Many crops seem less salt tolerant when grown in hot, dry environments than in cool, humid ones. High atmospheric humidity alone tends to increase the salt tolerance of some crops, with high humidity generally benefiting salt-sensitive more than salt-tolerant crops. A strong interaction between salinity and ozone, a major air pollutant, has been found for alfalfa, bean, and garden beet.

4.4.1.d Specific solute effects

Unlike most annual crops, trees and other woody perennials may be specifically sensitive to chloride, which is taken up with soil water, moves with the transpiration stream, and accumulates in the leaves. Crop, varietal, and rootstock differences in tolerance to chloride depend largely upon the rate of transport from the soil to the leaves.

TABLE 4.4.2

Maximum permissible chloride concentration in the soil saturation extract or plant leaves if leaf injury is to be avoided for various fruit crop rootstocks and varieties.

Crop	Rootstock or variety	Maximum permissible chloride concentration: Soil saturation extract mol/m^3	Plant leaf analysis kg of Cl/kg of dry leaf
	Rootstocks		
Citrus	Rangpur lime,	25	
Citrus spp.	mandarin		0.7
	Rough lemon, tangelo, sour orange	15	
	Sweet orange, citrange	10	
Stone fruit	Marianna,	25	0.3
Prunus spp.	Lovell, Shahl,	10	
	Yunnan	7	
Avocado *Persca americana*	West indian, mexican	8	0.25 to 0.50
Grape	Salt Creek, 1613-3	40	0.5
Vitis spp.	Dog Ridge	30	
	Varieties		
Grape	Thompson Seedless, Perlette	25	0.5
Vitis spp.	Cardinal, Black Rose	10	
Olive *Olea europaea*			0.5
Berries	Ollalie blackberry,	10	
Rubus spp.	Indian Summer raspberry	5	
Strawberry	Lassen,	8	
Fragaria spp.	Shasta	5	

In general, the slower the chloride absorption, the more tolerant the plant would be to this solute.

Leaf injury symptoms appear in chloride-sensitive crops when leaves accumulate about 0.3 to 0.5% chloride on a dry-weight basis. Symptoms develop as leaf burn or drying of leaf tissue, typically occurring first at the extreme tip of older leaves and progressing back along the leaf edges. Excessive leaf burn is often accompanied by defoliation. Chemical analysis of soil or leaves can be used to confirm probable chloride toxicity. The maximum permissible concentrations of chloride in the soil saturation extract or in plant leaves for several sensitive crops are given in table 4.4.2.

Boron, although an essential minor element, is phytotoxic if present in excess. Toxicity arises from high boron concentrations in well waters or springs located near geothermal areas or geological faults. Concentrations as low as 0.5 g/m^3 in the irrigation can be detrimental to sensitive crops (table 4.4.3). Few surface waters contain enough boron to cause toxicity. Sensitivity to boron is not limited to woody perennials, but to a wide variety of crops. Boron toxicity symptoms typically appear as yellowing, spotting, or drying of leaf tissue at

TABLE 4.4.3

Relative tolerance of trickle irrigated crops to boron[a].

Tolerant[b] (4.0 g/m^3 to 2.0 g/m^3)	Semitolerant[b] (2.0 g/m^3 to 1.0 g/m^3)	Sensitive[b] (1.0 g/m^3 to 0.3 g/m^3)
Asparagus	Sunflower	Pecan
Date palm	Potato	Walnut
Garden beet	Cotton	Jerusalem artichoke
Broadbean	Tomato	Navy bean
Onion	Radish	Plum
Turnip	Field pea	Pear
Cabbage	Olive	Apple
Lettuce	Pumpkin	Grape
Carrot	Bell pepper	Kadota fig
	Sweetpotato	Persimmon
	Lima bean	Cherry
		Peach
		Apricot
		Thornless blackberry
		Orange
		Avocado
		Grapefruit
		Lemon

[a] Relative tolerance is based on the boron concentration in irrigation water at which boron toxicity symptoms were observed when plants were grown in sand culture. It does not necessarily indicate a reduction in crop yield.
[b] Tolerance decreases in descending order in each column between the stated limits.

the tip and along the edges of older leaves. The damage gradually progresses interveinally toward the midleaf. A gummosis or exudate on limbs or trunks is sometimes noticeable on boron-affected trees such as almond. Many sensitive crops show toxicity symptoms when boron concentrations in leaf blades exceed 250 mg/kg, but not all sensitive crops accumulate boron in their leaves. Stone fruits (e.g., peach, plum, almond) and pome fruits (pear, apple, and others) do not accumulate boron in leaf tissue so that leaf analysis is not a reliable toxicity indicator.

A wide range of crops has been tested for boron tolerance in sand cultures. The results of these tests are summarized in table 4.4.3. The crops have been grouped according to their relative tolerance to boron in the irrigation water. These data are based on the boron level at which toxicity symptoms were observed and do not necessarily indicate corresponding reductions in yield. Establishment of the relationships between boron concentration and crop yield must await further research.

Although the concentrations of some solutes in saline soils may be several orders of magnitude greater than concentrations of some essential nutrients, plant nutritional disturbances are not common in salt-affected soils. In some instances, however, if the proportion of calcium to sodium becomes either extremely high or low, nutritional imbalances can occur that reduce crop yield below that expected from osmotic effects alone. Bean plants cannot tolerate excess calcium even though it is accumulated in the plant tissue. In contrast, calcium deficiency under some saline conditions results in blossom-end rot of tomatoes, internal browning of lettuce, and reduced corn growth. Because soil salinity in the field normally involves a mixture of salts, the effects of specific solutes on crop nutrition tend to be minimized so that the osmotic effect usually predominates.

4.4.2 *Irrigation water quality*

The suitability of any water for trickle irrigation depends on the concentration and composition of the soluble salts present. Water quality for salinity evaluation is normally based on three criteria: (1) salinity, the general effects of dissolved salts on crop growth that are associated with osmotic stress; (2) sodicity, the effect of an excessive proportion of sodium that induces a deterioration of soil permeability; and (3) toxicity, the effects of specific solutes that damage plant tissue or cause an imbalance in plant nutrition. Irrigation water, however, has no inherent quality independent of the specific conditions under which it is to be used. Thus, only a generalized guide for evaluating irrigation water quality is presented because site-specific conditions may alter the suitability of a particular irrigation water.

4.4.2.a Composition of irrigation water

The predominant cations in irrigation water are calcium, magnesium, sodium, and potassium; the typical anions are bicarbonate, sulfate, and chloride. Significant concentrations of other solutes, such as nitrate, carbonate, and minor elements are not common in natural waters. Nitrate, present in some localized areas, is beneficial to crop production unless it causes excessive vegetative growth or delays crop maturity. Appreciable amounts of carbonate occur only infrequently when the pH of the water exceeds 8.5. Trace elements (molybdenum, cadmium, selenium, arsenic, chromium, vanadium, beryllium, etc.) are not common in natural waters, but when present in only minute concentrations, can severely limit production of certain crops. Any high concentration of minor elements (copper, zinc, etc.) may be a problem in waste waters used for irrigation.

Concentration, the amount of substance per unit volume, is typically reported in the SI (Système International d'Unités) metric system as moles per cubic meter of solution (mol/m^3); alternately as grams per cubic meter (g/m^3) or milligrams per liter (mg/L). The units of g/m^3 or mg/L are numerically equivalent to parts per million (ppm) and will replace ppm as conversion to SI units occur. Traditionally, ionic concentration has been expressed as milliequivalents per liter of solution (me/L). Ion concentration in me/L can be determined by multiplying its concentration in mol/m^3 by its ionic valence. To convert from mol/m^3 to g/m^3, multiply by the ion's atomic weight. The valence and atomic weight of common ions are given for convenience in table 4.4.4. The total salt concentration (C) is merely the sum of the concentration of each ion present.

The electrical conductivity (σ) of an irrigation water is often measured to estimate total salt concentration. For convenience, measurements are reported in SI metric units of decisiemens per meter (dS/m) or the traditional units of millimhos per centimeter (mmho/cm); the values are equal numerically. The relationship between salt concentration (g/m^3) and electrical conductivity can be approximated by $C = 640\sigma$.

Example 4.4.3

Problem: Calculate the mass (g) of calcium in a m^3 of irrigation water if its ionic concentration is 4 me/L.

Solution: To convert from me/L to g/m^3, divide by the valence of calcium (2) and multiply by its atomic weight (40.1). Thus, a m^3 of the water would contain 80.2 g of calcium.

Example 4.4.4

Problem: Determine the total salt concentration of a water sample in g/m^3 that contains 4 me/L of magnesium, 8 me/L of sodium, and 12 me/L of chloride and estimate its electrical conductivity.

Solution: As in example 4.4.1 convert the concentration of each ion to g/m^3 and obtain the sum. The total concentration is 48.6 for magnesium plus 184 for sodium plus 423.6 for chloride, a total of 656.2 g/m^3. Its electrical conductivity is 1.03 dS/m (656.2/640).

TABLE 4.4.4

Determinations normally required to evaluate irrigation water quality along with their symbols and units of measure.

Determination	Symbol	Valence	Unit of measure	Atomic weight
Total salt content				
(1) Electrical conductivity	σ	—	dS/m[a]	—
(2) Concentration	c		mol/m^3 or g/m^3	—
Sodium hazard				
(1) Sodium adsorption ratio[b]	SAR	—	$(mol/m^3)^{1/2}$	—
Constituents				
(1) Cations				
calcium	Ca	+2	mol/m^3	40.1
magnesium	Mg	+2	"	24.3
sodium	Na	+1	"	23.0
potassium	K	+1	"	39.1
(2) Anions				
bicarbonate	HCO_3	-1	"	61.0
sulfate	SO_4	-2	"	96.1
chloride	Cl	-1	"	35.3
Trace elements				
Boron	B	—	g/m^3	10.8

[a] dS/m = decisiemens per meter = 1 millimho/cm, referenced to 25°C.

[b] SAR = $[Na]/([Ca]+[Mg])^{1/2}$

To assess the sodium hazard of a water, the sodium adsorption ratio (SAR) must be calculated. The defining equation for SAR is

$$SAR = [Na]/([Ca] + [Mg])^{1/2} \tag{4.4.3}$$

where all ion concentrations [C] are in mol/m^3. All of the determinations that are normally required to assess irrigation water quality are summarized in table 4.4.4, along with their symbols and units of measure.

The quantities and types of salts present in irrigation water vary widely. The major sources of irrigation water are rivers and groundwaters, but the use of brackish and waste water will increase as supplemental irrigation as land disposal of municipal and industrial waste waters increase. Irrigation is usually practiced in arid climates, and the rivers associated with such areas generally contain more salt than the average river water. Exceptions include the Pacific Northwest and some regions of California where the salinity of the rivers is low for much of their lengths. For comparison, compositions of representative river waters are listed in table 4.4.5. Groundwaters are generally more saline than surface waters, and usually contain higher proportions of sodium, boron, and nitrate. Typical water analyses of irrigation wells are also given in table 4.4.5.

4.4.2.b Salinity assessment

Most irrigation waters do not contain enough salt to cause immediate injury to trickle-irrigated crops. However, the concentration of the soluble salts in the soil increases as water is removed by evaporation and transpiration, leaving the salt behind. Without leaching, salt accumulates in the soil with successive irrigations. With leaching, the accumulation of soluble salts in the soil is controlled by the fraction of the applied water that drains below the crop root zone (leaching fraction). Hence, salinity assessment must be made in view of the salt tolerance of the crop and the leaching fraction. The fraction of salt in the irrigation water that will precipitate is a further consideration if the leaching fraction is less than about 0.1.

A generalized appraisal of the salinity hazard of an irrigation water can be made based upon proper water management and the ability of the soil to meet or exceed the minimum leaching fraction required to maintain full crop production termed the leaching requirement (L_r). The leaching requirement as a function of the salinity of the applied water and crop salt tolerance is given in figure 4.4.2.

Example 4.4.5

Problem: The salinity of the irrigation water in an arid region without any measurable rainfall is 3 dS/m. A tomato crop is to be trickle irrigated; what is the leaching requirement? If the evapotranspiration requirement of the crop is 900 mm, how much irrigation is required to prevent any loss in yield from excess soil salinity?

Solution: From table 4.4.1, the salt tolerance threshold for tomato is 2.5 dS/m. Entering figure 4.4.2 with this value and a value of 3 dS/m for σ_i gives a L_r of 0.20. Assuming 100% irrigation efficiency the amount of irrigation water (D_i) = D_{et} + D_d, where D_{et} is the depth of evapotranspiration and D_d is the depth of drainage. The required value of $D_d = L_r D_i$; thus $D_i = D_{et}/(1 - L_r)$ or $D_i = 900/(1 - 0.2)$ = 1125 mm.

TABLE 4.4.5

Water quality of selected river and well waters.

Water source and location	Salinity: Electrical conductivity σ (dS/m)	Salinity: Salt content C (g/m^3)	Sodicity: SAR $[(mol/m^3)^{1/2}]$
		RIVER WATERS	
SAN JOAQUIN Friant Dam, California	0.04	20	0.4
SNAKE King Hill, Idaho	0.5	370	0.9
RIO GRANDE Falcon Dam, Texas	1.0	680	2.9
COLORADO Hoover Dam, Arizona	1.3	920	3.2
GILA Gillespie Dam, Arizona	4.4	2920	10.7
PECOS Artesia, New Mexico	8.6	6010	12.5
		WELL WATERS	
Miller Co., Georgia	0.3	150	0.1
Van Buren Co., Michigan	0.5	330	0.4
Adams Co., Nebraska	0.5	370	0.9
Scott Co., Kansas	0.7	450	0.9
Maricopa Co., Arizona	1.2	820	3.9
Reeves Co., Texas	4.4	2860	6.1
Yuma Co., Arizona	7.2	4540	13.2

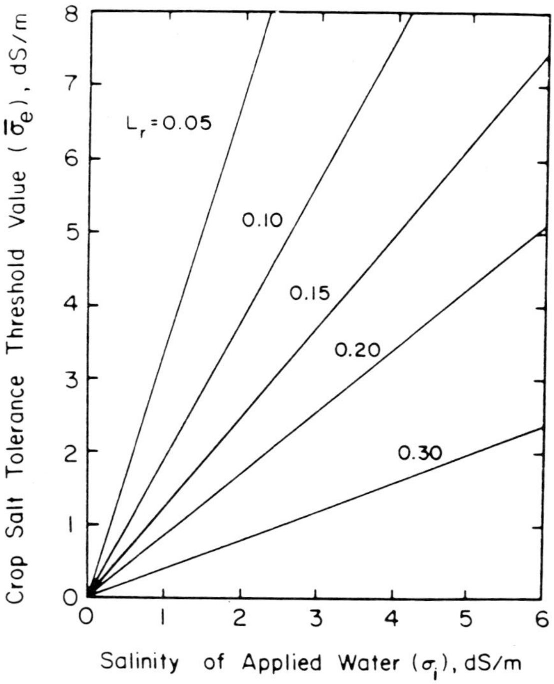

Fig. 4.4.2 Graphical solution for the leaching requirement as a function of the salinity of the applied water and the salt tolerance threshold value for the crop (after Hoffman and van Genuchten, 1983).

The salinity of applied water is the weighted average of the salt concentrations of the irrigation water and rainfall based on the amounts applied. The relationships in figure 4.4.2 allow the prediction of the minimum leaching fraction required for the particular water and crop combination for avoiding possible loss in yield.

4.4.2.c Sodicity assessment

Another consideration in evaluating irrigation water quality is the potential for soil structural deterioration caused by an excess buildup of sodium in the soil. When calcium and magnesium are the predominant cations adsorbed on the soil exchange complex, the soil tends to have a granular structure that is easily tilled and readily permeable. When the amount of adsorbed sodium exceeds about 10 percent of the total cation exchange capacity of the soil, however, soil mineral particles tend to disperse causing water penetration to decrease. Excess sodium becomes a problem when the rate of infiltration is reduced to the point that the crop is not adequately supplied with water or when the hydraulic conductivity of the soil profile is too low for adequate drainage. Sodium may also add to cropping difficulties because of crusting seed beds, temporary saturation at the soil surface, and/or possible disease, weed, oxygen, nutritional, and salinity problems. The sodium adsorption ratio (SAR) is influenced by the type of clay minerals present in the soil. In general, however, no permeability problem should be expected if the SAR is below 10 $(mol/m^3)^{1/2}$ providing the irrigation water is not too low in salt content.

Example 4.4.6

Problem: Assess the sodicity hazard of the well water in Kern County, California given the ion concentrations of 0.1, 0.0, and 7.3 mol/m^3 for calcium, magnesium, and sodium, respectively.

Solution: The sodium adsorption ratio, SAR, is given by the equation $[Na]/([Ca]+[Mg])^{1/2}$, and thus, SAR is $7.3/(0.1)^{1/2} = 23.1$ $(mol/m^3)^{1/2}$. This value is in excess of the guideline value of 10 $(mol/m^3)^{1/2}$ and water penetration problems from this water should be anticipated.

Water of very low salt content can aggravate a permeability problem because it allows a maximal swelling and dispersion of soil minerals and organic matter, and because it has a tremendous capacity to dissolve and remove calcium. This problem can arise if the electrical conductivity of the irrigation is less than 0.2 dS/m. Irrigating with water from the San Joaquin River (see table 4.4.5) may cause problems on soil with potential permeability problems. The addition of amendments to the dilute irrigation water in parts of the San Joaquin Valley of California is a common practice to improve infiltration. The addition of amendments of low solubility, such as gypsum, into waters to be applied through a trickle system is not recommended because of the serious threat of plugging emitters. To avoid this problem, amendments with low solubility should be applied directly to the soil.

4.4.3 *Management of salt-affected soils*

With trickle irrigation, salinity control is influenced by the quality and quantity of the applied water, the irrigation system and its management, drainage conditions, and agronomic techniques. These factors are often interrelated so that the solution to a salinity problem may not be obvious without proper diagnosis. The key to salinity control is a properly integrated total water management system.

4.4.3.a Irrigation management

The primary objective of any irrigation method is to supply water to the soil so that it will be readily available at all times for crop growth, but soil salinity is definitely an influencing factor. The soil salinity profile that develops as soil water is decreased through water uptake by roots or soil evaporation depends in part on the water distribution pattern inherent with the irrigation method. Distinctly different salinity profiles develop for various irrigation methods, and significantly different profiles can develop for each method within a given field because of differences in soil properties or in the management of the system.

Trickle irrigation systems provide water through perforated tubes or emitters spaced according to plant density. Their advantage is the maintenance of high soil water content in the root zone by frequent but small water applications. Plant roots tend to proliferate in the leached zone of high soil water content beneath the water sources. This allows water of relatively high salt content to be used successfully in many cases. Trickle irrigation is appropriate for orchards and for some widely spaced or valuable row crops.

The salinity profile under line sources, such as perforated or multi-emitter drip irrigation systems, has clearly recognizable lateral and downward components. The typical cross-sectional profile has an isolated pocket of accumulated salts at the soil surface midway between the line sources and a second, deep zone of accumulation, depending on the degree of leaching. Directly beneath the line source is a leached zone whose size depends on the irrigation rate and frequency, and the crop's water extraction pattern.

The distribution of chloride in the soil water is illustrated in figure 4.4.3 for a line source for a water having an electrical conductivity of 2.2 dS/m and a chloride content of 7.8 mol/m^3. While the salt distribution from line sources increases laterally and downward, the distribution from point irrigation sources such as microbasins and drip systems with widely spaced emitters increases radially in all directions below the soil surface. As the rate of water application increases, the shape of the salinity distribution changes. The salinity distribution in uniform, isotropic sand changes from elliptical (with the maximum movement vertical) to more circular as the rate of water application increases. In isotropic and layered soils, the horizontal rate of water movement increases relative to the vertical, resulting in relatively shallow salt accumulations. For tree crops irrigated with several drip emitters per tree, the wetting patterns may overlap, thereby reducing the level of salt accumulation midway between emitters under a tree.

Subsurface irrigation is the least practiced of the various methods of trickle irrigation where salinity is a hazard. The continuous upward water movement from a subsurface system results in rapid salt accumulation near the soil surface as water is lost by evapotranspiration. The extent of salt accumulation at shallow depths is a function of the depth of the subsurface system and the application rate; the more shallow the system and the larger the application rate, the smaller the amount of salts that accumulate above the system.

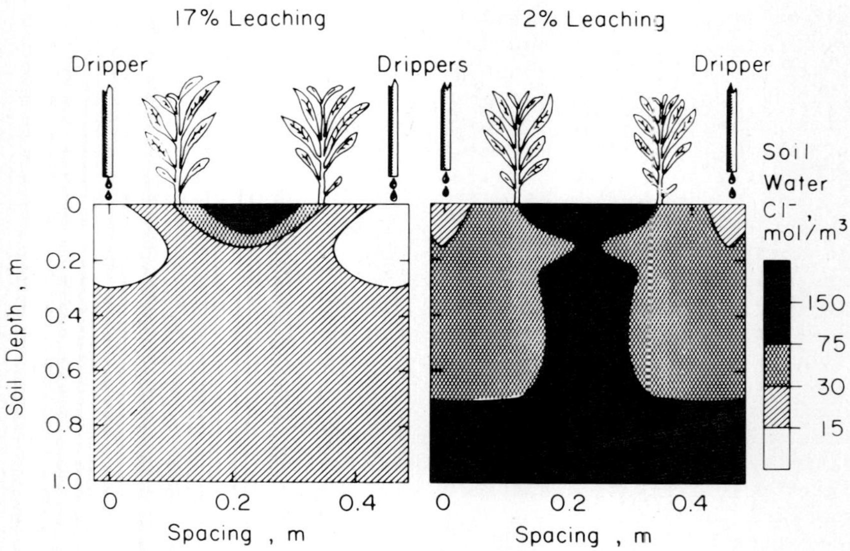

Fig. 4.4.3 Steady-state soil-water chloride profiles for 17 and 2% leaching under a line source trickle irrigation system. (after Hoffman et al., 1979).

Subsurface systems provide no means of leaching these shallow accumulations. Unless the soil is leached periodically by rainfall or surface irrigations, salt accumulations are a certainty.

4.4.3.b Drainage management

Without drainage, salination is a foregone conclusion and irrigated agriculture is bound to fail. Fortunately, most soils have some degree of natural drainage to facilitate leaching, but supplementary drainage is often required. The need for supplementary drainage may be lessened or even avoided by an efficient management of irrigation water.

For salinity control, the leaching requirement establishes the minimum drainage requirement. Additional drainage, however, may be necessary to remove excess precipitation, seepage from adjacent irrigated areas or from water conveyance structures, nonuniform irrigation applications, and other sources of water that are not directly related to leaching. The drainage requirement for salt removal under steady-state conditions is given by

$$D_d = L_r D_{et}/(1 - L_r) \qquad (4.4.4)$$

where D_d is the equivalent depth of drainage per unit land area, L_r is the leaching requirement, and D_{et} is the equivalent depth of evapotranspiration per unit land area. In terms of the electrical conductivities of the irrigation (σ_i) and drainage (σ_d) waters,

$$D_d = \sigma_i D_{et}/(\sigma_d - \sigma_i). \quad (4.4.5)$$

Crop salt tolerance dictates the selection of the leaching requirement in one equation and the permissible value of σ_d in the other.

The value of D_d refers only to the minimum amount of water that must leave the root zone at any point of a field. It is not the total amount of water that must be removed. Furthermore, the actual amount will vary significantly with the crop. Where a range of crops is grown in rotation, or where the cropping pattern changes over time, the effect of these changes on the drainage requirement must be considered. In planning a drainage system, the design should be based on the most demanding crop to be grown.

4.4.3.c Agronomic management

Many crop failures on salt-affected soils result from growing crops that have low salt tolerance. Approximately a ten-fold range in salt tolerance exists among agricultural crops (see table 4.4.1). In areas where only saline irrigation water is available, where shallow, saline water tables prevail, or where soil permeability is low, achieving non-saline conditions may not be economically feasible. In such areas, crops that produce satisfactory yields under saline conditions must be selected. Owing to the variations among cropping seasons and irrigation management, soils may tend to salinize during one crop and to desalinize during a following crop. Thus, by selecting an appropriate crop rotation, productivity can sometimes be maintained where, in the absence of a rotation, the soil would become unproductive.

Obtaining a satisfactory stand of trickle irrigated crops on saline soils or when using saline water is often a problem. Growers sometimes compensate for poor germination by planting two or three times as many seeds as normally required. In other cases, planting procedures are adjusted to ensure that the soil around the germinating seeds is low in salinity.

PROBLEMS

1. Describe the specific symptoms of salt-affected plants.

2. What is the anticipated yield loss for tomato if the average soil salinity is 6 dS/m? The average soil salinity should be below what value to prevent yield loss?

3. Discuss the types of plants most susceptible to chloride toxicity and describe the plant symptoms.

4. Calculate the sodium adsorption ratio of an irrigation water when the concentrations of sodium, calcium, and magnesium are 30, 10, and 8 me/L, respectively.

5. Describe the three major salinity factors in evaluating a water for irrigation purposes.

6. A Valencia orange orchard has an annual evapotranspiration of 1200 mm and receives 400 mm of rainfall annually that is effective in meeting part of the tree's water requirements. The only water available for irrigation has an electrical conductivity of 2 dS/m. What is the leaching requirement and how much irrigation water must be applied to prevent yield loss?

7. Discuss differences in the distribution of soil salinity for sprinkler compared to drip irrigation.

REFERENCES AND SUPPLEMENTAL READING

Bresler, E., McNeal, B.L. and Carter, D.L., 1982. Saline and Sodic Soils. Principles-dynamics-modeling. Advanced Series in Agricultural Sciences 10, Springer-Verlag, New York, 236 pp.

Hoffman, G.J., 1980. Guidelines for reclamation of salt-affected soils. Proc. of Inter-American Salinity and Water Management Tech. Conf., Juarez, Mexico, pp. 49-64.

Hoffman, G.J., 1981. Alleviating salinity stress. In Arkin, G.F. and Taylor, H.M. (Eds.), Modifying the Root Environment to Reduce Crop Stress, Monograph 4, Amer. Soc. Agric. Eng., St. Joseph, Michigan, pp. 305-343.

Hoffman, G.J., 1985. Irrigation water quality. In Handbook of Agriculture, Chapter 6a, Section F, Engineering Vol. 1, Production of Agriculture Products, CRC Press, Boca Raton, Florida (in press).

Hoffman, G.J., Rawlins, S.L., Oster, J.D., Jobes, J.A. and Merrill, S.D., 1979. Leaching requirement for salinity control. I. Wheat, sorghum, and lettuce. Agric. Water Manage. 2: 177-192.

Hoffman, G.J. and van Genuchten, M. Th., 1983. Efficient water management for salinity control. In Taylor, H.M., Jordan, W. and Sinclair, T. (Eds.), Efficient Water Use in Crop Production, Amer. Soc. Agron., Madison, Wisconsin, pp. 73-85.

Maas, E.V., 1985. Salt tolerance of plants. In Christie, B.R. (Ed.), Handbook of Science, CRC Press, Boca Raton, Florida (in press).

Maas, E.V. and Hoffman, G.J., 1977. Crop salt tolerance-current assessment. J. Irrig. and Drain. Div., Amer. Soc. Civil Eng. 103(IR2): 115-134.

SUBJECT INDEX

APPENDIX I. LIST OF SYMBOLS AND PREFIXES

Most engineering and scientific journals have gone from the "cgs" to the SI system (Le Systeme International d'Unites) for designating units of measurement. In the United States, conversion from the English to the metric unit has scarcely begun in industry and commerce, and to go one step further to the SI system is an additional hurdle to overcome for the general public. Even for the engineering and scientific communities, a change in units is not without problems because of the need to modify one's inherent thinking and feel for a given set of dimensions. The majority of reference tables and graphs are still in the English and metric units. Tables have been provided in the following Appendix section for converting from one unit to another.

Besides units, symbols used in equations or definitions have different connotations in the various fields. For example, "K" and "γ" are used extensively as a coefficient or constant and have different meanings for different fields. We have attempted to list the most commonly used symbols in this text, and when these symbols are used within the context of the particular discussion, hopefully, no problem should result.

SYMBOLS

SYMBOL	USE	UNITS
A	cross sectional area	square meter
A_d	fraction deficitly irrigated	decimal fraction
A_w	available soil water	percent
α	albedo	-
C	concentration	megagram per cubic meter mol per cubic meter milliequivalent per liter mol per liter parts per million
C_d	coefficient of discharge	-
C_d	concentration of drainage water	See C
C_f	concentration of fertilizer	milligram per liter

C_g	concentration of groundwater	See C
C_i	concentration of irrigation water	See C
C_s	Concentration of individual salts	See C
CV	coefficient of variation	-
CWSI	crop water stress index	decimal fraction
d	emitter diameter	millimeter
d_d	particle density	kilogram per cubic meter gram per cubic centimeter
d_i	amount of irrigation water	millimeter
d_w	water density	kilogram per cubic meter gram per cubic centimeter
D	dispersion coefficient	square meter per second
D	drainage water	millimeter
D	pipe diameter	meter inch
D	soil water diffusivity	square meter per second
D_d	amount of drainage water	millimeter
D_{et}	amount of evapotranspiration	millimeter
D_i	amount of irrigation water	millimeter
∇	vector gradient operator	reciprocal meter
∇^2	Laplacian operator	reciprocal square meter
ε	emissivity	-
e	number of emitters per plant	-
e_s	saturated vapor pressure	kilopascal

E_a	application efficiency	percent
E_d	distribution efficiency	percent
E_i	application efficiency	decimal fraction
E_r	water requirement efficiency	percent
E_{tc}	evapotranspiration	millimeter per day
E_{tr}	evapotranspiration of reference crop	millimeter per day
ET	evapotranspiration	millimeter per day
EU	emitter uniformity coefficient	percent
EU_a	absolute emission uniformity	percent
EU'	field emission uniformity	percent
f	friction coefficient	-
g	acceleration of gravity	meter per square second
gpm	flow rate	gallon per minute
G	soil heat flux	megajoule per square meter
γ	activity coefficient	-
γ	psychrometric constant	-
h	matric potential	kilopascal meter (water)
h	pressure head	meter (water)
h_i	head loss	meter (water)
H	hydraulic head, soil water	meter (water)
H	operating pressure, pipe	kilopascal pound per square inch
H_{max}	maximum pressure in line	kilopascal pound per square inch

H_{min}	minimum pressure in line	kilopascal pound per square inch
H_{var}	pressure variation	-
HDPE	high density polyethylene	-
HP	horsepower	kilowatt
ΔH	energy drop	meter (water)
$\overline{H}$	total specific energy	meter (water)
I_a	irrigation amount	millimeter
I_r	irrigation requirement	millimeter
J	flux of salt	mol per square meter-second
k	emitter discharge constant	-
K_{cb}	crop coefficient	-
K_{cs}	soil water stress coefficient	-
K_d	dissociation constant	-
K_g	soil cation exchange coefficient, Gapon	-
K(h)	unsaturated hydraulic conductivity	centimeter per day
K_{IAP}	ion activity product constant	-
K_k	soil cation exchange coefficient	-
K_p	first order rate reaction for salt precipitation	-
K_p	pan evaporation coefficient	-
K_s	solubility product constant	variable
K_s	saturated hydraulic conductivity	centimeter per day

K_{so}	soil evaporation factor	-
L	latent heat of vaporization	joule per kilogram
L	length	meter
L	volume	liter
L_e	equivalent pipe length	meter
L_r	leaching requirement	-
LF	leaching fraction	-
LM	maxiumum lateral soil water movement	centimeter
LDPE	low density polyethylene	-
LSI	Langelier saturation index	-
M	mass	megagram
μ	viscosity	newton-second per square meter
ν	kinematic viscosity	square meter per second
p	porosity	percent
psi	pressure	pound per square inch
P	precipitation	millimeter
P	line pressure	kilopascal pound per square inch
P_b	barometric pressure	kilopascal
P_D	amount of deficit irrigation	percent
Φ_T, ϕ_T	soil water potential	kilopascal meter (water)
π	osmotic potential	kilopascal meter (water)

ψ_m	soil water potential	kilopascal meter (water)
ψ_L	leaf water potential	kilopascal meter (water)
ψ_p	turgor pressure	kilopascal meter (water)
ppm	solution concentration	milligram per liter
PE	polyethylene	-
PVC	polyvinyl chloride	-
q	soil water flux	gram per square meter second
q_e	emitter flow rate	liter per hour
q'_e	normalized q_e at temperature T	percent
$\overline{q}$	mean emitter flow rate	liter per hour
q_{avg}	mean emitter flow rate	liter per hour
q_{max}	maximum emitter flow rate	liter per hour
q_{min}	minimum emitter flow rate	liter per hour
q_r	required emitter flow rate	liter per hour
q_{var}	emitter flow variability	percent
Δq	measured deviation of q_e	liter per hour
Q, Q_a	pipe flow rate	cubic foot per second liter per second gallon per second
Q	runoff water	millimeter
Q_s	water supply rate	liter per second
Q_f	fertilizer injection rate	liter per hour
r_o	radius of saturated zone	meter

R	black body radiation	watt per square meter
R	universal gas constant	megapascal-liter per mol-Kelvin joule per mol-Kelvin
R	solute retardation factor	-
ρ_b	bulk density of soil	megagram per cubic meter
ρ_w	density of water	megagram per cubic meter
R_e	Reynolds number	-
R_h	hydraulic radius	meter
R_i	energy drop ratio	-
R_n	net radiation	megajoule per square meter
R_s	daily solar radiation	"
RH	relative humidity	percent
s	standard deviation	-
S	soil storage water	millimeter
S_f	slope of energy line	-
SAR	sodium adsorption ratio	square root of millimol per liter square root of mol per cubic meter
S_p	amount of salt precipitation	mol
S_y	standard deviation	-
σ(EC)	solution electrical conductivity	decisiemen per meter millimho per centimeter
$\sigma_d(EC_d)$	σ of drainage water	"
$\sigma_e(EC_e)$	σ of soil saturation extract	"

$\sigma_i(EC_i)$	σ of irrigation water	"
T	irrigation setting	hour
T	temperature	Celsius Kelvin
T	time	second hour day
T	transpiration	millimeter millimeter per square meter-second
T_a	mean daily temperature	Celsius Kelvin
T_s	surface temperature	Celsius Kelvin
T_i	net irrigation time	hour per day
θ_g	soil water content, gravimetric	kilogram water per kilogram dry soil
θ_m	soil water content, gravimetric	kilogram water per kilogram dry soil
θ_v	soil water content, volumetric	cubic meter water per cubic meter soil
θ_r	residual soil water content	"
θ_s	saturated soil water content	"
U	wind speed	meter per second
U_s	uniformity coefficient for emitter flow	percent decimal fraction
U_c	uniformity coefficient for irrigation	percent decimal fraction
V	flow velocity	meter per second
V_c	scouring velocity	meter per second

V_e	CV with emitter plugging	-
V_g	volume of groundwater	cubic meter liter
V_i	volume of irrigation water	cubic meter liter
V_m	CV of emitter discharge (manufacturer's variation)	-
V_o	critical settling velocity	centimeter per second
V_q	CV of emitter discharge (operational)	-
V_{qs}	CV of submain unit	-
V_y	coefficient of variation	-
VM	maximum soil water movement	meter
VPD	vapor pressure deficit	kilopascal
VS	wetted soil volume	liter
W	required amount of irrigation water	cubic meter liter
x	emitter emission exponent	-
y	relative yield	percent
z	elevation	meter

PREFIXES

SYMBOL	DESIGNATION	FACTOR
T	tetra-	10^{12}
G	giga-	10^{9}
M	mega-	10^{6}
k	kilo-	10^{3}
h	hecto-	10^{2}
da	deka-	10^{1}
d	deci-	10^{-1}
c	centi-	10^{-2}
m	milli-	10^{-3}
μ	micro-	10^{-6}
n	nano-	10^{-9}
p	pico-	10^{-12}

APPENDIX II. UNITS CONVERSION TABLE

Units

The compilation of this text involved a mixed group of engineers and scientists with different public and private contacts and backgrounds. Attempts to standardize units to the SI system can become one of frustration and confusion since equipment literature has been based on either the English or metric system. Also, the mixed audience using this book may not be comfortable with another unit, and errors may be caused especially under field operational situations. Thus, we have aimed for clarity and consistency in the use of units in the text. The tables in this section were developed to help alleviate confusion.

To convert units from the vertical column to the horizonal one, multiply the column unit by the numerical value at the intersection of the two units.

LENGTH CONVERSION

FROM \ TO	micron	mm	cm	m	km	in	ft	mile
micron	1	1 E-03	1 E-04	1 E-06	1 E-09	3.937 E-05	3.281 E-06	6.214 E-10
mm	1 E+03	1	1 E-01	1 E-03	1 E-06	3.937 E-02	3.281 E-03	6.214 E-07
cm	1 E+04	10	1	1 E-02	1 E-05	.3937	3.281 E-02	6.214 E-06
m	1 E+06	1 E+03	1 E+02	1	1 E-03	39.37	3.281	6.214 E-04
km	1 E+09	1 E+06	1 E+05	1 E+03	1	3.937 E+04	3.281 E+03	.6214
in	2.540 E+04	25.40	2.540	2.540 E-02	2.540 E-05	1	8.333 E-02	1.578 E-05
ft	3.048 E+05	3.048 E+02	30.48	.3048	3.048 E-04	12	1	1.894 E-04
mile	1.609 E+09	1.609 E+06	1.609 E+05	1.609 E+03	1.609	6.336 E+04	5280	1

AREA CONVERSION

FROM \ TO	mm^2	cm^2	m^2	km^2	ha	in^2	ft^2	acre	$mile^2$
mm^2	1	1 E-02	1 E-06	1 E-12	1 E-10	1.550 E-03	1.076 E-05	2.471 E-10	3.861 E-13
cm^2	1 E+02	1	1 E-04	1 E-10	1 E-08	.1550	1.076 E-03	2.471 E-08	3.861 E-11
m^2	1 E+06	1 E+04	1	1 E-06	1 E-04	1.550 E+03	10.76	2.471 E-04	3.861 E-07
km^2	1 E+12	1 E+10	1 E+06	1	1 E+02	1.550 E+09	1.076 E+07	2.471 E+02	.3861
ha	1 E+10	1 E+08	1 E+04	1 E-02	1	1.550 E+07	1.076 E+05	2.471	3.861 E-03
in^2	6.452 E+02	6.452	6.452 E-04	6.452 E-10	6.452 E-08	1	6.944 E-03	1.594 E-07	2.491 E-10
ft^2	9.290 E+04	9.290 E+02	9.290 E-02	9.290 E-08	9.290 E-06	144	1	2.296 E-05	3.587 E-08
acre	4.047 E+09	4.047 E+07	4.047 E+03	4.047 E-03	.4047	6.273 E+06	4.356 E+04	1	1.562 E-03
$mile^2$	2.590 E+12	2.590 E+10	2.590 E+06	2.590	2.590 E+02	4.014 E+09	2.788 E+07	640	1

VOLUME CONVERSION

FROM \ TO	cm^3	L	m^3	in^3	ft^3	pint	qt	gal	ac-ft
cm^3	1	1 E-03	1 E-06	6.102 E-02	3.531 E-05	2.113 E-03	1.057 E-03	2.642 E-04	8.107 E-10
L	1 E+03	1	1 E-03	61.02	3.531 E-02	2.113	1.057	.2642	8.107 E-07
m^3	1 E+06	1 E+03	1	6.102 E+04	35.31	2.113 E+03	1.057 E+03	2.642 E+02	8.107 E-04
in^3	16.39	1.639 E-02	1.639 E-05	1	5.787 E-04	3.464 E-02	1.732 E-02	4.329 E-03	1.329 E-08
ft^3	2.832 E+04	28.32	2.832 E-02	1.728 E-03	1	59.84	29.92	7.481	2.297 E-05
pint	4.732 E+02	.4732	4.732 E-04	28.87	1.671 E-02	1	.5000	.1250	3.837 E-07
qt	9.464 E+02	.9464	9.464 E-04	57.74	3.342 E-02	2	1	.2500	7.674 E-07
gal	3.785 E+03	3.785	3.785 E-03	2.310 E+02	.1337	8	4	1	3.069 E-06
ac-ft	1.233 E+09	1.233 E+06	1.233 E+03	7.523 E+07	4.354 E+04	2.606 E+06	1.303 E+06	3.258 E+05	1

MASS CONVERSION

FROM \ TO	mg	g	kg	metric ton	oz	lb	short ton	grain
mg	1	1 E-03	1 E-06	1 E-09	3.527 E-05	2.205 E-06	1.102 E-09	1.543 E-02
g	1 E+03	1	1 E-03	1 E-06	3.527 E-02	2.205 E-03	1.102 E-06	15.43
kg	1 E+06	1 E+03	1	1 E-03	35.27	2.205	1.102 E-03	1.543 E+04
metric ton	1 E+09	1 E+06	1 E+03	1	3.527 E+04	2.205 E+03	1.102	1.543 E+07
oz	2.835 E+04	28.35	2.835 E-02	2.835 E-05	1	6.250 E-02	3.125 E-05	437.5
lb	4.536 E+05	453.6	.4536	4.536 E-04	16	1	5.000 E-04	7.000 E+03
short ton	9.072 E+05	9.072 E+05	9.072 E+02	.9072	3.200 E+04	2000	1	1.400 E+07
grain	64.80	6.480 E-02	6.480 E-05	6.480 E-08	2.285 E-03	1.429 E-04	7.143 E-08	1

CONCENTRATION CONVERSION

FROM \ TO	mg/L	g/L	kg/m^3	lb/in^3	lb/ft^3	lb/gal	grain/gal
mg/L	1	1 E-03	1 E-03	3.613 E-08	6.243 E-05	8.345 E-06	5.842 E-02
g/L	1 E+03	1	1	3.613 E-05	6.243 E-02	8.345 E-03	58.42
kg/m^3	1 E+03	1	1	3.613 E-05	6.243 E-02	8.345 E-03	58.42
lb/in^3	2.768 E+07	2.768 E+04	2.768 E+04	1	1.728 E+03	2.310 E+02	1.617 E+06
lb/ft^3	1.602 E+04	16.02	16.02	5.787 E-04	1	.1337	9.354 E+02
lb/gal	1.198 E+05	1.198 E+02	1.198 E+02	4.328 E-03	7.480	1	6.998 E+03
grain/gal	17.12	1.712 E-02	1.712 E-02	6.185 E-07	1.069 E-03	1.429 E-04	1

FLOW RATE CONVERSION

FROM \ TO	L/sec	m^3/min	m^3/hr	gal/min	gal/hr	ft^3/sec
L/sec	1	.0600	3.600	15.85	9.510 E+02	3.532 E-02
m^3/min	16.67	1	60	2.642 E+02	1.585 E+04	.5886
m^3/hr	.2778	1.667 E-02	1	4.404	2.642 E+02	9.812 E-03
gal/min	6.309 E-02	3.785 E-03	.2271	1	60	2.228 E-03
gal/hr	1.051 E-03	6.309 E-05	3.785 E-03	1.667 E-02	1	3.713 E-05
ft^3/sec	28.32	1.699	1.020 E+02	4.489 E+02	2.693 E+04	1

PRESSURE CONVERSION

FROM \ TO	dyne/cm^2	kg/m^2	Pa	cm Hg	ft H_2O	bar	atm	lb/in^2	lb/ft^2
dyne/cm^2	1	1.010 E-02	.1000	7.501 E-05	3.346 E-05	1 E-06	9.869 E-07	1.450 E-05	2.089 E-03
kg/m^2	98.07	1	9.807	7.356 E-03	3.281 E-03	9.807 E-05	9.678 E-05	1.422 E-03	.2048
Pa	10.00	.1020	1	7.501 E-04	3.346 E-04	1 E-05	9.870 E-06	1.450 E-05	2.088 E-02
cm Hg	1.333 E+04	1.360 E+02	1.333 E+03	1	.4460	1.333 E-02	1.316 E-02	.1934	27.85
ft H_2O	2.989 E+04	3.046 E+02	2.989 E+03	2.241	1	2.989 E-02	2.949 E-02	.4335	62.42
bar	1 E+06	1.020 E+04	1 E+05	75.01	33.46	1	.9869	14.50	2.089 E+03
atm	1.013 E+06	1.033 E+04	1.013 E+05	76.00	33.90	1.013	1	14.70	2.116 E+03
lb/in^2	6.895 E+04	7.031 E+02	6.895 E+03	5.171	2.307	6.895 E-02	6.804 E-02	1	144
lb/ft^2	4.788 E+02	4.882	47.88	3.591 E-02	1.602 E-02	4.788 E-04	4.725 E-04	6.944 E-03	1